**Technology and employment
in industry**

Technology
and employment
in industry

A case study approach

Edited by A. S. Bhalla
Foreword by Amartya Sen

International Labour Office   Geneva

ISBN 92-2-101238-7 (paperback)
ISBN 92-2-101241-7 (hardback)

First published 1975

ILO publications can be obtained through major booksellers or ILO local offices in many countries, or direct fom ILO Publications, International Labour Office, CH-1211 Geneva 22, Switzerland. A catalogue or list of new publications will be sent free of charge from the above address.

Printed by Imprimeries Réunies, Lausanne, Switzerland.

FOREWORD

This collection of case studies, edited by Dr. A. S. Bhalla, together with the preceding examination of some of the conceptual and measurement issues relevant to the problem of technological choice, represents a contribution to the economics of technological possibilities open to developing countries. Most of the studies are concerned with identifying and analysing alternative techniques of production, and examining their implications for specific policy decisions. The motivation is practical and the focus is down to earth. The intention is not to suggest a grand strategy for technological planning, but to discover factual relations in the detailed technical experience of specific countries—as regards can making in Kenya, Tanzania and Thailand, jute processing in Kenya, textile manufacturing in the United Kingdom, sugar processing in India, manufacturing cement blocks in Kenya, running engineering industries in Colombia, metalworking in Mexico, extracting and processing copper and aluminium in the United States, Zambia, Zaire and Chile.

A NEGLECTED FIELD

It is not unfair to describe this book as being concerned with the details of technological economics. It is precisely these "details" that make all the difference between success and failure in technological planning. Nevertheless, this has been a much neglected field of research, and it is worth inquiring why this is so.

One reason, certainly, is the lack of glamour in such detailed work; the mention of can making in Tanzania does not make the air electric with expectation. The number of economists who will be absorbed by a study on manufacturing cement blocks in Kenya is likely to be remarkably lower than the number who will dig deep into a general article by the same author on the conflict between output maximisation and employment maximisation in technological choice. The system of recognition and acclaim that applies in the world of economics does not particularly encourage the preparation of works concerned with details, however important they may be for practical planning.

EFFECTIVE DEMAND AND SECTORAL BALANCES

But this, I suspect, is not the whole story. The shift in the focus of interest from micro-economic problems to issues of aggregative economics that took place around the 1930s has not yet fully worked itself out. That shift was related to two important, real world phenomena: the Great Depression and the emergence of central planning in the Soviet Union. The former led to the Keynesian *General Theory* and to the related developments in macro-economics. The latter had a profound impact on the study of inter-sectoral relations, of which "input-output analysis" is one example. The study of inter-sectoral relations exposed the limitations of "partial equilibrium" approaches, and while this was a major advance in many ways it also led to the practice of glossing over the details of technological complexities (e.g. ruling out substitution possibilities, or assuming that there is only one "primary" factor of production, as in alternative interpretations of the Leontief system).

Both approaches, while technologically weak, were very successful when used for explaining, and for predicting as well as for planning. Keynesian macro-economics, with its simple concern with aggregate "effective demand" regulating the level of employment and prices—"full employment" being flanked on one side by unemployment (if there is too little effective demand) and on the other by inflation (if there is too much of it)—changed the focus of government policy in the advanced capitalist economies, and helped considerably to avoid a recurrence of a depression like that of the 1930s. Only recently, with the persistence of the so-called "stagflation" and related phenomena, has it become clear that the role assigned to aggregate effective demand in the Keynesian system is exaggerated and that the macro-economics of the one-commodity world is a somewhat oversimplified concept.

The success of central planning in the USSR and in Eastern Europe owed not a little to the socialist planners' awareness of the importance of inter-industry relations, and the switch from partial equilibrium piecemeal planning indeed contributed to achieving structural transformation and to generating economic growth.

The technological constraints on Soviet planning were, however, substantially eased by the fact that the kind of technology the Soviet planners were seeking was not altogether different from the technology prevailing in the advanced capitalist countries. The *relations* of production were going to be different, but not so much the *means* of production. In terms of the supply of "primary" factors of production, i.e. of labour, land and natural resources, the Soviet economy was not very far removed from the American economy, and the deficiency consisting in a low level of accumulated capital was soon to be tackled by a massive programme of growth and accumulation. Soviet and Western technology, never very far apart, have increasingly converged.

THE SEARCH FOR TECHNOLOGY

The situation was quite different when China began its socialist planning. In terms of the supply of labour, land and natural resources, the Chinese faced a problem altogether different from both that of the Western world and that of the Soviet Union. The Chinese planners were very much concerned about the evolution of new technologies. Whether engaging in the massive use of mobilised labour for modern construction or undertaking the daring experiment of making steel in small units, they were to a large extent sailing on an uncharted ocean. The focus on the details of technological possibilities was inevitable. On the technological plane, if one of the major Soviet contributions was the development of inter-sectoral balances, one of the primary Chinese contributions has been focusing on technological complexes appropriate to a labour-abundant economy.

The spectacular economic growth of Japan in the period following the Second World War has been another important influence. Many factors contributed to Japanese growth, but the development of technologies suited to Japanese conditions, or the adaptation of existing technologies to these conditions, were certainly among them. The Japanese resource situation had some very special features, including an exceptionally broad educational base that already at the time of the Meiji restoration was more extensive at primary level than in Europe, and in recent years far more extensive also at the levels of higher and technical education. The choice of skill-intensive appropriate technology played a significant part in Japanese postwar growth.

AN INFORMATIONAL DICHOTOMY

The concentration on aggregative economics—reflecting the influence of the important events of the 1930s—has begun to lessen in recent years, partly as a result of the above-mentioned developments and of related ones. The emphasis on aggregate effective demand and on inter-sectoral interdependences had an important effect on the development of both economic theory and practice, but the simplification of technological complexities occurring in such aggregative analysis—macro as well as sectoral—also had certain drawbacks. For the developing countries in particular, a shift in emphasis to the study of technological "details" is long overdue. For example, a study of Indian planning experience reveals a repeated lack of precise technological information in oversimplified sectoral planning models, contributing to the remarkable gap between expectations and achievements.

A persistent problem has been the fundamental dichotomy as regards information: while the central body knows a great deal about planning objectives and over-all balances, it knows rather little about technological details, and while the operating units are familiar with the technological details, they do not know a great deal about over-all balances. [1]

The solution that has sometimes been sought in economic theory is recourse to iterative planning, the idea being that, through a sequence of revisions of tentative decisions, the same optimal plan could emerge as would have done if there had been no such informational dichotomy. For various reasons, however, such iterative models have not been very much used in practice, and the problem of the informational dichotomy has remained largely unresolved.

Faced with this problem, planning strategies have evolved typically on one side or the other of the informational dividing line. Those focusing on aggregative or sectoral planning models have concentrated on over-all balances while making do with rather crude technological information. Those focusing on, say, project evaluation and cost-benefit analysis have concentrated on the technological "details"—broadly defined—and tended to gloss over the interdependence—or general equilibrium—aspects of planning. That each type of strategy is essentially incomplete is clear enough, but without using a fully fledged iterative model capable of encompassing both aspects, it has been a matter of judgement which part of the information base to concentrate on, and which to leave to crude guesswork. It is fair to say that there has been some switch in recent years from the sectoral balance focus to the technological details focus, partly connected with the real world developments already discussed.

---

[1] See for example E. Malinvaud: "Decentralised procedures for planning", in E. Malinvaud and M. O. L. Bacharach (eds.): *Activity analysis in the theory of growth and planning* (London, Macmillan; New York, St. Martin's Press, 1967), and G. Heal: *The theory of economic planning* (Amsterdam, North-Holland Publishing Co., 1973).

A DIFFERENT PARTITIONING OF INFORMATION?

The set of studies presented in this book make a contribution to the technological "details" side of the dichotomy, and are perhaps best viewed as an attempt to extend precise technological knowledge that is crucial for practical development planning. However, an important question has also been implicitly raised about the appropriateness of the particular form of the informational dichotomy assumed to exist in the literature of iterative planning. Detailed technological information in terms of local labour conditions, the resource situation, transport facilities, etc., may well be more easily accessible to the man on the spot, but does he really know very much about the potentially relevant techniques used in other economies but not yet locally? Certainly, if he learns more about the experience of other countries, he may well be in a better position than the man at the centre to judge the local technological possibilities in the light of local conditions. But there is a need to feed information on technological possibilities also to decentralised units. Technical knowledge has to come from several quarters. The studies here presented thus attempt to tackle the important question of the partitioning of information on technology, over and above the informational dichotomy assumed to exist in the standard literature on iterative planning.

It need hardly be emphasised that only a small part of this important question has been dealt with in this book, but that is not surprising given the enormity of the problem.

Amartya Sen

CONTENTS

EDITOR'S INTRODUCTION[1]

The issue of technological choice is a hardy perennial. Yet, in spite of much theoretical debate and some empirical work, all that could hitherto be said with some certainty as a basis for the establishment of general policy in the matter was that the technical choice was wide in agriculture and construction, but very limited in manufacturing. The reason for remaining at that level of generalisation has been the lack of representative information on "intermediate" or "appropriate" techniques actually in use, collected on the basis of the study of practical decisions made by entrepreneurs at the plant or firm level. The present volume, which is the product of the technology and employment project of the ILO World Employment Programme[2], is designed to contribute to filling this gap.

The World Employment Programme is the main ILO contribution to the International Development Strategy for the Second United Nations Development Decade. The major aim of this Programme can be summed up quite briefly, namely to identify particular policies and measures to improve the employment situation in the developing countries, and to assist in the implementation of such policies and measures. Four principal means of action have been developed under the Programme—comprehensive and exploratory country missions, regional employment teams for Africa, Asia and Latin America and the Caribbean, country employment teams, and an action-oriented research programme.

---

[1] A. S. Bhalla, the editor of this volume, is project manager of the technology and employment project under the ILO World Employment Programme, and has had many years of experience as a development economist. He has written extensively on development planning, technological choice, and employment and manpower planning, and has contributed to various professional journals.

[2] For details of this and related research activities, see ILO: *World Employment Programme: A progress report on its research-oriented activities* (Geneva, 1973).

In the research carried out under the technology and employment project as well as others, particular stress is laid on country-based case studies, with the co-operation of local research institutions, government agencies and individuals, the aim being to put forward practical suggestions for policy formulation based on a better knowledge of actual conditions. The studies in this volume that relate to jute processing and can making in Kenya and the capacity of the engineering industry in Colombia provide concrete examples of the local co-operation mentioned above. These studies were undertaken as a result of the ILO-organised comprehensive employment strategy missions to the two countries.

The employment missions to Colombia, Sri Lanka, Kenya, Iran and the Philippines [1] all attached considerable importance to the possibility that more labour-intensive technologies could be used or designed for agriculture, industry, construction and services. The Kenya mission went further than others in emphasising the need for an appropriate, over-all technology policy covering choice of products, choice of techniques, labour-intensive technical change, and research and development.

In fairness to those missions it must be acknowledged that the limited time available to them while they were in the field did not allow them to carry out empirical micro-studies to collect representative information on the effects on output, employment and income distribution of alternative technologies used in each sector. However, some such studies, e.g. on can sealing and road construction in Kenya, were attempted during the field phases of the missions. Nevertheless, the ILO's major effort to obtain representative information on alternative technologies on the basis of a sufficiently wide sample is being undertaken under the technology and employment project which was established in 1972 after the United Nations Advisory Committee on the Application of Science and Technology to Development (ACAST) endorsed a proposed action programme at its 15th Session, held in Geneva in November 1971. As may be seen from some of the papers included in this volume, the engineering and economic aspects of technological choice in different economic

---

[1] See:

ILO: *Towards full employment: A programme for Colombia*, prepared by an inter-agency team organised by the International Labour Office (Geneva, 1970);

Idem: *Matching employment opportunities and expectations: A programme of action for Ceylon*, report and technical papers in two volumes (Geneva, 1971) (at the time the mission took place, Sri Lanka was known as Ceylon);

Idem: *Employment, incomes and equality: A strategy for increasing productive employment in Kenya* (Geneva, 1972);

Idem: *Employment and income policies for Iran* (Geneva, 1973);

Idem: *Sharing in development: A programme of employment, equity and growth for the Philippines* (Geneva, 1974).

sectors are being examined under the project to provide a sounder basis for policy making than exists at present. The project is supported by a generous grant from the Swedish Government.

The results obtained from the project are being linked with consultancy and other advisory forms of technical assistance to developing countries. In addition, an increasing effort is being made to disseminate the results of these various case studies as widely as feasible among policy makers. The present volume, which is a synthesis of results of the case studies in industry, is but one step in this direction.

This volume is divided into two parts, one dealing with conceptual issues and questions of measurement and the other containing empirical case studies; while the latter account for the bulk of the book's contents, it was also felt necessary to raise some conceptual and measurement issues which, although previously discussed in publications on the subject, still remain largely intractable, yet of practical importance. Moreover, even though the real world situations covered in the case studies shed more light on the subject than anything that can be said on the current conceptual situation, some underlying conceptual basis is required and the limitations of the existing state of the art need to be identified. For one thing, the concept of labour intensity is often used in a wide variety of different ways to suit different objectives. This volume therefore opens with an analysis of the limitations of this concept. It is explained that even though the concept is ill defined, it can be operationally meaningful provided its users recognise that different indicators of labour intensity are only shorthand expressions and can be expected to give only a partial picture.

Much has also been written on the empirical production functions: more and more work on the subject is coming to light. However, the estimates of the elasticity of substitution are too aggregative to be meaningful for policy making in a large variety of production conditions in developing countries. This is shown in Chapter 2, in which it is argued that if the concept of elasticity is to be useful, it would be desirable to have additional disaggregated information obtained through sample surveys of firms and plants. Until a few years ago, a survey of the question of the elasticity of substitution would have been regarded as tantamount to a survey of purely academic work published on the subject, containing very little of interest to a planner wishing to choose a technique. The studies in this volume, however, demonstrate that in practice the elasticities of substitution between capital and labour are positive, so that factor pricing policy is quite relevant to the problem of making appropriate technical choices. On the other hand, they also demonstrate that there are important considerations, beyond factor prices, which need serious attention if action on factor prices and technical choices is to have a sizable effect. These non-price considerations are particularly brought out in the studies on can production, jute and sugar processing, textiles, and the metalworking and engineering industries.

Chapter 3, on indirect employment effects, shows the relevance of this concept for industrial planning, the choice of industries and technical choice both within and outside the input-output framework. In the latter case, indirect employment effects could be considered in industrial project evaluation, and for estimating the value of public works as an instrument of employment policy.

Part II contains eight empirical studies, most of which were undertaken on the basis of field surveys; most of them, especially those on jute processing, sugar processing, cement blocks and tin cans, involved extensive field work, interviews and primary data collection. They deal with a very wide variety of cases. Two studies deal with consumer goods industries, i.e. textiles and sugar processing; two others relate to the capital goods sector, namely the metal-machining and engineering industries; two deal with packaging industries, namely jute processing and can making; there is a study dealing with the production of cement blocks for construction purposes; another, on extractive industries (copper and aluminium), could be regarded as relating to intermediate goods. The case studies on jute processing, sugar processing, metalworking and cement blocks relate to particular countries, whereas those on the copper and aluminium industries and the production of tin cans are inter-country comparative studies. In addition, two chapters relate to Latin America, two to Asia and three to Africa.

The studies were undertaken at different levels of aggregation. The one on the copper and aluminium industries (Chapter 11) utilises both primary and secondary data in explaining the scope for capital-labour substitution. The two Zambian copper mines provided data the results of which are compared with those of the United States and Chile at a higher level of aggregation. Unlike most empirical studies of this kind, it has the merit of taking into account the influences of different degrees of capacity utilisation in industry on a number of variables entering into the production function analysis.

The next study in this class of international comparisons is that on the production of cans in Kenya, Tanzania and Thailand. For that study, extensive data were collected from factories in the three countries during the field phases of research work. Some very important and new features of this study include the attention devoted to the influence of the organisation and growth of markets on technical choice, the role of product quality and specifications, supervision and skill requirements, the risk perceptions of private decision makers and the need for adequate information about the existing technical alternatives, and the relationship of labour intensity to management attitudes. The fact that in Kenya and Tanzania the manufacture of cans is carried out by a subsidiary of a large international company, whereas in Thailand there are both large and small companies serving a rather segmented market, brings out interesting differences in the behaviour of local and foreign firms respectively in making decisions regarding machinery choice.

Chapter 6, on the textile industry, is different from the others in the sense that it does not deal specifically with any developing country; instead, it draws heavily on a study on the United Kingdom textile industry which considered technological alternatives, in products and processes, that exist in current practice in that country. However, this study is of considerable relevance in analysing the technical choice facing the less developed countries in regard to the textile industry. This is so because most of the techniques considered by a developing country have generally evolved in a developed country. In view of the early establishment of the Lancashire textile industry, the United Kingdom is a supplier of textile machinery, both new and used, to many developing countries.

The study on textiles has something in common with the one on jute processing in Kenya contained in Chapter 5. Both deal, by and large, with the question of the economic feasibility of using second-hand machinery in industry in developing countries. However, different approaches are followed by these two studies. The Kenyan study is based on primary data collected from a single factory that used both new and second-hand machines to produce maize bags; the study on textiles, on the other hand, utilises published data on new and second-hand machines compiled by the United Kingdom Textile Council. The differences between the two go further. The Kenyan study covers the organisation of the second-hand machinery market in the United Kingdom in addition to the technical and economic performance of the jute processing industry in Kenya. The study on the textile industry, on the other hand, examines the alternative hypothetical values of factor prices, especially wage rates, to determine the break-even wage at which machines bought new and second-hand would give the same total unit costs of production.

The study on sugar processing in Chapter 7 demonstrates, on the basis of field work in India, that the choice of techniques is complicated by product differences, government intervention in the market structure and the social and economic objectives of the particular society. It is one of the few studies in the whole set which explicitly attempts to apply social cost-benefit analysis for economic evaluation of alternative technologies. A simplified version of the guidelines for project evaluation [1] of the United Nations Industrial Development Organization (UNIDO) is used, and it is shown that the simpler technology, which is capital-saving and employment-generating by comparison

---

[1] UNIDO: *Guidelines for project evaluation*, Project Formulation and Evaluation Series, No. 2 (New York, United Nations, 1972).

with the modern sugar mills, is at least competitive with them, and clearly preferable if planners prefer an increase in rural income to an equal increase in urban income. Where the eradication of poverty is an overwhelming national objective, as in India, this may well be so.

Chapter 8, on cement blocks in Kenya, is similar to the previous chapter in one major respect; it carries the issue of product choice and its link with that of technical choice much further, and is perhaps the first empirical attempt in the published work on technological choice to come to grips with this issue on the basis of hard facts instead of purely intuitive reasoning. Three other significant features of this study include the attention devoted to the relationship between skill intensity and physical capital intensity, the process instead of the product approach to the technology issue, and the emphasis on urban-rural differences in technical choice, largely owing to differences in scale and product requirements. The latter two aspects receive particularly extensive treatment in that chapter. It also deals with the manufacture of different types of materials for the building industry, and thus constitutes important links between manufacturing proper on the one hand and housing and construction on the other.

Two chapters (9 and 10), dealing with Colombia and Mexico respectively, relate to the capital goods sector. The study of the engineering industry in Colombia attempts to examine the extent to which indigenous Colombian capacity exists for supplying machines and equipment to agriculture, manufacturing and the construction sector. While such capacity does exist, there is very little interaction between producers and users of machinery, with the result that the equipment being produced in Colombia does not necessarily answer the needs of Colombian users. This deficiency is further reinforced by the fact that in most cases Colombian engineering industry copies imported designs, often with little modification and adaptation.

The study on technological choice in metalworking (Chapter 10) makes a strong plea for process or task-level analysis. It is argued that only a high level of disaggregation such as that can satisfactorily reveal possibilities of substitution. Moreover, private enterprises make decisions largely at that level of disaggregation. Unlike other empirical studies, this micro-analysis relies very largely on basic engineering data on metal-machining tasks. Little is said, however, about the ways in which microdata could be aggregated for purposes of making possible generalisations.

The final chapter in this volume attempts to draw together the main conclusions of the various case studies and brings out the key issues and the implications for policy making. Some of the conclusions may be worth mentioning here. First, the studies demonstrate quite clearly that substitution possibilities exist in industry in both core and ancillary operations. This conclusion, based on empirical evidence, is important, since it has often been assumed that there is no choice of techniques in manufacturing industry. Secondly, the range of available techniques can be widened by re-designing or copying older designs and blueprints with local engineering adaptations, or through local manufacture of equipment. This comes out quite clearly in the study on can production. Thirdly, quite often the use of capital-intensive techniques, where more labour-intensive ones could have been used equally efficiently, is due not to the fact that there are no other technical possibilities in industry—there are—but to imperfect knowledge and inappropriate selection systems.

Selection processes are inappropriate because private decision makers do not know of the existence of other technical possibilities in the form of different types of equipment. Even subsidiaries of international firms, with superior organisation and management, often lack adequate information. One of the reasons for this may be that the technological decisions for these firms are made by staff at the head office of the parent company, who are more conversant with the technology used in already advanced countries than with that suited to the conditions of developing countries.

What lessons can we draw from all this? One solution is an international technological information system on alternative technologies, as proposed by the Advisory Committee on the Application of Science and Technology to Development and more recently by the Committee on Science and Technology for Development. In fact, the Economic and Social Council of the United Nations at its 57th Session, held in Geneva in the summer of 1974, adopted a resolution on the "role of an international technological information system in the transfer and assessment of technology and in the indigenous growth of appropriate technologies in developing countries". This resolution calls upon the United Nations system to undertake a "feasibility study on the progressive establishment of an international information exchange system for the transfer and assessment of technology: such a system should serve the actual needs of potential users of the information...".

The conclusions of several studies in this volume, in particular of the one on can production in East Africa and Thailand, fully support the spirit of the above-mentioned resolution—that is, the developing countries must have knowledge of the various technologies available, together with analyses of economic costs and engineering specifications. However, the studies in this volume also caution against placing too much emphasis on international information centres. It is important that the detailed information that is assembled is actually brought to the attention of entrepreneurs and decision makers in the developing countries as well as in the developed ones. To achieve this objective, equal emphasis needs to be placed on the gathering at the national level of information sufficiently relevant and reliable to reduce the entrepreneur's risks.

Finally, attempts to correct factor price distortions may be less successful unless information about existing technological alternatives is more widely disseminated than has hitherto been the case. Therefore appropriate incentives constitute only one element of action to ensure appropriate technological decision making.

PART I

ISSUES OF CONCEPT AND MEASUREMENT

THE CONCEPT AND MEASUREMENT OF LABOUR INTENSITY

1

by A. S. Bhalla [1]

   The problem of how to measure the intensity of labour has plagued theorists from the time of Adam Smith and the labour theory of value. Wages are a notoriously poor measure of labour intensity and many other factors have to be allowed for before there would be even a positive relation between hard work and high wages. . . . In any event, the question "has an increase in output come about because of a rise in efficiency or because of a rise in the intensity of labour?" deserves serious attention in view of the Marxist tendency to attribute an increase in product under capitalism to an increase in labour intensity. . . . [2]

   Intensity of labour per unit of time implies working hard or not for any given length of time. Another more popularly understood connotation of the degree of labour intensity is the quantity of labour or labour time per unit of output or per unit of investment. We concern ourselves with this latter interpretation, rather than with the intensity of effort.

   In spite of the long controversy over technological choice, which has now entered into its third decade, the concept of labour intensity still remains ambiguous and ill defined. In publications on economic development, labour intensity is often inferred indirectly from the degree of capital intensity measured in terms of capital-labour (K/L) or capital-output (K/V) ratios. This indirect approach to the measurement of labour intensity implies that production functions are of such form that an increase in K/L and K/V satisfactorily reflects a decline in the degree of labour intensity. [3] A few other measures such as the labour-output ratio and the share of wages in value added have also been considered. We shall critically examine these in this chapter.

---

   [1] International Labour Office. The author is grateful to C. G. Baron and J. Gaude for comments and to H. Pack and Amartya Sen for many valuable suggestions for improvement on an earlier draft.

   [2] Kenneth E. Boulding: "Some difficulties in the concept of economic input", in National Bureau of Economic Research, New York: *Output, input and productivity measurement*, Studies in Income and Wealth, Vol. 25 (Princeton University Press; London, Oxford University Press, 1961).

   [3] See Chapter 2 below.

A practical planner clearly needs an operational and useful shorthand indicator of labour intensity to choose particular industrial projects, e.g. to rank the labour intensity of various textile or maize-milling projects and techniques. For lack of adequate project-level data, economic analysts are often forced to use industrial aggregates instead of the detailed information on plants and establishments. However, the real conditions are reflected much more by microdata than by aggregates, which conceal useful information about differences in the labour intensity of projects and techniques. Therefore a distinction between different levels of aggregation needs to be borne in mind.

The importance of project-level analysis and data is well brought out in the following citation:

Perhaps the type of reference data most seriously needed in today's industrial planning and programming machinery is that which helps planners to handle their economic calculations at a level not too many steps removed from individual industrial plants and at the same time on a basis that is sufficiently comprehensive to retain the standpoint of over-all national and regional economy. [1]

In the case studies in Part II of this volume, wherever possible an attempt has been made to provide plant-level primary data collected by the authors on the basis of factory visits and sample surveys. This effort is intended to supplement the efforts of UNIDO and the Organisation for Economic Co-operation and Development (OECD) to provide hard facts on industrial projects and establishments. [2]

In this chapter, we argue that the existing studies on labour or capital intensity often test macro-economic arguments on the basis of aggregated data for industries and not on that of plant-level data. We shall attempt to compare alternative indices of labour intensity using both plant data and industry data in order to examine whether there is any major difference in the results. For purposes of illustration, ranking by different indices of labour intensity will be carried out at four levels of aggregation, namely country, sector, industry and establishments.

LABOUR-INTENSIVE versus CAPITAL-INTENSIVE DICHOTOMY [3]

A dichotomy between capital-intensive and labour-intensive investments, industries or techniques (so familiar in writings on choice of technology) is

---

[1] UNIDO: *Profiles of manufacturing establishments*, 3 vols. (New York, United Nations, 1967-71), Vol. I, p. 2.

[2] See UNIDO: *Profiles of manufacturing establishments*, op. cit., and OECD: *Manual of industrial project analysis in developing countries* (Paris, 1968), Vol. I, Annex I, on "Industrial profiles".

[3] For an analysis of situations in which a combination of capital-intensive and labour-intensive techniques are used instead of one or the other see Amartya Sen: *Choice of tech-*

both confusing and inappropriate. Under conditions of factor price flexibility and smooth neo-classical production functions, the technical choice is not merely dualistic but pluralistic. The factor proportions are variable so that any amount of capital can be associated with any amount of labour, output increasing with an increase in the input of either factor, although a limit to the absorption of labour may be set by the level of wages relative to capital or by the subsistence wage.

What is more realistic is an optimal degree of total factor intensity (a combination of different inputs, including capital and labour) determined by relative factor prices and the state of technological knowledge. Private investors, who are invariably the majority element in decision making in market economies, are motivated by cost minimisation. From a private point of view, it is only an optimisation of output per unit of all inputs that would lead to cost minimisation. Thus what enters into the cost calculations of entrepreneurs is the total factor productivity, not so much the capital intensity or labour intensity of productive operations, nor the maximisation of labour productivity. Neither the capital-intensive nor the labour-intensive methods need be optimal from the point of view of cost minimisation.

If we assume that market prices are poor guides to resource allocation and that in labour surplus economies the social cost of labour is nil [1], the correct degree of labour intensity would be high, at least in absolute terms. In theory, it may be higher still if we assume a negative shadow price of labour on the grounds that unemployment is a social evil and that any extra employment is a benefit in itself [2] (irrespective of the additional cost of employment in terms of additional consumption which may reduce savings and potential investment). Therefore, strictly speaking, any positive labour cost will give us techniques which by reference to the extreme degree of labour intensity as defined above, would all be capital-intensive. Even apart from this conceptual weakness, another question of policy significance is: what should determine the acceptable level of labour intensity against the maximum limit? If the objective is not simply to maximise employment as such but to increase *productive* employment up to a point at which additional employment generates

---

niques: *An aspect of the theory of planned economic development* (Oxford, Blackwell, 3rd ed., 1968), Ch. IV; R. Krishna: "A model of the unemployment trap, with policy implications", in ILO: *Fiscal measures for employment promotion in developing countries* (Geneva, 1972).

[1] As in fact has been assumed in the debate on investment criteria for developing countries. See A. E. Kahn: "Investment criteria in development programs", in *Quarterly Journal of Economics* (Cambridge, Mass.), Vol. LXV, No. 1, Feb. 1951; W. A. Lewis: *The theory of economic growth* (London, Allen and Unwin, 1955), and Jan Tinbergen: *The design of development* (London, Oxford University Press, 1958).

[2] Cf. A. K. Sen: "General criteria of industrial project evaluation", in UNIDO: *Evaluation of industrial projects*, Project Formulation and Evaluation Series, Vol. I (New York, United Nations, 1968).

some extra output, then the optimum degree of labour intensity can be chosen only on the basis of some criterion of efficiency and cost minimisation. An efficient technique will be measured by a ratio of output to input, so that the higher the ratio the more efficient is the process or technique of production.

If $X = f(x, y, z)$ where X is output and x, y, z are inputs, there may be a set of technologically efficient alternatives to produce X, viz.—

(a) with more of x, and less of both y and z; or

(b) more of y and less of both x and z; or

(c) more of z and less of x and y.

If a process of production called for more of x without reducing inputs of y and z, it would be technically inefficient. While the efficiency criterion eliminates the inefficient techniques, it does not help in choosing among the other equally efficient possibilities. [1] Therefore the concept of technical efficiency needs to be supplemented by other criteria, such as optimality, to choose the best from a set of efficient techniques.

Given the set of physically efficient points, if one knows the shadow prices of all inputs a unique economically efficient method can be chosen. If this turns out not to be the most labour-intensive one but if the most labour-intensive one is chosen anyway on the basis of some other criterion, output will be given up somewhere else in the economy. In the light of our objective of obtaining maximum productive employment, for example, one might argue that a technically efficient possibility with the highest labour intensity (per unit of output or per unit of investment) would qualify for selection. Nevertheless if the choice of the most labour-intensive method means output forgone it might make sense to choose instead the economically efficient technique and simply "waste" labour, that is, use more than is technically necessary, dividing a given amount of work among all labourers.

A gap between the maximum and the optimum degree of labour intensity, even under the assumption of a zero shadow price of labour, arises from the fact that a positive social cost of labour can be compatible with its zero shadow price. Additional employment even under conditions of excess labour supply implies additional purchasing power, leading to more consumption at the expense of investment. A reduction in investment and thereby in the future rate of growth represents a cost to society of employing extra labour. [2] In fact

---

[1] For a lucid analysis of efficiency and optimality in choice of techniques see A. K. Sen: "Choice of technology: A critical survey of a class of debates" in UNIDO: *Planning for advanced skills and technologies*, Industrial Planning and Programming Series, No. 3 (New York, United Nations, 1969).

[2] The ILO comprehensive employment mission to Sri Lanka stated that rough estimates suggested that employment of otherwise unemployed unskilled labour meant an expenditure

much of the debate on the choice of technology centres on this sub-optimality of savings and investment and the effect of the selected factor proportions on these variables. [1]

The foregoing implies that a distinction between a positive approach and a normative one is relevant. If one adhered to the former, only such indicators of labour intensity as pertain to efficiency (e.g. the labour coefficient or the capital coefficient) would qualify in determining choice of a particular technique. On the other hand, if a normative approach is adopted, employment maximisation may well be an objective in itself, in which case the relevant index or area of information may simply be the capital-labour ratio which links one input to another.

However, too much should not be made of the concept of productive efficiency as a basis for technological choice. Some techniques which are inefficient may still be regarded as socially desirable, especially if one introduces such notions as people's preferences in relation to different modes of production and to the systems of organisation under which different techniques operate. A technique which qualifies as inefficient under a normal economic test may be used in family-based production because it is preferred by those applying it. [2] Moreover, given the alleged discontent with work that characterises industrial processes, those which are more pleasant or interesting might be chosen if the efficiency calculus is supplemented by a "humanitarian" one. Of course, such a choice, if generalised, would imply a lower material standard of living for the population in question. [3]

WHY DEFINE LABOUR INTENSITY?

If the concept of labour intensity is ill defined and ambiguous, it is logical to ask: why define labour intensity at all? The large body of economists

---

of some 60 to 70 per cent of their extra income on food, which at the margin was imported or diverted from exports of coconuts (see ILO: *Matching employment opportunities and expectations*, op. cit., Report, p. 82).

[1] See A. K. Sen: "Some notes on the choice of capital-intensity in development planning", in *Quarterly Journal of Economics*, Vol. LXXI, No. 4, Nov. 1957; A. S. Bhalla: "Galenson-Leibenstein criterion of growth reconsidered", in *Economia Internazionale* (Genoa), Vol. XVII, No. 2, May 1964 and Vol. XIX, No. 4, Nov. 1966; and Koji Taira: "A note on the analytical properties of Galenson-Leibenstein investment criterion", in *Bulletin of the Oxford University Institute of Economics and Statistics* (Oxford, Blackwell), Vol. 27, No. 2, May 1965.

[2] This point is brought out quite well in Amartya Sen (on behalf of the ILO): *Employment, technology and development* (Oxford, Clarendon Press, 1975), and in the same author's *The concept of efficiency*, a lecture delivered to the Annual Conference of the Association of University Teachers of England, Coventry, 26 March 1973.

[3] This point is due to H. Pack.

writing on the subject have used indicators of labour intensity or capital intensity in order to study—

(a) the variations in factor proportions and functional relationships between partial factor inputs and outputs;

(b) the allocation of investment and its effects on the choice of production techniques, and vice versa; and

(c) consistency or conflict between output and employment objectives.

The growing unemployment in the less developed countries has brought the need for an operationally meaningful indicator or indicators to the forefront. It is necessary to examine the employment potential of production and invest-ment with the existing techniques and the effect of varying techniques and the product mix in sectors and individual industries, and especially projects on the absorption of the surplus factor, i.e. labour. There is another operation which calls for some index of labour intensity, namely tracing the effect of international trade specialisation on labour absorption in the developing countries through the export of products that are labour-intensive and use techniques which employ a large amount of labour per unit of investment or output. [1] It is assumed that the less developed countries have a comparative cost advantage in the export of these products.

To summarise, the reasons for our interest in labour intensity for policy making and for related economic analysis revolve around the following central issues, which obviously overlap to some extent:

(a) the determination, through time series as well as cross-section analysis at both the macro-economic and the micro-economic levels, of the degree of factor substitution (between capital and labour) and the elasticity of employ-ment with respect to output in manufacturing [2];

(b) the classification of industries and projects, to the extent that this is feasible, according to the degree of labour intensity; planners need to know which particular projects and industries rank high in the labour intensity scale in order to determine priorities in investment allocation that will reflect more accurately the objectives of raising employment (direct and indirect [3]) as well as output;

---

[1] Other points to be borne in mind in connection with changing the composition of output in favour of labour-intensive goods and services are that *(a)* income can be redistributed through taxation schemes and subsidies, proper credit facilities, import-substituting and export incentives measures, and *(b)* the demand for labour-intensive goods in export markets is income-elastic.

[2] See Chapter 2 below.

[3] See Chapter 3 below.

(c) the investigation into whether labour intensity, in whatever way it may be defined or measured, remains invariant to scale of production or not. (A test of this hypothesis is of great policy significance: if returns to labour-intensive techniques were higher at lower scales and volume of production than at higher ones, then a shift in policy from the establishment of a few large enterprises towards that of a large number of small enterprises would, other things being equal, lead to greater labour absorption.[1])

Some conceptual issues

Before reviewing the various indicators used to link labour input and output, or labour input and capital, some of the conceptual aspects need to be examined in relation to the published work on the choice of techniques, investment allocation and factor proportions.

First, the use of the concept of labour intensity has been concerned with the sub-optimality of savings in the economy, investment allocation and technological choice in a general equilibrium framework more than with the partial equilibrium formulation in which the choice of technique or a degree of labour intensity pertains to a particular industry or firm. In the latter case, unlike the former, no difference to the over-all inoptimality of savings may be made since the firm or industry in question accounts for a very small part of the economy as a whole. The application of the macro-economic theoretical arguments about over-all rates of growth, investment allocation and optimality of savings etc. to industry data are quite numerous. [2] However, it is not clear that conceptually it is appropriate to test macro-economic theories with micro-economic data or micro-economic theories with macro-economic data. Logically, it would seem fallacious to carry over arguments worked out at a macro-economic level to the evaluation of an industry at the micro-economic level.

Similarly, crucial propositions, e.g. that the degree of capital intensity is positively correlated with the rate of growth of output, are often not specified in terms of the degree of aggregation to which they relate. In other words, does this proposition apply to a situation of an industry, a sector or the economy as a whole? If it applies to an industry, how does one select and define it?

---

[1] See Chapters 8 and 10 below.

[2] G. Ranis: *Industrial efficiency and economic growth: A case study of Karachi*, Monograph No. 5 (Karachi, Institute of Development Economics, 1961); A. S. Bhalla: "Investment allocation and technological choice: A case of cotton spinning techniques", in *Economic Journal* (London, Macmillan), Vol. LXXIV, No. 295, Sep. 1964; idem: "Choosing techniques: Hand pounding versus machine-milling of rice: An Indian case" in *Oxford Economic Papers* (Oxford, Clarendon Press), Vol. 17, No. 1, Mar. 1965; J. Gouverneur: *Productivity and factor proportions in less developed countries: The case of industrial firms in the Congo* (Oxford, Clarendon Press, 1971); and J. C. Sandasera: *Size and capital-intensity in Indian industry* (University of Bombay, 1969).

Are inter-industry or inter-sectoral data for a single economy appropriate for an empirical test, or is the appropriate methodology the one which makes use of macro-economic data for a number of countries? Such methodological questions have not received as much attention as they deserve.

Thirdly, most of the studies consider only the final stage of production and the techniques used at that stage. Intermediate stages and different processes of production at each stage tend to be ignored. Labour-intensive final products would require that intermediate inputs to the manufacture of those products should also be more labour-intensive than capital-intensive final products. This may or may not be so in practice. In principle, both labour-intensive and capital-intensive final products could be produced with either labour-intensive or capital-intensive intermediate inputs: in practice, however, some industries such as fruit canning, which is relatively labour-intensive (in terms of labour input per unit of output), use refined sugar, which is a capital-intensive input. On the other hand there may be industries of a luxury type (e.g. cigars and cigarettes), with a backward linkage to agriculture, which might use labour-intensive inputs. Thus an aggregate index of labour intensity (which is often taken into consideration in evaluating an industry) is the weighted average of the labour intensity at each stage of production. In practice, if one stage of production has higher technical requirements for labour input per unit of output or of capital, the aggregate labour intensity would be higher than otherwise.

Fourthly, related to the above issues is the fact that the majority of empirical studies [1] on labour intensity or capital intensity consider only the direct effects on employment or reinvestible surpluses of the adoption of different kinds of technology. However, as Chapters 3 and 6 below show, the indirect employment effects of investment in certain industries and techniques can often be extremely important, sometimes even more so than the direct employment effects. If account is taken of these also, it is probable that the ranking of alternative techniques might be reversed. This would be equally true if computations of employment (direct and indirect) were made not only for the final stage of production but also all the way backwards, consideration being given also to labour inputs in the intermediate input supplying stages. An interesting question is whether industries or projects that have greater indirect than direct effects should be promoted. In some cases some capital-intensive industries such as fertilisers may curiously be supported even on employment grounds, since agriculture can be more labour absorbing through the provision of fertiliser inputs. However, the empirical studies cited in Chapter 3 below

---

[1] Chapter 6 in this volume on "The choice of technique and employment in the textile industry", is an exception, however.

demonstrate that the industries which have great indirect employment effects are not necessarily the capital-intensive ones.

Finally, most of the works dealing with the factor proportions approach the labour intensity issue purely from the supply side, that is, taking the demand for a particular product as given and considering whether it can be produced either with more labour and less capital or with more capital and less labour. Interest in the related demand-oriented aspect of technology, namely change in the product or industry mix through changes in the consumption pattern, and the effects of this product substitution on technical choice, is only of a very recent origin. [1] Much of the discussion on technological choice in the past excluded the question of product substitution by assuming identical physical attributes and quality. In practice, however, consumer needs are met by a range of different products among which particular choices are made depending on consumer purchasing power. Thus an intra-industry demand shift, say from drip-dry nylon shirts to more labour-intensive cotton shirts, widens the range of technical choice and can lead to an over-all increase in the labour intensity of a given industry. [2]

INDICATORS OF LABOUR INTENSITY

In general, different indices can be used to measure labour intensity, be it in agriculture, manufacturing or construction. A distinction is therefore needed before we consider the degree of labour intensity by industrial projects and aggregates. Apart from the usual difficulty of finding capital data, the ratio of capital to labour (K/L) which is used as a measure of capital intensity (and its inverse, that of labour intensity), leaves out of account efficiency considerations which are better examined through input-output coefficients, e.g. labour-output ratios, capital-output ratios, or the share of wages in value added. For example, at any level of aggregation a production function does not describe what output will be achieved for given levels of different inputs: rather a production function describes the maximum level of output that can be achieved if the inputs are efficiently employed. [3]

---

[1] See F. Stewart: "Technology and employment in LDCs", in *World Development* (Oxford), Vol. 2, No. 3, Mar. 1974. See also Chapter 8 below for an interaction between product choice and technical choice in a specific context.

[2] Cotton shirts are considered more appropriate and labour intensive in two ways. First, cotton shirts are more labour-using in consumption (they require ironing); secondly, they are made of natural instead of synthetic fibres, which tend to be more capital-intensive. However, the use of labour in consumption is good only if no value is attached to leisure.

[3] See Franklin M. Fisher: "The existence of aggregate production functions", in *Econometrica* (New Haven, Conn.), Vol. 37, No. 4, Oct. 1969.

The various indices of labour intensity that are often used are not mutually exclusive. The labour coefficient and capital coefficient represent efficiency in input use, which with technical progress implies a reduction in the size of the coefficients. We shall examine the different indicators one by one below.

The labour coefficient (L/V)

The labour coefficient defines a functional relationship between the input of labour and the output of a commodity. Some ambiguity in the concept of labour intensity also arises from a confusion between the elasticity of employment with respect to capital and the elasticity with respect to output. The two indices need not give identical results. An increase in the rate of capital utilisation (increase in the number of shifts) raises the labour absorption from a given amount of capital (that is, there is a decline in capital-labour ratio), but this will not necessarily raise the input of labour per unit of output. This would occur under decreasing returns to scale when the marginal returns from labour are lower on the second and third shifts than the average returns.

Given the following technical relationship,

$$L/V = f(L/K) \qquad \qquad \ldots (1)$$

a rise in the cost of labour in relation to capital is likely to raise the capital-labour ratio (K/L), lower the labour coefficient (L/V) or raise its inverse, the average output per worker (V/L). The labour intensity of different techniques can be illustrated in terms of V/L, as in figure 1. The vertical axis measures output, and the horizontal axis the amount of labour employed. The slopes of the lines OA, OB and OC measure output per worker. The degree of labour intensity is the highest in technique II since it absorbs the largest amount of labour, OL_1. Technique III is less labour intensive since it employs a lower amount of labour and generates a smaller output. Technique I, with the highest output and the highest productivity of labour, is the superior technique in the sense that it requires a lower amount of one factor per unit of output, with the other factors remaining constant.

Hal B. Lary [1] has used value added per employee (V/L), that is the inverse of the labour coefficient, as an index of labour intensity. A number of merits are claimed for this indicator, viz.—

(a) it reflects the flow of capital services rather than a stock, and is therefore more relevant to the theory of production functions;

---

[1] Hal B. Lary: *Imports of manufactures from less developed countries* (New York, National Bureau of Economic Research, 1968).

Figure 1. Labour intensity of different techniques.

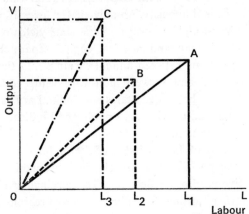

—.—.—. = superior technique (I).
————— = labour-intensive or less capital-intensive technique (II).
– – – – – = capital-intensive or less labour-intensive technique (III).

Note: The figure indicates purely technical relations. Technique III (OB) is therefore ineffi-cient because of the machine design, not because labour is inefficiently used or more generally because there is "X" inefficiency.

(b) it incorporates the contribution of human capital and of skill differences [1]; and

(c) it by-passes the difficulty of measuring physical capital.

The treatment of value added per employee as an index of factor intensity simply implies that labour productivity is a composite index of the contribu-tions of both capital and labour, which can be easily separated as follows:

If
$$V = wL + pK \qquad \qquad \dots (2)$$

then wage value added $= \dfrac{wL}{V} =$ index of labour intensity measuring the contribution of human capital;

and non-wage value added $= \dfrac{pK}{V} =$ index of capital intensity measuring the contribution of physical capital.

Although Lary recognises the limitations of the separation between wage and non-wage value added for measuring capital intensity, he oversimplifies by assuming that some of the inputs covered in the non-wage value added but not reflecting the share of physical capital have a systematic relation to different

---

[1] There is a bit of circularity in Lary's reasoning. Although he claims this merit of his approach, for purposes of his analysis, he assumes homogeneity of labour. To quote: "The higher the value added per employee, the more capital-intensive the industry on both accounts combined *[i.e. human as well as physical capital]*; the lower the value added per employee, the more labour-intensive the industry" (Lary, op. cit., p. 22).

21

industries, so that they "enhance the usefulness of total value added per employee as a guide to capital intensity, broadly viewed, even though the division of the total into its wage and non-wage components yields only an imperfect measure of inputs of human and physical capital separately considered". [1] On the basis of correlations between annual earnings and the ratio of skilled labour to the workforce, and between non-wage value added and physical assets across industries in the United States and India, Lary assumes that these components of value added provide reasonably good indications of the inputs of human and physical capital.

The underlying basis of Lary's index of factor intensity is the theory of competitive factor shares and the neo-classical production function. It is assumed that relative earnings reflect relative marginal products of factors, so that identical average earnings for industries across regions within a country or across countries imply the same average level of human capital per worker. [2] Apart from this simplifying assumption the following limitations severely restrict the usefulness of V/L as an index.

Imperfections in product markets also account for productivity (V/L) differentials which cannot be attributed to differences in technical requirements of factor inputs. Some firms and industries are more monopolistic than others. They may charge very high prices for their products, which results in very high value added per employee (V/L), which would largely reflect an element of monopoly rent rather than a high contribution of labour or capital. The extent of concentration may be particularly large in firms which are subject to control by multinational enterprises. Differences in the relative importance of these enterprises in different industries would also influence inter-industry differences in wage rates as well as profit margins. [3] Secondly, industries have both new plants and old plants operating side by side. The newer plants tend to have higher value added per employee (V/L) than older plants. Thus variations in the proportions of old to new plants over time caused by technological advance would also account for differences in value added per employee across industries. Although these variations should indicate technological differences, they are vitiated by the random variations in such components of value added as inventory charges, depreciation charges and profit rates. Thirdly, differences in the nature of employment, part-time or seasonal, among firms and industries and countries may further distort comparability of V/L. Variations in the scale of operation of firms would further influence the size of the labour coefficient. Assuming a positive correlation between

---

[1] Lary, op. cit., p. 35.

[2] As a corollary of this, V/L or productivity differentials across industries and countries subsume that all firms are on the same neo-classical production function.

[3] See Chapter 11 below.

scale and capital intensity, an increase in output might tend to lower the labour coefficient (L/V). This decline will be further aggravated by capital-using technical progress.

We observed earlier that a realistic analysis of capital intensity or labour intensity must also consider the requirements of indirect and intermediate inputs in the process of production. By excluding raw material inputs, value added per employee (V/L) as an index of labour intensity gives only an indication, if that, of substitution possibilities between direct inputs of capital and labour. If the objective is to consider the possibilities of substitution between labour and non-labour inputs (e.g. intermediate products and land), then estimates of gross output would be more relevant than estimates of value added. [1]

The share of wages in value added (wL/V)

Apart from Lary (who does not, however, apply this index), some other authors [2] have used wages as a percentage of value added as an index of labour intensity to rank industries. For lack of adequate data the share of wages paid to production workers alone, which would in principle be a better approximation since employees not directly engaged in the production process introduce complications, could not be considered. Indices of labour intensity based on wage shares also suffer from the simplifying textbook assumptions, viz. functioning of a perfect factor market. In practice, the influence of wage legislation and of trade unions distorts factor prices and the share of wages in value added, which has thus no relevance to the predictable technical relationship between factor inputs and output. Moreover, within a country or between countries differences in the shares of unincorporated enterprises and owner-operators also tend to vitiate inter-industry comparisons of the shares of wages in value added. In some manufacturing industries, the contribution of self-employed non-wage-earning labour to value added may be much larger than that in others.

In principle, this indicator is a true measure of labour intensity not only under the restrictive assumptions of perfect competition in factor and product markets but also under those of scale independence and of the elasticity of substitution of labour for capital being greater or smaller than unity. If the elasticity is unity (as in the case of the Cobb-Douglas production function),

---

[1] A recent study of Philippine manufacturing concludes that "long-run substitution possibilities between primary and intermediate inputs are far greater than is normally assumed". See Jeffrey G. Williamson: "Relative price changes, adjustment dynamics, and productivity growth: The case of Philippine manufacturing", in *Economic Development and Cultural Change* (University of Chicago Press), Vol. 19, No. 4, July 1971.

[2] For example Carlos F. Diaz Alejandro: "Industrialization and labor productivity differentials", in *Review of Economics and Statistics* (Cambridge, Mass., Harvard University Press), Vol. XLVII, No. 2, May 1965.

the relative shares of wages and profits in value added will always remain the same. If it is less than unity, then as the K/L ratio increases the share of wages rises. If it is greater than unity, as capital intensity (K/L) rises the wage share decreases. Thus a process that permits substitutability and could thus potentially be run labour-intensively may be observed to have low wage share if the elasticity is greater than unity and it is in fact operated in a capital-intensive manner.

The wage share clearly tells us nothing about factor intensity when elasticity is unity. Even if the elasticity is not unity, it is a poor indicator so long as the capital-labour ratio and the elasticity are not known beforehand. One can argue that if there is a need for prior knowledge about K/L ratio there is no need to bother at all with the wage share as an indicator of labour intensity. If two or more techniques are in use, we can infer from the wage share the one which is more capital intensive if the elasticity of the isoquant along which the substitution is taking place is known, but this inference cannot be made as between processes in different industries.

Presumably it is for some of these reasons that Diaz Alejandro states that "indices of labour intensity based on wage shares are clearly inferior to those based on capital-labour ratios...". However, he is rather unclear as to the precise reasons why this index is clearly inferior to others. He states that the index assumes uniform wage rates in all industries and that it may give rise to statistical biases; but these objections to the above index may apply also to other indices considered in this paper.

Furthermore, the share of wages may not be independent of scale differences. In principle, given the factor price ratios and technical knowledge, the scale effect would tend to lower the share of wages in value added. This seems to be borne out empirically in the Diaz Alejandro study on Argentina. The study which measures the size of firms by the average number of production workers per establishment shows that the share of wages is small in many industries which have a large average size (see table 1).

Finally, the use of a technique unsuited to the size of the local market in a developing country would also tend to lower the share of labour, since the existence of unused capacity raises K/V more than L/V.

The capital coefficient *(K/V)*

The definition of labour intensity or capital intensity in terms of the capital coefficient refers to the income aspect, which cannot be ignored since it is one of the elements of labour productivity.

$$V/L = \frac{V}{K} \cdot \frac{K}{L} \qquad \qquad \dots (3)$$

Table 1. Share of wages in value added and the average size of establishments in Argentina

Industry	Wages and salaries as percentages of value added	Average productivity of production workers (US = 100)	Average number of production workers per establishment	Average size of establishment (US = 100)
Cigarettes	6.7	82.9	383	14.5
Distilled liquors	11.6	67.1	35	13.3
Petroleum refining	14.5	104.5	384	137.2
Other petroleum and coal products	16.1	76.2	47	46.3
Pulp mills	16.2	63.9	194	25.2
Tyres and inner tubes	19.2	73.6	561	82.9
Matches	20.6	55.1	193	114.8
Motorcycles, bicycles and parts	22.2	66.2	25	1.2
Printing ink	26.8	54.0	26	58.4
Canned and dehydrated fruits and vegetables	27.5	47.8	26	15.5
Electrical machinery except radio and television equipment	29.0	46.0	45	8.9
Fish	29.1	79.3	27	29.6
Toys	35.4	28.0	31	13.4
Bread and other bakery products	36.4	21.8	13	12.3
Paperboard containers and boxes	36.9	25.7	21	17.2
Lime	37.2	23.8	22	18.2
Metals, excluding machinery	38.5	31.9	38	12.5
Macaroni and spaghetti	39.8	22.0	22	25.2
Elevators and moving stairways	41.6	17.2	54	51.7
Tobacco, excluding cigarettes	42.1	27.5	56	34.9
Musical instruments	42.8	24.4	25	9.1
Printing and publishing, except newspapers and magazines	43.8	27.4	29	42.9
Glass and glass products	47.6	27.2	82	2.5
Explosive substances	49.0	29.7	159	105.3

Source: Carlos F. Diaz Alejandro: "Industrialisation and labour productivity differentials", op. cit.

This simple identity shows that V/L grows faster if either V/K or K/L increases with the second factor remaining unchanged.

One of the first writers to use the capital and labour coefficient for measuring relative inputs of capital and labour in different industries was Leontief. [1] In particular, the relative factor intensity of different American industries

---

See W. Leontief: *Studies in the structure of the American economy* (London, Oxford University Press, 1953); idem: "Factor proportions and the structure of American trade: Further theoretical and empirical analysis", in *Review of Economics and Statistics*, Vol. XXXVIII, No. 4, Nov. 1956. See also V. V. Bhatt: "Capital-output ratios of certain industries: A comparative study of certain countries", in *Review of Economics and Statistics*, Vol. XXXVI, No. 3, Aug. 1954.

was measured by the ratio of the capital coefficient and the labour coefficient, that is $\left(\dfrac{K}{V}\bigg/\dfrac{L}{V}\right)$, thus giving a ratio of the value of stock of capital per worker (K/L). [1]

The capital-output ratio represents a stock-flow concept relating the value of the stock of capital to the value of the flow of output at a given time. This stock-flow concept seems to be more limited than a concept measuring the flow of capital services to the flow of output. This is so because capital goods vary in durability. Let us compare K/V ratios in two different industries: in one case capital loses its practical value after the current period, whereas in the second it continues contributing to output in subsequent periods. It may be erroneous, therefore, to attribute the output of the latter industry entirely to capital: the use of more capital per unit of output in the second case does not necessarily imply that it is more capital-intensive than the first industry. Apart from considering differences in the durability of capital, time patterns of output yields need also to be taken into account to make empirical comparisons meaningful. [2] The concept of annual investment or acquisition cost to daily output used in Chapter 8 below by Frances Stewart attempts to relate the flow of capital services to the volume of output, thus taking account of the above objection. [3]

There are several other practical limitations in the use of the conventional capital-output ratio. It is a poor indicator of the labour or capital intensity of an industry since changes in the denominator and numerator need not necessarily take place in response to technological factors. Increases in output depend on—

(a) the application of better methods to old plants without the use of any additional capital (that is, the case of technical progress);

(b) fuller utilisation of old plant in response to demand fluctuations;

(c) changes due to the introduction of multiple shifts; and

(d) changes of the kind for which a certain relationship between capital and output may be assumed as given by technical factors. [4]

---

[1] More of this latter indicator below.

[2] See E. Borukhov: "The capital-output ratio, factors-intensity and the input of capital", in *Economia Internazionale*, Vol. XIX, No. 2, May 1966.

[3] See Chapter 8 below for the use of this indicator, and others, namely investment-labour and labour-output ratios.

[4] See W. B. Reddaway: *The development of the Indian economy* (Homewood, Ill., Irwin; London, Allen and Unwin, 1962), Appendix C.

Unless the contribution of the above non-technical factors can be isolated, inter-industry comparisons of capital-output ratios cannot be a very meaningful index of capital intensity or labour intensity.

Industries vary in their production functions, scale economies and capacity utilisation, besides their levels of technology. Even within a given industry, K/V may vary between extensions of existing factories and the completely new units, being lower in the former than in the latter. Variations between countries at a given point of time do also occur, but not necessarily owing to differences in the degree of capital intensity of the methods of production. Instead, K/V may simply be an increasing function of the amount of investment, with the result that differential investment policies across countries and not the labour intensity of industries would be reflected in variations of K/V. In other words, the size of K/V may depend less on the technological factors and more on the rate of growth of investment. The definition of capital intensity in terms of the K/V ratio, on the other hand, implies that the existing capital stock, which has been accumulated over the years, and its ratio to output are independent of the above size relation. [1]

The capital-labour ratio *(K/L)*

The capital-labour ratio is one of the most commonly used indicators of labour intensity, especially in studies relating to employment implications of technological choice. This index, which reflects essentially the degree of mechanisation, is closely related to the capital coefficient. The latter reflects a degree of labour or capital intensity, depending on whether a high or a low K/L is associated with it. If K/V is large (or capital productivity is low) and K/L is also large, then the labour intensity of a given investment is also low. This situation, where an increasing K/L is associated with a rising K/V (or diminishing average productivity of capital), obtains under conditions of constant returns to scale. However, the experience of less developed countries tends to contradict this neo-classical assumption. The techniques with a higher K/L yield a lower output per unit of capital (V/K) or a higher K/V. Of course, these situations in the less developed countries can be reconciled with the neo-classical theory by assuming increasing instead of constant returns to scale. However, contrary to what the traditional production theory would suggest, returns to mere capital-intensive processes are not invariant to scale. At higher scale of operations returns to more capital-intensive operations (higher K/L) may be greater than those to labour-intensive ones (lower K/L ratios).

---

[1] This objection, however, refers to the aggregate capital coefficient. At a disaggregated level this problem may not arise.

The use of K/L and K/V for measuring capital intensity or labour intensity in isolation could lead to the identification of a given industry or technique as labour-intensive and capital-intensive at the same time. This inconsistency has been illustrated elsewhere by advocating that the triple objectives of output, employment and reinvestment from a given total investment need to be considered in conjunction rather than in isolation. [1]

Given a certain level of technology, there is a certain maximum amount of labour that can be employed on a given piece of capital equipment. Although this employment level can be varied by varying the utilisation of the capacity of equipment beyond full capacity level, an increase in labour may simply result in a negative marginal productivity of labour (over-staffing is a case of sharing a given output among a larger number of workers). Labour hoarding amounts to a subsidy during the period over which it is not technically required.

Variations in capacity utilisation across industries affect the K/L ratios, which may not be easy to identify empirically. The ratios derived from statistical data are not comparable unless it can be determined that the industries in comparison use the same number of shifts. In practice, censuses of manufactures rarely provide any information on the number of shifts or the percentage of utilised capacity.

In addition to capacity utilisation, the K/L ratio may be influenced by the nature of the bias of technical change and by relative factor prices which vary across industries. For example, while neutral technical change will leave K/L unaffected, capital-using technical change (in the Hicks sense) will raise it. Similarly, K/L will be rigid under extreme values of elasticity of substitution.

In fact, influences of these factors on the K/L ratio can be isolated, though not most accurately, within a production function framework.

Further complications in empirical comparisons of K/L ratios arise owing to the heterogeneity of both capital (K) and labour (L). If K includes investment in human capital, the relative proportions of fixed and human capital may vary across industries. Some industries may be more skill-intensive and capital-light than others. In general, capital-labour substitution implies substitution of fixed capital for unskilled labour assuming complementarity between fixed capital and skilled labour. To the extent to which capital substitutes for skills, as may occur in the case of automation, the K/L ratio may well conceal the substitution of physical capital for human capital.

Similar difficulties arise from differences in working capital requirements across industries. A lower degree of capital intensity need not necessarily mean a lower degree of mechanisation, which is in fact measured by the amount of fixed equipment employed per person. The K/L ratio will represent capital

---

[1] See A. S. Bhalla: "Investment allocation and technological choice", op. cit.

intensity only if it is a ratio of the stock of investment in fixed capital and in working capital to the flow of labour working with it. [1] In some cases, the degree of mechanisation and of capital intensity might move in opposite directions. For instance, given the extent of mechanisation, multiple-shift working will reduce the capital intensity of investment. [2]

RANKING OF PROJECTS AND INDUSTRIES

Projects and industries could be ranked according to the different indices of labour intensity. However, since none of these indices are by themselves pure ones in reflecting labour intensity, unless these were purged of non-technological factors they would be of limited usefulness for purposes of comparison across a heterogenous mass of industries. If we are able to rank firms and industries by their pure labour intensity (be it in terms of input coefficients or K/L ratios), we could have an ordering of the degree of inter-industry or intra-industry labour intensity. The key factor here is the purity of indicators, which to a large extent depends on the assumptions about techniques of production. Purity of indicators can be improved at least to some extent by the use of project or plant data. This may be particularly true in the case of industries in which only a few projects or plants constitute the bulk of the industry. Unless details of labour intensity were obtained for the enterprises, the evaluation of industry could be misleading.

In theory and practice, there are considerable difficulties in ranking projects and industries according to labour intensity. Much depends on the substi-tutability of products, differences in scale and variations in relative prices and their effects on substitution. The actual prices may not necessarily be relevant; even if they are, they may at best be relevant only for marginal changes. One can argue that the problem of pricing and valuation can be overcome by considering labour intensity in terms of physical quantities (e.g. labour per unit of physical product). Nevertheless, for purposes of comparison a common value denominator may be essential even though imperfect.

---

[1] See Amartya Sen: "Choice of techniques of production: with special reference to East Asia", in Kenneth Berrill (ed.): *Economic development with special reference to East Asia* (London, Macmillan; New York, St. Martin's Press, 1964). The ILO-organised comprehensive employment mission to the Philippines, which estimated the book value and replacement cost of capital including inventories, demonstrates the sizable importance of inventory capital (see ILO: *Sharing in development: A programme of employment, equity and growth for the Philippines*, op. cit.).

[2] Even in the absence of multiple-shift work, such economic activities as retail distribution may require more working capital in relation to fixed capital. See A. S. Bhalla: *Economic efficiency, capital-intensity and capital-labour substitution in retail trade*, Discussion Paper No. 94 (Yale University Economic Growth Center, 1970).

Table 2.   Ranking of manufacturing sector by size of employment and labour intensity

Country	Size of employment	Indicators of labour intensity			
		Labour-output ratio (L/V)	Share of wages in value added (wL/V)	Capital-labour ratio (K/L)	Capital-output ratio (K/V)
Argentina (1963)	7	1	2	.	.
Brazil (1959)	3	4	4	3	6
Mexico (1955)	6	5	5	4	4
Iran (1963)	.	6	8	7	8
Pakistan (1960)	2	8	6	5	1
Philippines (1961)	4	3	3	2	5
Singapore (1966)	5	2	7	1	3
Sri Lanka (1968)	.	7	5	6	7
Thailand (1963)	1	9	1	8	2

Note: Ranking is done in such a way that the increase in numbers indicates an increasing degree of labour intensity and an increase in size of employment.

Sources: United Nations: *The growth of world industry, 1953-65, National tables* (New York, 1967); *The growth of world industry, 1968 edition* (New York, 1970) Vol. I; International Monetary Fund: *International Financial Statistics* (Washington, DC), Vol. XIII, 1960; and ILO: *Year Book of Labour Statistics*, 1960 and 1970 (Geneva).

If the object of ranking projects and industries is to explore their employment-generating capacity, the share of employment in the total, that is absolute size, is as relevant as the ratios that reflect labour intensity.

In table 2, for purposes of illustration, the manufacturing sectors of a few developing countries (for which data are easily available) are ranked according to four indices of labour intensity considered in the previous section. Table 3 shows a ranking for two-digit manufacturing industries in Brazil for 1959 and table 4 for individual establishments in Mexico.

If we assume the same efficiency across all establishments and industries, and uniform price vectors, then rankings according to capital coefficient, capital-labour ratio, labour coefficient and the share of labour in value added should be identical. In tables 2 and 3, this does occur in some cases either at the aggregative level of manufacturing sectors or at the level of individual industries or establishments. For example, it is interesting to note in table 2 that different indices give fairly closely comparable rankings in the cases of Brazil, Mexico and Sri Lanka. This is much less so at a disaggregated level in the case of Brazilian industries and Mexican establishments. In table 3 only a few industries, namely basic metals and furnitures and fixtures, show close rank ordering by different indices. Similarly, in table 4 establishments in the vegetable oils and fats and ammonium sulfate industries show closely similar rankings.

On the basis of the available evidence, it is difficult to explain the precise reasons for cases of close ranking at the two different levels of aggregation.

Table 3. Ranking of manufacturing industries by size of employment and labour intensity in Brazil in 1959

Industry	Size of employment	Indicators of labour intensity			
		(L/V)	(wL/V)	(K/L)	(K/V)
Food	17	8.5	3	6	4
Beverages	6	6.5	7	3.5	3
Tobacco	1	3	2	10.5	18
Textiles	18	18	15	14.5	5
Clothing, footwear and made-up textiles	13	16.5	13	18	17
Wood and cork products	11	15	11	10.5	6
Furniture and fixtures	7.5	16.5	18	17	14
Paper and paper products	5	5	5	7.5	12
Printing and publishing	9	11	16	10.5	11
Leather and leather products	3	12.5	10	14.5	13
Rubber products	2	1	1	3.5	16
Chemical and chemical products ⎱ Petroleum and coal products ⎰	14	2	4	2	10
Non-metallic minerals	15	14	9	10.5	8
Basic metals ⎱ Metal products except machinery and transport ⎰	16	8.5	8	7.5	7
Manufacture of machinery	10	10	17	5	1
Electrical machinery and appliances	7.5	6.5	12	7.5	9
Transport equipment	12	4	6	1	2
Other manufacturing	4	12.5	14	16	15

Note: Ranking is done in such a way that an increase in numbers indicates an increasing degree of labour intensity and an increase in size of employment.

Source: United Nations: *The growth of world industry, 1953-1965, National tables*, op. cit.

One can only speculate that in the countries, industries and establishments concerned actual conditions conform closely to the above assumptions, namely same efficiency in input use, uniform price vectors, absence of scale effect, etc. A non-similar ranking in establishments and industries may occur owing to different degrees of market imperfections, and variations in returns to scale. More empirical evidence at the level of the firms is desirable to unravel the various influences that are concealed at the level of an industry.

It is worth noting that in both tables 2 and 3 the four indices of labour intensity can be classified into two sub-sets. The first sub-set includes the labour coefficient, the share of wages in value added and the capital-labour ratio, whereas the second consists only of the capital coefficient. It is in fact within the former sub-set that the ranking is quite close. With few exceptions, the ranking of industries by capital coefficient is quite different from that given by the other three indices.

Table 4.   Ranking of establishments by size of employment and labour intensity in Mexico

Establishment	Year	Size of employment	Indicators of labour intensity			
			(L/V)	(wL/V)	(K/L)	(K/V)
Cheese, butter, cream, etc.	1966	5	18	4	20	22
Canned vegetables, fruit and fruit pieces	1966	12	9	13	14	21
Wheat flour and by-products	1966	16	20	12	12	3
Cane sugar refinery	1965/6	17	11	18	5	4
Synthetic fibre fabrics	1965/6	22	6.5	22	16	10.5
Shirts, trunk hoses, etc.	1966	6	12.5	11	21	17.5
Plywood and sawn timber	1966	13	19	14	17	14
Newsprint and printing paper for books	1967	10	8	3	2	1
Dried skins for ladies' handbags	1966	4	16	7	13	15
Natural and synthetic rubber tyres	1966	9	2	10	18	16
Ammonium sulfate and single superphosphate	1966	11	6.5	5	3	5.5
Vegetable oils, fats and margarine	1967	1	1	1	1	9
Paints and synthetic resins	1966	2	4.5	2	8	19
Soap, detergents and glycerine	1966	20	17	20	11	8
Bricks and other construction materials	1966	8	14	8	9	7
Rolled, drawn, forged and cast metal products	1966	15	12.5	17	10	10.5
Production of refined antimonial lead	1966	7	21	9	19	13
Steel angles, bolts, nuts, steel wire, rods, etc.	1966	21	15	21	4	2
Tractors, seeders and other agricultural implements	1966/7	14	10	15	7	5.5
Electric lamps	1966/7	19	22	16	22	20
Automobiles and trucks	1967	18	3	19	6	17.5
Production of motor cycles and bicycles	1966	3	4.5	6	15	10.5

Note: Ranking is done in such a way that an increase in numbers indicates an increasing degree of labour intensity and an increase in size of employment.

Source: UNIDO: *Profiles of manufacturing establishments*, op. cit., Vol. III.

CONCLUDING REMARKS

We have observed that the concept of labour intensity, like that of productive efficiency, is not without its limitations. For one thing, ambiguity about it arises from the various shades of meaning attached to it. The multiplicity of interpretation occurs on account of the different objectives one has in view for using the concept.

In spite of conceptual and measurement difficulties, we find that the concept is of some operational significance for planning and policy making. The ranking of industrial projects by different indices of labour intensity would be more useful if those indices could be purged of impurities. Labour intensity together with the absolute size of employment can be helpful in defining priorities for the choice of projects from the point of view of employment planning. In addition to examining indices of labour intensity and size of employment, it would also be useful to look at the range of elasticity of substitution.

However, in practice there is perhaps no such thing as a true or pure index of labour intensity. The different indicators examined in this chapter relate factor inputs and outputs, and represent shorthand expressions which capture only a part of the picture. Given these limitations, these indicators can serve a useful purpose for analysis, planning and policy making, as will be shown in the subsequent chapters. Their usefulness can be enhanced by analyses at the level of the firms and projects rather than at the level of industries.

CAPITAL-LABOUR SUBSTITUTION POSSIBILITIES: A REVIEW OF EMPIRICAL EVIDENCE

2

by J. Gaude[1]

There is a growing body of evidence that the growth of industrial employment is lagging behind the growth of output in less developed countries. Rates of investment have been quite high, but labour absorption has been held constant, or has even declined in some cases. In this connection it is commonly suggested that either an inappropriate choice of technology or a lack of such choice, or both, prevent less developed countries from absorbing labour more quickly. We are primarily concerned here with this hypothesis in so far as the value of the elasticity of substitution tends to reflect technological alternatives.

Two main issues are to be examined in this study. The first is the appropriateness of the elasticity of substitution as a guide for policy making. In this connection the concept of the elasticity of substitution is defined and its implications for policy making are examined. The second part of this chapter is a critical examination of some empirical estimates of the elasticity of substitution in the manufacturing and other sectors of the less developed countries. The reasons for the separate treatment of cross-section and time series estimates are also discussed.

THE ELASTICITY OF SUBSTITUTION

The elasticity of substitution is a measure of the ease with which two inputs such as capital and labour can be substituted for each other. Let us assume that an economy experiences capital-deepening (an increase in the

[1] International Labour Office. The author would like to thank A. S. Bhalla, C. G. Baron, and Professors H. Pack and P. Strassmann for their comments on an earlier draft, and L. Massenet for research assistance.

amount of capital in relation to the amount of labour). If substitution is comparatively easy (the elasticity of substitution is comparatively high), then (competitive) firms can be induced to absorb the increase by a comparatively small rise in wages and a comparatively small fall in the interest rate as entrepreneurs seek profit maximisation. On the other hand, if substitution is relatively difficult (the elasticity of substitution is relatively low), then (competitive) firms will absorb the increased capital (in relation to labour) only after they have bid down the interest rate and bid up wages by comparatively greater amounts. Clearly the wage increases can reduce the rate of growth of employment, the extent to which this happens depending on the wage change (relative to capital price) and on the size of the elasticity of substitution, which appears to be an important component of the static flexibility of the economy in responses to changes of relative factor prices. Other things being equal, the higher the elasticity of substitution between capital and labour the greater is the possible rate of growth of product because the relatively fast-growing primary factor can be substituted more easily for the relatively slow-growing primary factor. [1] Consequently, a high value of the elasticity of substitution would reflect important policy implications for governments of less developed countries that are faced with unemployment problems.

Let us assume, for example, that a typical less developed country seeks a shift to efficient factor combinations with a higher labour-capital ratio at an industrial or sectoral level, whether on average or only at the margin. Let us assume further that the relevant estimated values of the elasticity of substitution are expected to be low. Then there is little point in investigating the effects of fiscal or financial measures on factor prices the changes in which will induce little or no change in sectoral or industrial factor proportions, even if all entrepreneurs are profit maximisers. [2] It follows that knowledge of the values of the elasticity of substitution in industrial sectors and sub-sectors can be useful for policy makers for changing the market signals to ensure greater labour absorption. This leads us to the measurement of the

---

[1] J. R. Behrman: "Sectoral elasticities of substitution between capital and labor in a developing economy: Time series analysis in the case of postwar Chile", in *Econometrica* (New Haven, Conn., Econometric Society), Vol. 40, No. 2, Mar. 1972; and M. Brown: *On the theory and measurement of technological change* (Cambridge University Press, 1965).

[2] S. N. Acharya includes three other restraints which could impede the potential shifts to efficient combinations with more employment per unit of capital, namely: (*a*) the possibilities of changing the output mix in favour of labour-intensive "sectors", "industries" or "products"; (*b*) the responsiveness of producing units to price signals; and (*c*) the success of action to change prevailing factor prices in the desired direction. See S. N. Acharya: *Fiscal/ financial intervention, factor prices and factor proportions: A review of issues*, Staff Working Paper No. 183 (Washington, DC, International Bank for Reconstruction and Development, Development Economics Department, 1974).

elasticity of substitution, which is a technical parameter characterising a production function. [1] Estimation of the elasticity of substitution implies an analysis of the production function and its translation into an estimating form. [2] The estimating form of this concept correlates labour productivity and the real wage rate. The former is explained by a number of variables such as technical progress, scale of output and changes in the wage rate. For example, the expansion of output would tend to lead to a further division of labour and to a higher labour productivity because of internal and external economies. [3] Also, an increase in the real wage rate tends to raise the wage-rental ratio, and consequently to induce entrepreneurs to use more capital and less labour, thus leading to an increase in labour productivity. The higher the elasticity of substitution the larger the effect of a change of real wage rate on labour productivity arising from substitution of capital for labour for a given scale of operations.

---

[1] For details see Brown, op. cit., and M. I. Nadiri: "Some approaches to the theory and measurement of total factor productivity: A survey", in *Journal of Economic Literature* (Menasha, Wis.), Vol. VIII, No. 4, Dec. 1970.

[2] The most convenient variant of the production function is the CES (constant elasticity of substitution) function (see K. J. Arrow, H. B. Chenery, B. S. Minhas and R. M. Solow: "Capital-labor substitution and economic efficiency", in *Review of Economics and Statistics*, Vol. XLIII, No. 3, Aug. 1961). This is written as follows:

$$V = \gamma e^{\lambda t} (\delta K^{-\rho} + (1 - \delta) L^{-\rho})^{-\nu/\rho} \qquad \ldots (1)$$

where V: real value added; K: employed capital; L: employed labour; δ: the distribution parameter; γ: the efficiency parameter; ρ: the substitution parameter; ν: the scale parameter; λ: the (disembodied) rate of neutral technical progress.

If the *ex-post* demand for labour in a non-competitive setting is

$$\frac{\partial V}{\partial L} = N (w/p),$$

where $N = (1 + \Sigma^{-1}) (1 + \eta^{-1})^{-1}$; w/p: real wage rate; Σ: supply elasticity of labour; η: demand elasticity for output, the estimating equation (1) can be derived by substituting the equation above into (1), solving for V/L and taking logarithms in both sides:

$$\text{Log } V/L = b_0 + b_1 \log V + b_2 \log (Nw/p) + b_3 t + u \qquad \ldots (2a)$$

with: $b_0 = \sigma \log [\nu (1 - \delta) \gamma^{-\rho/\nu}]^{-1}$; $b_1 = (1 - \nu^{-1}) (1 - \sigma)$;
$b_2 = \sigma = (1 + \rho)^{-1}$; $b_3 = \lambda \nu^{-1} (1 - \sigma)$;
$u =$ disturbance variable.

Equation (2a) will provide us with the general framework for most of the time series studies discussed below. Obviously, in the cross-section case, time t clearly does not appear, and equation (2a) becomes:

$$\log V/L = b'_0 + b'_1 \log V + b'_2 \log (Nw/p) + u' \qquad \ldots (2b)$$

while the coefficients b'_i contain the same (implied) structural parameters as the b_is.

Between the two points of time, the general equation (2a) can be written:

$$(\dot{V/L}) = \nu^{-1} \lambda (1 - \sigma) + \sigma (\dot{w} + \dot{N} - \dot{p}) + [(1 - \nu^{-1}) (1 - \sigma)] \dot{V} + u'' \qquad \ldots (2c)$$

where the dot above each letter denotes the percentage change of the appropriate variable.

[3] It has been argued that external economies follow from a change in any of the four factors: *(a)* structure of final demand, *(b)* the "state of arts", *(c)* supply of labour and other resources, *(d)* the existing organisation of the industry. These four factors are responsible for the market size, technological change, quality and quantity of productive factors and the indivisibility of productive units respectively.

Economic problems

The economic background underlying the concept and estimating equations can be viewed as follows. First, the capital-labour ratio is assumed not to vary with wages and labour productivity. This implies that the elasticity of substitution is independent of factor proportions [1] and scale. Such an assumption is generally not supported by either casual empiricism or the few available micro-studies of production processes. [2] Secondly, the effect depends on the type of analysis (cross-section or time series) that is considered. In the latter case, it is difficult to isolate the effects of economies of scale without technical progress from economies reaped through technical progress. Besides, as Kotowitz notes [3], the meaning of economies of scale is very vague when dealing with highly aggregative relations. In other words, the hypothesis of constant returns to scale is stronger in cross-section than in time series analysis. However, it should be noted that if there are increasing returns to scale at the level of the firm as well as of the industry, then the assumption of equality of the wage rate and the marginal value product of labour is very likely to be invalid. All factors cannot be paid the marginal value product without exceeding the total value added. Increasing returns to scale are therefore compatible with a profit maximisation model only if there are imperfections in product or factor markets. [4] Finally, the CES function cannot incorporate returns to scale which vary with factor proportions [5]; applied to production at either industrial or sectoral level, it assumes perfect malleability of factor combinations. A more realistic assumption is that factor combinations, once chosen, can seldom be adjusted quickly without any loss of efficiency.

---

[1] The empirical usefulness of this hypothesis derives from the ease of substitution between inputs, which directly depends on the manner in which factors are combined, with important implications for the input decisions of the firm. See H. H. Tsang: "Economic hypotheses and the derivation of production functions", in *The Economic Record* (Melbourne), Vol. 49, No. 127, Sep. 1973.

[2] See for instance G. K. Boon: "Substitución de capital y trabajo, comparaciones de productividad e insumos primarios y proyectados", in *Demografía y Economía* (Colegio de México), Vol. VII, 1973, No. 3; C. A. Knox Lovell: "Capacity utilization and production function estimation in postwar American manufacturing", in *Quarterly Journal of Economics*, Vol. LXXXII, No. 2, May 1968; idem: "Estimation and prediction with CES and VES production functions", in *International Economic Review* (Philadelphia and Osaka), Vol. 14, No. 3, Oct. 1973.

[3] Y. Kotowitz: "On the estimation of a non-neutral CES production function", in *The Canadian Journal of Economics* (Toronto), Vol. I, No. 2, May 1968; idem: "Capital-labour substitution in Canadian manufacturing, 1926-39 and 1946-61", ibid., Vol. I, No. 3, Aug. 1968.

[4] M. S. Feldstein: "Alternative methods of estimating a CES production function for Britain", in *Economica* (London), Vol. XXXIV, No. 136, Nov. 1967; R. K. Diwan: "Alternative specification of economies of scale", ibid., Vol. XXXIII, No. 132, Nov. 1966.

[5] The relationships between the elasticity of substitution, the scale elasticity and the Verdoorn coefficient have been analysed by J. M. Katz: *Production functions, foreign investment and growth* (Amsterdam, North-Holland Publishing Co., 1969) and C. St.J. O'Herlihy: *The capital-labour substitution elasticity in developing countries*, unpublished working paper (Geneva, ILO, 1970).

Attempts to fit production functions to less developed countries take value added as the measure of output. Consequently the treatment of intermediate input is concealed. [1] There is an implicit assumption that the share of the intermediate input in gross output is nearly constant, which amounts to saying that the elasticity of substitution between intermediate input and value added is zero. The omission of materials from the production function often affects the elasticity of substitution between capital and labour. [2] Furthermore, capital may be substituted more efficiently for intermediate input than for labour.

Labour and capital are aggregates of elements that are basically heterogeneous, with divergent characteristics: they differ in their longevity, productive qualities, mobility, etc. Following Kotowitz, treating labour as homogeneous input introduces two types of bias into the measure: *(a)* improvements in the quality of labour over time are ignored; and *(b)* changes in the quality of labour over the business cycle are ignored. [3] If the length of time is fairly short, the first type of bias will be small. This heterogeneity appears to be the main cause and consequence of technical progress in an economy. The estimated value of the elasticity of substitution strongly depends on the manner of grouping heterogeneous units of capital, labour or output [4], which is closely related to the level of aggregation. If the definition of sectors is too aggregative, the estimates of the elasticity of substitution based on time series data become intra-sectoral substitutions of one kind of product for another. This change in output mix may or may not be due solely to changes in factor prices; in both cases, the meaning of the measured elasticity is unclear. Finally, the two-factor CES production function does not permit adequate treatment of the role of land, education, entrepreneurship [5], and labour mix or capital mix in the production processes.

---

[1] See Acharya, op. cit.; Frances Stewart: "Choice of technique in developing countries", in *Journal of Development Studies* (London, Frank Cass), Vol. 9, No. 1, Oct. 1972, and reprinted in C. M. Cooper (ed.): *Science, technology and development* (London, Frank Cass, 1973); and Boon, op. cit.; also Chapter 1.

[2] See Chapter 11.

[3] Kotowitz, op. cit., uses per cent unemployment of the civilian labour force (and its square) as proxy variable to represent departures of demand from its expectations.

[4] For a more comprehensive treatment of these problems see A. A. Walters: "Production and cost functions: An econometric survey", in *Econometrica*, Vol. 31, No. 1-2, Jan.-Apr. 1963; idem: *An introduction to econometrics* (London, Macmillan, 1968), and Nadiri, op. cit.

[5] The contribution of the entrepreneur's education to an increase in a firm's output and productivity can be caused partly by better choice of inputs and partly by more effective use of labour. Some technological change might be induced because some entrepreneurs decide to innovate and experiment while others wait till the economic profitability of these innovations is established and technological change becomes less risky. For a statistical treatment of these concepts within a production function framework, see J. Gaude: *Analyse des relations entre la main-d'œuvre, l'échelle et la technologie dans le petit commerce à Lima (Pérou)* (Geneva, ILO, 1973; mimeographed), and E. J. Mitchell: "Explaining the international pattern of labor productivity and wages: A production model with two labor inputs", in *Review of Economics and Statistics*, Vol. L, No. 4, Nov. 1968.

Estimation problems

There are a number of problems connected with the estimation of the elasticity of substitution. First, the relationship between (real) wages and employment is wholly explainable by the productivity of labour and not by substitutability against labour [1] unless the estimate of the elasticity is shown to be significantly different from unity. The classical procedure of estimation assumes an exogenous determination of the wage rate. If the wage rate is not exogenously determined, the CES procedure yields a biased and inconsistent estimate of the elasticity even if returns to scale are constant and the wage rate is equal to the marginal product of labour. The exogenous determination of the wage rate is likely to be a more relevant assumption for an international sample within an industry than for an aggregate time series within a single country or an inter-industry sample. [2]

Secondly, the data are assumed to represent points on the production frontier, that is, all firms observed are assumed to have adjusted fully to the prevailing factor prices. This is quite unrealistic: for instance, for cross-section data the discrepancy between actual and optimal combinations may reflect differences in managerial quality across firms; the analogous problem for time series data is the implicit assumption of full adjustment within the observation period.

Thirdly, except for a few rare cases [3] the studies do not make adjustments for under-utilisation of capacity. In time series analysis, much of the variation in value added may be attributed to different rates of capacity utilisation over a business cycle. Besides, in a cross-section analysis, if most firms are producing below capacity we should assess the apparently increasing returns to scale: conversely, if most firms are (mistakenly) over-producing, a function fitted to the firms gives rise to a scale elasticity of less than unity. Consequently, depending on estimation procedures, under-capacity or over-capacity of most firms, on average, can lead to a biased estimate of the elasticity of substitution.

Fourthly, it is assumed that the same technological alternatives are available, and that a single production function exists for all countries or for all sectors in a country. This would appear to be a particularly difficult condition to satisfy for a cross-section analysis of firms in less developed countries where transmission of technological change is generally slow, and one could expect to find different technologies being used simultaneously in the same sector. [4]

---

[1] D. R. Fusfeld, review of *Profile of Michigan: Economic trends and paradoxes*, by Stephen P. Sobotka with T. A. Domencich, in *American Economic Review* (Menasha, Wis.), Vol. LIV, No. 4, Part I, June 1964.

[2] Feldstein, op. cit.

[3] Behrman, op. cit., and Chapter 11 below.

[4] Acharya, op. cit.

In other words, the estimates of the elasticity relate not to the substitution possibilities of a unique production function, but rather reflect the "varying responses to market conditions of firms producing with different vintages of technology". [1] With time series data on aggregate relationships an analogous problem is that of correctly specifying the nature of technological change, which is an important—indeed the most important—element in the statistical interpretation.

Fifthly, if time series data are used undeflated from an inflationary situation, variations in the rate of inflation bias the estimate of elasticity towards unity. [2] Even in cross-section, if undeflated data are used and if prices are positively correlated with nominal wages, a bias towards unity occurs. [3]

Finally, as one might expect, there are marked multicollinearities in the time series. The attempt to eliminate the trend by moving average techniques or by linear (or exponential) terms tends to introduce spurious relationships and to magnify the errors in measurement of the variables. It might be thought that this problem of multicollinearity would be avoided in cross-section samples, but this is not so. With intra-industry cross-sections, for instance, size of industry, in terms of labour, capital and output, is the dominant relationship; consequently technologies are similar and the capital-labour and capital-output ratios are more or less the same. The important conclusion, therefore, is that one does not avoid multicollinearity by the use of cross-section data. [4]

The empirical evidence below seems to indicate that the parameters of the production function are highly sensitive to slight changes in the data, measurement of variables, and methods of estimation.

A CRITICAL SURVEY OF ESTIMATES OF THE ELASTICITY OF SUBSTITUTION

We begin by saying a few words about the reason for separating cross-section studies from time series studies. Secondly, we describe why and how particular studies have been chosen.

Most of the time series estimates of elasticity are lower than unity, while the cross-section estimates are generally higher than the time series estimates

---

[1] M. Roemer: *The neoclassical employment model applied to Ghanaian manufacturing*, Economic Development Report, No. 225 (Cambridge, Mass., Harvard University Development Research Group, 1972).

[2] O'Herlihy, op. cit.

[3] M. Nerlove: "Recent empirical studies of the CES and related production functions", in M. Brown (ed.): *The theory and empirical analysis of production* (New York, National Bureau of Economic Research, 1967).

[4] Walters: *An introduction to econometrics*, op. cit.

and close to unity. In fact, the time series data actually reflect a dynamic adjustment due to a combination of factors such as changes in relative prices, technical change and external shocks, which are generally excluded in cross-section data. The time series estimates are often biased because of simultaneity between the inputs and their prices, mis-specification of adjustment lags between the inputs and the output, and the dominance of cyclical conditions, e.g. under-utilisation of capacity. Furthermore, in a competitive market there is no reason for relative prices to differ among firms. Differences in observed input proportions tend to reflect differences in intra-firm managerial ability, and consequently the individual production function is not identified. [1] Besides, cross-section estimates tend to be higher than time series estimates because the former reflect long-run equilibrium states whereas the latter are dependent on annual changes where short-run reactions may prevail. In other respects, in a cross-regional analysis, positive correlations are to be expected between the size of operation, technological change and the resulting labour productivity on the one hand, and the average wage on the other. This implies a possible upward bias of the cross-section estimates of the elasticity of substitution. [2] It follows from the above that both conceptual and statistical reasons justify our decision to separate cross-section estimates from time series estimates.

The studies presented have been selected according to two criteria—diversity of the developing countries where the analysis was conducted and diversity of estimating methods or sampling centred on the manufacturing sector. We shall compare selected results by sub-dividing them into four kinds of estimates—cross-section, time series, pooling of cross-section and time series, and others.

Cross-section estimates

Five studies providing cross-section estimates of elasticity of substitution between one or two points of time have been selected. In chronological order, these are a study at the country level by Reynolds and Gregory for Puerto Rico, units of observation being undifferentiated industries; a study by Eriksson of five Latin American countries with the same specification as Reynolds and Gregory; a study at a cross-regional level for Argentina between 1946 and 1954 by Katz; a study of the Philippines manufacturing industries by Sicat; and a study of Nigerian industry by Oyelabi. [3]

---

[1] Nadiri, op. cit., and Walters: "Production and cost functions: [An econometric survey", op. cit.

[2] O'Herlihy, op. cit.

[3] L. G. Reynolds and P. Gregory: *Wages, productivity and industrialization in Puerto Rico* (Homewood, Ill., Irwin, 1965), Ch. 3, pp. 82-103; J. R. Eriksson: "Wage policy and

As can be seen, the specifications are fairly different, as is the testing of hypotheses. Whereas the estimation of the elasticity of substitution is at the very root of the analysis in Sicat's and Katz's studies, Eriksson [1] is concerned with it only incidentally, while Reynolds and Gregory do not mention it anywhere.

Since differences in specification and level of aggregation might lead to biased estimates of the elasticity of substitution and hence to unjustified comparisons, we shall—

(1) compare Katz's (Argentina) results with those of Sicat (Philippines) and Oyelabi (Nigeria), in both cases using equation (5), though at a different level of aggregation (at the regional level for Argentina and at the country level for the Philippines and Nigeria);

(2) compare Katz's results for Argentina with those of Reynolds and Gregory for Puerto Rico, in both cases using equation (3), at the regional level for Argentina and at the country level for Puerto Rico;

(3) examine the Eriksson model; and

(4) also discuss the Katz model (see equation (3b)).

The results obtained by Katz, Sicat and Oyelabi are compared in table 5.

1. As can be seen when comparing the 1960 estimates by Sicat with the 1954 estimates by Katz, the elasticities of substitution are fairly comparable for electrical machinery (about 1.0), tobacco (about 1.60), leather products (about 0.90), chemical products (about 1.0), printed and published materials (about 0.85) or non-metallic minerals (about 1.25). The paper industry tends to exhibit high flexibility since the elasticity is higher than unity in the three countries. On the other hand one can observe fairly strong divergences in the three countries for textiles, rubber and metal products. Such deviations can be

---

economic development in Latin American countries", in *Tijdschrift voor Sociale Wetenschappen* (Ghent), Vol. 14, No. 4, 1969; Katz, op. cit.; G. P. Sicat: "Capital-labor substitution in manufacturing in a developing economy: The Philippines", in *The Developing Economies* (Tokyo), Vol. VIII, No. 1, Mar. 1970; J. A. Oyelabi: "Tests of factor substitution in Nigeria's manufacturing sector", in *Eastern Africa Economic Review* (Nairobi), Vol. 3, No. 1, June 1971. The estimating equations used in each case are:

$(\dot{V}/L = a + b\,(\dot{w}/p) \quad \dots (3a)$ (Reynolds and Gregory)
$(\dot{V}/L) = -a + (1 - b)\,\dot{V} - c\,(\dot{w}) \quad \dots (4)$ (Eriksson)
$(\dot{V}/L) = a + b\,(\dot{w}/p) \quad \dots (3b)$ (Katz)
$\log (V/L) = b_0 + b_1 \log (w/p) \quad \dots (5)$ (Katz)
$\log (V/L) = b_0 + b_1 \log (w/p) \quad \dots (6)$ (Sicat)
$\log (V/L) = b_0 + b_1 \log (w/p) + b_2 \log (K/L) \quad \dots (7)$ (Sicat)
$\log (V/L) = b_0 + b_1 \log (w) \quad \dots (8)$ (Oyelabi)

[1] Eriksson is mainly concerned with the explanation of the demand for labour variation, while Reynolds and Gregory endeavoured to estimate the demand elasticity for labour (through the link between the employment forgone—differences between changes in output and employment under the linearity hypothesis—and change in wages).

Table 5.   Cross-section estimates by industries

Industries [1] (abridged titles, author's grouping)	ISIC No.	Country, author, date of source data and equation number					
		Philippines (Sicat)		Argentina (Katz)			Nigeria (Oyelabi)
		1960		1946	1954	1946-1954	?
		(6)	(7)	(5)	(5)	(3b)	(8)
A. Food	20	1.37	1.567	1.35	0.87	1.15	
Beverages	21	1.11	2.14				
Tobacco	22	1.57	1.541	1.76	1.73		
B. Textiles	23	0.44		0.98		0.77	1.348
Footwear and apparel	24	0.60					
Leather products	29	1.01	1.338	0.87	0.71	0.81	
C. Wood and cork	25	0.86	0.464	0.93		1.06	0.747
Furniture and fixtures	26	1.43	2.075				
D. Rubber products	30	1.58		0.92			0.977
Chemical products	31	1.09	0.902	0.90	1.01		1.802
E. Basic metals	34	0.94	1.441				
Metal products	35	1.36	3.073	0.87	0.47	0.17	0.978
Non-electrical machinery	36	1.06	2.707	0.46 [2]	0.68 [2]	0.62 [2]	
Electrical machinery	37	0.87		0.45	1.11	0.79	
Transport equipment	38	0.325	2.006				
F. Paper	27	1.25	1.339	1.49	1.63		1.432
Printing, publishing, etc.	28	0.79	neg	0.87	0.91	0.79	1.044
Non-metallic minerals	33	1.35	4.313	1.19	1.20	1.17	

[1] Major groups of the *International Standard Industrial Classification of All Economic Activities*, Statistical Papers, Series M, No. 4, Rev. 1 (New York, United Nations, 1958).   [2] Aggregation of industries 36 to 38 in the Classification.

partly attributed to the level of aggregation for a cross-regional analysis of the Katz type, which is more heterogeneous than a cross-establishment one of the Sicat type even at a country level. As a matter of fact, as Katz wrote, whereas industrial categories are defined in the same way in all regions, the product mix is not exactly comparable because of specialisation. [1] Secondly, the year used by Sicat, i.e. 1960, was one of slight recession in the Philippines while 1954, the year for Argentina, was one of expansion. [2] These differences can consequently be accounted for by the differential effects of the full capacity operation in Argentina in 1954 as compared with partial capacity utilisation for the Philippines in 1960.

---

[1] Katz, op. cit., p. 46.

[2] Katz does not say anything about the trend around 1946 in Argentina.

In other respects, Sicat shows some estimates for developed countries which can be compared with the estimates for the Philippines. It is generally contended that the elasticity of substitution is likely to be higher for less developed countries than for developed countries owing to slow supply response. However, as Sicat[1] noted, the average elasticities by industries directly obtained for the Philippines do not appear to be any smaller than those obtained for American manufacturing using the same level of aggregation. Williamson's results in some industries, obtained through a method of pooling cross-section and time series observations and other specifications of the estimating function[2], tend to confirm this argument. However, Sicat's or Williamson's comparisons are likely to be meaningless for the level of aggregation is different and thus the product mix is more heterogeneous.

Finally, Sicat compares the estimates of Liu-Hildebrand[3] cited by Nerlove[4] with his results for the Philippines obtained through a more general production function. The estimates obtained under this more sophisticated specification are higher than those found by standard methods. Besides, the correlation coefficient between the two Sicat estimates is 0.54, which would suggest strong deviations of the elasticity of substitution. For instance, the elasticity for wood and cork and for electrical machinery is the same when calculated by standard methods but differs when estimated by this more general production function.

2. For manufacturing as a whole, Katz's results for Argentina (0.82) are fairly different from the estimates of Reynolds and Gregory (1.138) for Puerto Rico; however, both estimates are not significantly different from 1. Besides, the elasticity of substitution for local industries in Puerto Rico is less than that of export industries—which would imply that the more dynamic export sector can be induced to absorb a greater amount of labour by a comparatively small rise in wages than is the case for local industries. [5]

3. In Eriksson's model, the elasticity of substitution tends to be less than unity for every country except Colombia. [6] If one compares these results with those of Reynolds and Gregory for the period 1954-58 (the same assumptions prevail except that in the latter case the rate of increase of output is omitted),

---

[1] Sicat, op. cit., p. 30.

[2] See equations (17) to (19) below.

[3] T. C. Liu and G. H. Hildebrand: *Manufacturing production functions in the United States, 1957* (Ithaca, NY, New York State School of Industrial and Labor Relations, 1965).

[4] Nerlove, op. cit.

[5] O'Herlihy, op. cit.

[6] There is a marked tendency for economies of scale in Argentina and Brazil while these countries experience the highest rate of (neutral) technological change.

there are reasons for thinking that the value of the elasticity of substitution is similar in Colombia, Mexico and Puerto Rico. [1]

4. The Katz study, which is the most comprehensive, relates to both cross-section and time series data relating to Argentina after the Second World War. Selecting the cross-sectional sub-set of the study, three different groups of elasticity of substitution estimates were made up from 15 industries for 1946 and 10 industries for 1954 (equation (5)), and from 10 industries with changes in the cross-regional census data between 1946 and 1954 (equation (3b)). Where the most restrictive assumptions held (equation (5)), the industries have corresponding estimates for each year; only three out of ten industries had estimates that differed substantially from each other—food, metal products, and electrical machinery. The observed range of estimates is 0.46 (vehicles and machinery) to 2.02 (petrochemicals) in 1946, and 0.47 (metal products) to 1.73 (tobacco) in 1954; however, the true range (statistically) is between 0 and 1.50.

When estimates based on intercensal changes are considered, a degree of imperfect competition is assumed to exist but not to change over the period. Katz correctly shows that the absence of this variable tends to bias the estimated elasticity coefficient upwards. In fact the change of hypothesis does not significantly alter the results, as it does for stone, glass and ceramics, knitting mills, printing, leather products, and wood and furniture. Katz does not provide any explanation for this.

As regards the stability of elasticity of substitution through time, Katz maintains that in spite of the fact that the elasticity of substitution did not remain stable from 1946 to 1954, the pattern of change of this parameter is "consistent" across industries, which means that the ranking of industries with respect to their elasticity of substitution in 1946 tends to be similar to their ranking with respect to elasticity in 1954. The industries that tended to exhibit a lower than average σ in 1946 maintained a lower than average value of that parameter in 1954 (except for foodstuffs), and vice versa for industries in which the elasticity of substitution was higher than average (except for machinery and electrical equipment). However, if the σ values for 1946 are

---

[1] The expected value of the elasticity of substitution in the Reynolds and Gregory model is: $E(b) = \sigma + c.r(\dot{w}, \dot{V})$ where c is the expected regression coefficient of $\dot{V}$ in a regression of $\dot{V}$ and $\dot{w}$ on $\dot{V}/L$; $r(\dot{w}, \dot{V})$ is the expected value of the regression coefficient of $\dot{w}$ on $\dot{V}$; b is the elasticity of substitution estimated by Gregory and Reynolds; σ would have been the "true" elasticity if Gregory and Reynolds had added the variable $\dot{V}$ (as is the case in the Eriksson model). Therefore, the omission of the $\dot{V}$ variable in the Puerto Rican case tends to lead to an over-estimate if one could expect a significant correlation coefficient between $\dot{w}$ and $\dot{V}$. Conversely, the existence of the $\dot{V}$ variable in the Eriksson model tends to lead to under-estimates if $\dot{w}$ and $\dot{V}$ are positively correlated: that is the case for Argentina (0.44) and, to a lesser degree, for Colombia (0.32) and Mexico (0.31).

(statistically) equal to those of 1954, and if in particular the exogenous variables that do not enter into each estimated relation are not correlated with the logarithm of the real wage (or the proportionate change of the real wage), the elasticity of substitution is bound to be the same. In such industries as printing and publishing or knitting mills, leather products, stone and glass, Katz obtains a comparable elasticity of substitution. [1]

Time series estimates

We have selected five time series studies which measure the elasticity of substitution. A serious limitation of these studies is the suggestion that the elasticity of substitution remains fairly constant over a wide range of capital-labour ratios, and that it is not altered by technology. With the exception of Bruno, this assumption is made by all the authors. Three studies (out of five) assume that the industries contain many firms operating under perfect competition in the factor and product markets. [2] This assumption appears reasonable for industries defined by two or more digits in the International Standard Industrial Classification, but is clearly violated in total manufacturing, which is a sufficiently large sector of the economy to have feedback effects on wages. These five studies [3] are a study of Kenya by Harris and Todaro at the country level of aggregation; the Katz estimates for the Argentine manufacturing industry; Diwan and Gujarati's three models corresponding to different specifications of the production function for Indian manufacturing industries; Behrman's estimates for Chilean economic sectors; and the Bruno model of the constant marginal share (CMS). [4]

---

[1] Katz wrote in these respects that, in 8 industries out of 15, σ is significantly different from unity at a 20 per cent level of confidence in 1946 and in 6 out of 10 in 1954. We are not convinced by such a low level of confidence. We think that, on the contrary, the level of confidence must be as severe as possible due to the set of hypotheses implicitly made.

[2] As each industry employs a small fraction of the labour force, it can be assumed that the price of labour for each industry is given and is only marginally affected by the state of demand in any industry.

[3] J. R. Harris and M. P. Todaro: "Wages, industrial employment and labour productivity: The Kenyan experience", in *Eastern Africa Economic Review*, Vol. I, No. 1, June 1969; Katz, op. cit.; R. K. Diwan and D. N. Gujarati: "Employment and productivity in Indian industries: Some questions of theory and policy", in *Artha Vijñāna* (Poona), Vol. 10, No. 1, Mar. 1968; Behrman, op. cit.; M. Bruno: "A note on the implications of an empirical relationship between output per unit of labour, the wage rate and the capital-labour ratio" (Stanford University, July 1962; mimeographed), and idem: "Estimation of factor contribution to growth under structural disequilibrium", in *International Economic Review*, Vol. 9, No. 1, Feb. 1968.

[4] The estimating equations in the above cases are:

$$(\dot{V}/L) = a + b\dot{w} \qquad \qquad \ldots(9)$$

$(\dot{V}/L)$ = percentage change of value added—in money terms—per employee (or African); $\dot{w}$ = percentage change of average money wage of employees (or Africans). Neither output
(footnote concluded overleaf)

The Diwan and Gujarati model (equation (14)) estimates changes in market structures. In the product market, the authors elaborate a simple growth process where the firm (or industry) plans a "target" output and makes policy decisions to achieve it. Growth follows if "planned" output is always kept higher than "achieved" output by the same ratio (namely θ/λ, where $\theta < \lambda$, which yields the percentage of "planned" output actually being achieved). The Behrman model first tests the degree of flexibility of the Chilean economy in response to changes in international markets and then examines the implications between aggregate elasticity of substitution, the sectoral elasticities and the relative growth of each sector.

Bruno elaborates a new production function which first links labour productivity linearly to the real wage rate and then again the marginal product of labour to the real wage rate. The first equation implies that marginal factor shares are constant, and the second one that market conditions cannot be perfect. [1] The model is estimated for manufacturing and the entire private sector in Israel from 1953 to 1964. The values of the elasticity of substitution he found are fairly small, though depending on labour productivity. On average, they are significantly less than unity and comparable for both sectors.

---

nor wages were deflated for price changes because of the lack of an adequate deflator. Equation (9) is estimated from Kenyan data for "all industry and commerce" and per employee only for "manufacturing", "trade and commerce" and "other services" for the period 1955-66 excluding the change 1963-64 because of the break in series.

Katz

$$\log (V/L) = a + b \log (w/p) + ct \qquad \text{...(10)}$$
$$\log (V/L) = a' + b' \log (w/p) + c't + d' \log V \qquad \text{...(11)}$$

where V/L = real value added per worker, w/p = average real wage per worker, t = trend variable. Equation (10) was estimated for two periods: 1943-53 and 1954-61, corresponding to two "technological epochs", while the more general equation (11) was only computed from 1954-61.

Diwan and Gujarati

$$\log L = \log A' - \sigma \log (w/p) + \alpha \log V \qquad \text{...(12)}$$
$$\log L = \log A'' + \alpha_1 \log (w/p) + \alpha_2 \log V + \alpha_3 \log L_{-1} \qquad \text{...(13)}$$
$$\log L = \log A^* + \beta_1 \log (w/p) + \beta_2 \log (w/p)_{-1}$$
$$+ \beta_3 \log V$$
$$+ \beta_4 \log V_{-1} \qquad \text{...(14)}$$

where L = employment (production and non-production workers); w/p = real wage rate; V = net value added.

Behrman

$$\log CV/L = a + b \log \tilde{w}/p + c \log (CV/L)_{-1} + dt \qquad \text{...(15)}$$

where CV = capacity of real value added (defined by "the trend through the peaks" method); L = secular trend in labour force; w/p = expected ratio of nominal wage (including employers' social security payments) to nominal product as represented by a weighted average of past values of $\tilde{w}/p$.

[1] By eliminating real wage in both equations, and resolving the resulting partial differential equation, the following production function is obtained.

$$V = \gamma e^{\lambda t} K^\alpha L^{1-\alpha} - mL \qquad \text{...(16)}$$

The elasticity of substitution is not a fixed value but depends on the average labour productivity.

Another interesting aspect of Bruno's model is that it successfully evaluates the degree to which the labour market is imperfect.

Tables 6 and 7 show time series estimates for different developing countries with respect to a set of manufacturing industries and to the average for all manufacturing. In the latter case, it is found, as expected, that the estimates of the elasticity of substitution for each country are significantly lower than unity. At the industrial level, estimates for the chemical industry turn out to be very low in India and Argentina. On the other hand, the results for the paper and pulp industry tend to show rather considerable differences between the two countries.

It is impossible to compare the results of these studies since the level of aggregation (from three-digit or four-digit industries to total manufacturing) and the specifications adopted are different. However, the elasticity of sub-stitution, whatever the specifications, tends to lie below unity in manufacturing as a whole, with practically the same value in Kenya and Israel. This does not mean that a similar estimate would necessarily be obtained from the same specifications. In fact, what is important is the appropriateness of the set of hypotheses in a country. For instance, the instantaneous optimality condition (e.g. the industry adjusts its capital-labour ratio within a year of a change in average wages) assumed in the Kenya study is certainly a more restrictive assumption in less developed countries than in developed countries partly because of organisational rigidities. Even in the long run the assumption of constant returns to scale is certainly more appropriate in a developed country like Canada [1] than in a developing one. A first problem would be to evaluate the consequences of the relaxing of successive hypotheses on the elasticity of substitution in order to ascertain why and how the elasticity of substitution is sensitive to these changes. This is the purpose of Diwan and Gujarati's study for Indian industries. Unfortunately, the increasing complexity of the estimating equations (and hence their greater realism) was achieved at the cost of increasing multicollinearity; it follows that the estimates of elasticity of substitution are certainly biased. In the simplest model (equation 12 above), 17 out of 28 industries have an elasticity not significantly different from zero. Of the 11 remaining ones, 6 lie between zero and unity, while in 4 cases the elasticity amounts to unity.

It is impossible to disentangle the effects on σ differentials due to the changing of hypotheses from those associated with higher multicollinearity. However, Diwan and Gujarati's main contribution concerns their treatment of imperfect

---

[1] Kotowitz, op. cit.; idem: "Technical progress, factor substitution and income distribu-tion in Canadian manufacturing, 1926-39 and 1946-61", in *Canadian Journal of Economics*, Vol. II, No. 1, Feb. 1969.

Table 6.   Time series by industry

Industry	Long-run	Short-run			
	India (Diwan and Gujarati)	India (Diwan and Gujarati)		Argentina (Katz)	
	1946-58	1946-58		1943-53	1954-61
	Equation 13	Equation 12	Equation 13		
Cotton textiles	0.245	0.040	0.129		
Woollen textiles	0.032	0.061	0.022	0.26	0.34
Jute textiles	0.505	0.443	0.229		
Paper and pulp	1.295 [1]	0.939 [1]	0.649 [1]	0.21	0.48
Chemicals	0.066	0.058	0.058	0.03	0.32
Stone, glass, etc.	0.302 [2]	0.043 [2]	0.200 [2]	negative	0.54
Electric lamps	negative	negative	negative		
Electric fans	0.851	0.661	0.549	0.10 [3]	0.53 [3]
General and electrical engineering	0.449	0.380	0.402		

[1] Paper and paperboard.   [2] Glass and glassware.   [3] Machinery and electrical equipment.

Table 7.   Time series and pooled estimates for all manufacturing

Kind of series		Country (and author)	Period	Value
Time series	Short-run	Argentina (Katz)	1943-53	0.26
			1954-61	0.43
		Chile (Behrman)	1945-65	0.21
		Israel (Bruno)	1953-64	0.754 [1]
	Long-run	Chile (Behrman)	1945-65	0.760
Pooled cross-section and time series	Short-run	Philippines (Sicat)	1957-63	0.424 [2]
				0.213 [3]
			1957-65	0.422 [4]
	Long-run		1957-63	1.588 [2]
				0.820 [3]
			1957-65	2.510 [4]

[1] Author's own estimate.   [2] Equation 17.   [3] Equation 18. Arithmetic mean of the two estimates obtained when w includes the capital gains in r on the one hand and when w does not on the other.   [4] Equation 19. Arithmetic mean of the two estimates obtained with one-year lag for value added per worker on the one hand, and with two-year lag for gross output per worker on the other.

market conditions. It is interesting to compare their approach with that of Bruno. [1]

Diwan and Gujarati versus Bruno on market imperfections

Diwan and Gujarati assume that initially the wage rate is less than its marginal product. Therefore, "a rate of growth of the wage rate higher than

---

[1] In his "Estimation of factor contribution to growth under structural disequilibrium", op. cit.

the rate of growth of marginal productivity would reduce market imperfections". Imperfections in the labour market are assumed to decrease with an increasing rate of growth of the wage rate.

If the value of the elasticity of change in market imperfections with respect to the rate of growth of wages is negative, it tends to decrease labour exploitation (i.e. the positive difference between the marginal product and the wage rate). The authors found a tremendous range of this elasticity (-30.08 to 47.85) due mainly to multicollinearity, whereas reasonable values should be between -1 and 0. Consequently it is impossible to draw any conclusion about the degree of market imperfections from this procedure.

On the other hand Bruno's approach to market imperfections is quite different. He specifies the following equation:

$$\partial V/\partial L = pw + q,$$

where p and q are assumed to be fixed institutional parameters.

The p coefficient is called an "imperfection parameter" and the q coefficient a "disequilibrium parameter" ($p = 1$ and $q = 0$ would imply that both labour and product markets are in competitive equilibrium). The following values for Israel were found:

$$p = \begin{cases} 0.978 \text{ (manufacturing)} \\ 0.935 \text{ (total private sector)} \end{cases}$$

$$q = \begin{cases} -1.154 \text{ (manufacturing)} \\ -0.707 \text{ (total private sector)} \end{cases}$$

In both cases, labour receives more than its marginal product for any value of the wage rate, though the relative importance of the discrepancy becomes increasingly small over time. Bruno attributes these results to the fact that strong trade unions had a major influence on wage determination during the 1950s. He states that if real wages are rigid and cannot be brought down to the marginal productivity of labour, one way of increasing employment is to bring the marginal productivity of labour up to the wage rate and invest as much as one can from domestic savings and foreign aid. [1]

It is rather hard to compare Bruno's treatment of market imperfections directly with that of Diwan and Gujarati. The specifications of the production functions are not identical and the level of aggregation is quite different—the firm level for India and the aggregate private sector for Israel. We can, however, mention the following points.

First of all, the conditions assumed move in opposite directions in the two cases: the workers' exploitation assumption in India (i.e. marginal product

---

[1] Bruno: "Estimation of factor contribution to growth under structural disequilibrium", op. cit., p. 53.

greater than wages) contrasts with trade unions' pressure in Israel which results in wages higher than the marginal product.

Theoretically, if Indian entrepreneurs expect a long-term decrease of workers' exploitation (i.e. the labour market to become more perfect), they will be inclined to substitute capital for labour since the latter would be relatively more expensive. The balancing factor would be provided by the indirect increase in the marginal productivity of labour. The higher the elasticity of substitution, the faster and larger will be the influence of the input-price ratio on the capital intensity.

The picture looks quite different in the case of Israel. Under the working conditions, labour receives more than its marginal product. Therefore, if entrepreneurs, in the long run, expect a diminishing gap between wages and marginal product, they will be inclined to substitute capital for labour in order to minimise their losses. Because of the low values of estimated elasticity of substitution, a similar process is likely to occur in India.

As a consequence, both models would yield the same results though the conditions with regard to market imperfections are diametrically opposed in the two cases.

Let us now see the long-term effect of a change in market conditions towards a better equilibrium, using Bruno's model.

Other things being equal, a once-for-all increase of q (which corresponds to an increase of the workers' marginal product relative to the real wage rate, since q was found to be negative and p nearly equal to unity) should induce entrepreneurs to substitute labour for capital since the use of labour would become more profitable. This is likely to be feasible under Bruno's model because an increase of q is associated with an increase in the elasticity of substitution for a given value of the productivity of labour, independently of the other coefficients except for the share of capital.

The Katz study: a synthesis of time series and cross-section results

We now present the time series estimates under two specifications and two periods of time for some Argentine industries. When equation (10) (see above) is calculated through two time periods, 1943-53 and 1954-61, the value of the elasticity of substitution, for a given (neutral) technical progress, appears to be very fluctuating between corresponding industries. As expected, the time series estimates are consistently lower than the cross-section estimates for the 18 industries.

However, we disagree with Katz's finding that 10 industries out of 13 had a higher elasticity of substitution in 1954-61 than in 1943-53. If we adopt a rigorous test of significance imposed by the number of hypotheses made (constant returns to scale, homogeneity of labour, neutral technical progress), four elasticities decrease, four others increase and five are more or less constant; on the whole, however, the elasticity significantly increases.

Besides, when the same period (1954-61) is used for two different functions (equations 10 and 11 above), the constancy of the scale of operations tends to reduce the elasticity estimate in almost all industries. First, the (probable) high collinearity between time and output does not allow for testing the "true" effect of the scale of operation on the elasticity of substitution. Secondly, the fact of regressing V on V/L, if small changes in employment are associated with large changes in output, is fairly tautological in time series analysis since it amounts to regressing V on itself. [1]

In a more general way, the sources of labour productivity growth in Argentine manufacturing are examined in relation to the elasticity of substitution computations. More specifically, in order to explain the sources of productivity growth, Katz associates for 14 industries the cross-section value of the elasticity of substitution for 1946 with the changes in labour productivity for the period 1946-54.

He found a good rank correlation coefficient (R =0.88) [2], which tends to imply that above-average increases in labour productivity tended to be obtained in the industries where the elasticity of substitution between capital and labour was highest, and lower than average changes in labour productivity appeared in industries with lower than average elasticity of substitution. [3] This argument, together with the fact that wage-rental ratio increased in Argentina by approximately 60 per cent between 1946 and 1954, induces Katz to assert that "capital intensity must have grown more than average in sectors where the elasticity of substitution was found to be greater than unity and much above the average elasticity of substitution of the whole manufacturing sector". [4] However, there is room for argument. First, the cross-section 1946 estimates can in fact be subdivided into three or four sets according to the more or less rigorous level of confidence adopted, i.e. different from 0, not different from 0.5, not different from 1, higher than 1. Can one successfully test a four-points regression? Secondly, our proper rank correlation coefficient between changes in labour productivity for the period 1946-54 and the elasticity of substitution obtained through changes in a cross-regional sample for the same period is about − 0.65. Why such a difference? Thirdly, the rank correlation coefficient between

---

[1] On the basis of the data in table 2.1 (Katz, op. cit., p. 24) the manufacturing employment index (1954 = 100) would have been 114.6 in 1961 while the corresponding output index would have stood at 149.5

[2] Table 5.2 in Katz, op. cit., p. 88.

[3] Ibid., pp. 88-89.

[4] Ibid., p. 89.

changes in labour productivity for that period and the time series elasticity estimate for the 1943-53 period is very low, which implies that high deviations from the labour productivity per industry can be associated with a similar elasticity of substitution. Conversely, industry differences in possibilities of substitution do not tend to explain the sources of labour productivity growth for the corresponding period. [1]

Pooling of time series and cross-section estimates

We present separate estimates of elasticity of substitution computed with time series and cross-section analysis. The studies dealing with pooled data are more complicated in their specifications than the previous ones. Williamson's two papers [2] present three production function specifications for a grouping of two-digit industries in Philippine manufacturing. [3]

The three models, though homogeneous for original data and period employed, are different in specifications. For manufacturing as a whole, the long-run values of elasticity of substitution differ substantially—about 0.80

---

[1] Besides, referring to the 1954-61 period, the main determinants of labour productivity growth are (disembodied) technical progress and returns to scale. Evidently, these factors do not affect the curvature of isoquants and, thus, tend to differ from those found for the previous period of 1946-53. However, since it is mentioned (Katz, op. cit., p. 87) that the Cobb-Douglas function is a misleading tool ($\sigma = 1$ in any case), one could logically argue that the CES function with an explicit time variable tends to be a misleading tool also, since it assumes a neutral technological change. Thus, a non-neutral technological change of the kind considered by Bruno (op. cit.) or Kotowitz (op. cit.), or any specification assuming a variable elasticity of substitution, could at least noticeably modify the weighting of the determinants of labour productivity for both periods.

[2] J. G. Williamson: "Capital accumulation, labor saving, and labor absorption once more", in *Quarterly Journal of Economics*, Vol. LXXXV, No. 1, Feb. 1971; idem: "Relative price changes, adjustment dynamics, and productivity growth: The case of Philippine manufacturing", op. cit.

[3] The first equation is:

$$\log L = \log A + b_0 \log (w/p) + b_1 \log V + b_2 \log L_{-1} + b_3 t \qquad \ldots (17)$$

where: L = number of employed (production line) workers at time t; w/p = average annual earnings of production workers deflated by output price; V = value added deflated by output price. It is estimated from a cross-section of manufacturing industries in 1957-63. However, this equation does not embody capital formation as an explanatory variable, as done in Nadiri's model (1968) which provides us with both short-run and long-run elasticities of substitution. Its estimating equation is:

$$\log L = \log B + c_0 \log \widetilde{w} + c_1 \log K + c_2 \log L_{-1} + c_3 t \qquad \ldots (18)$$

$\widetilde{w} = Pw/r$; r = user's cost of capital at time t; w = real wage rate; P = the output price; K = physical stock of capital (not otherwise specified). The last model structure is an application to Philippine manufacturing of H. Kaneda: "Substitution of labor and non-labor inputs and technical change in Japanese agriculture", in *Review of Economics and Statistics*, Vol. XLVII, No. 2, May 1965. The estimating equation is:

$$\log y' = \log C + d_0 \log w' + d_1 \log y'_{-1} \qquad \ldots (19)$$

$\log y'$, $\log w'$, $\log y'_{-1}$ are variables measured from their (logarithmic) sectoral mean values of period t.

(equation 18), 1.60 (equation 17), 2.70 (equation 19). By sectors, the results can be highly different according to the model used. For instance, for sectors A (processed foods, beverages, tobacco) and B (textiles, footwear and apparel, leather products), equation 18 gives estimates slightly lower than equations 17 and 19. For sector C (wood and cork and furniture and fixtures) in the case of two models the estimates are nearly similar, while with equation 19, values are substantially higher. Conversely, for other sectors (D, E, F) [1], the three models give results statistically equivalent; the estimates are not different from 0 in the two first groups and from 1.5 in the last group.

Heavy industries (sectors D and E) tend to have, on average, rather lower— in fact insignificant—elasticities of substitution than light industries (sectors A, B, C), and that fact supports the hypothesis that the elasticities of substitution tend to decrease as the share of capital rises. However, these lower or insignificant estimates obtained for heavy industries could also mean that σ is very small or that the data are not very good in accounting for the inter-industry annual fluctuations in demand for labour.

Other estimates

We now present two studies which differ from the preceding ones in their methods of estimation. Clague [2] is mainly concerned with directly estimating the elasticity of substitution in less developed countries by using capital-labour ratios and factor prices for Peru and the United States. The main conclusion is that elasticities are strikingly low, whatever the panel (the estimates are on average lower for actual plant figures than for engineering estimates). The smallness of these values does not seem to be due to the set of assumptions that were adopted. Moreover, the results are not very sensitive to alternative assumptions (for instance, the range is: 0.176 - 0.243 for leather products, or 0.446 - 0.578 for cement industry). It should be mentioned that Pack [3], although using an analogous method, arrived at estimates of elasticity of substitution significantly higher than 1; the data used involve firm-level observation in a number of less developed and more developed countries. However, the results cannot be directly compared in so far as the level of aggregation is not the same and consequently industry-wide averages

---

[1] D: rubber products, chemicals; E: basic metals, metal products, non-electrical and electrical machinery, transport equipment; F: paper, printing, publishing, etc.; non-metallic minerals.

[2] C. K. Clague: "Capital-labor substitution in manufacturing in underdeveloped countries", in *Econometrica*, Vol. 37, No. 3, July 1969.

[3] H. Pack: "The employment-output trade-off in LDCs: A micro-economic approach", in *Oxford Economic Papers*, Vol. 26, No. 3, Nov. 1974.

(Clague's sample basis) do not allow for an independent estimate of the production frontier. Of course, several objections can be made against directly calculating the elasticity of substitution from observed factor proportions. As Clague himself stated, under-estimation may be due to the optimality assumption of observed capital-labour ratios in each country. On the other hand, an upward bias can arise from the fact that Peruvian capital-labour ratios may be lower than the United States ones because "Peruvian workers, being less skilled, cannot handle as many machines as American workers". Nevertheless, this study is one of the rare examples of the direct estimation of elasticity of substitution on the basis of plant data.

Bassoco and Norton [1] also try to estimate directly the elasticity of substitution through a multi-product production function embodied in a linear programming model based on Mexican agricultural data. The two inputs used are medium-term capital (agricultural machinery, which is normally a direct substitute for labour) and non-homogeneous labour composed of hired and farm labour (including family workers). The cost of labour represents the total net income received by hired and farm labour per man-year worked. Total net income consists of wage payments and producers' profits, that is, economic rents received by farmers thanks to their own labour and to their rights to land and water. The cost of machinery is simply estimated by the rate of interest. In order to induce movement along the isoquant [2], increases in nominal wages (a component of the cost of labour) are stipulated. The estimates of the elasticity of substitution range from 1 to more than 3. The lowest estimates are derived from the assumed constancy of producers' profits, while the largest are obtained by assuming that the total value of production at endogenous prices is constant. [3] When nothing is assumed to be changed, the estimate turns around 1.4. The authors convincingly show that the elasticity of substitution obtained from the assumed constancy of producers' profit is under-estimated because there tend to be compensating changes in production levels and product prices. As can be seen, therefore, the size of estimates tends to show significant possibilities of substitution between capital and labour even if these estimates are based only on one element of fixed capital and even though the measurement of the cost of capital is much less elaborate than the cost of labour.

---

[1] L. M. Bassoco and R. D. Norton: *A quantitative agricultural planning methodology*, paper presented to the Sixth International Conference on Input-Output Techniques, Vienna, 25 April 1974.

[2] The isoquant is obtained from an envelope of production functions defined over multiple factors, in particular land and irrigation supplies.

[3] This case permits the measurement of a pure substitution effect as far as a unit isoquant is concerned.

CONCLUSIONS

The studies described above were attempts to evaluate possibilities of substitution between capital and labour by using the concept of the elasticity of substitution. Most of them estimated the elasticity indirectly within the production function framework under a large variety of simplifying assumptions. As a matter of fact, the most popular specification is to assume that labour productivity changes, between regions and over time, can be explained by wage-induced changes. In particular, no serious attempt was made to test other assumptions that are more realistic in an underdeveloped setting, such as dependency of the capital-labour ratio on labour productivity through technological change and economies of scale, under-utilisation of inputs (especially fixed capital), market imperfections, the relation between the level of aggregation and the size of the elasticity of substitution, and the effect of the passage of time. All these assumptions should be rigorously tested in order to assess their relative effect on the "apparent" value of the elasticity of substitution. Moreover, highly aggregate estimates tend to be meaningless because of the large variety of production conditions in a developing setting. Thus the heterogeneity of results is closely linked to that of assumptions. A production function taking into account the whole set of hypotheses relevant to developing countries is very difficult to estimate directly by traditional methods of estimation because of the lack of quantitative data and problems related to multicollinearity between explanatory variables. Whatever its weaknesses, the direct estimation of the elasticity of substitution avoids such restrictive assumptions to a large extent.

Two major conclusions related to the utility of the elasticity of substitution for policy making may be distilled from the foregoing discussion. The first concerns the usefulness of comparing estimates of the elasticity of substitution drawn from different sample bases and levels of aggregation and different equations reflecting varying assumptions about the production function. The second major question, which is strongly related to the first, is whether the elasticity of substitution can in any way be measured, so as to constitute an adequate indicator of actual economic behaviour. As regards the first major question identified here, comparability of estimates of the elasticity of substitution (as between the same sector in two countries, or two different sectors in the same country, etc.) requires very strict comparability of both the treatment of data and the estimation procedures. This restriction is not often taken into account as fully as is necessary. Turning now to the second major question, the difficulties of comparison that have been mentioned render estimates of the elasticity of substitution a doubtful indicator on which to base economic policy formulation. Its utility could be greatly reinforced if support for prima facie estimates could be obtained from empirical surveys, field investigations, and interviews with entrepreneurs.

These ideas of how to treat the elasticity of substitution concept in practice deserve to be spelt out more fully. Suppose, for example, that an economic planner in a typical developing country x requires information about the value of elasticity (σ) for policy purposes, but that an estimate of elasticity for all manufacturing may be derived only from pooled data for a sample of countries in which country x is not included. The traditional argument would be that developing country x more or less follows the average behaviour and that an estimate of the elasticity based upon such a sample can therefore be usefully applied in that country. As we saw, however, the elasticity is likely to be sensitive in particular to the degree of aggregation and if the output mix of country x is significantly more or less concentrated than that of the average country in the sample, then, other things being equal, the sample estimate of elasticity cannot be useful, or at least not directly; in fact if the output mix in x is more dispersed than the output mix in the sample, we would expect σ to be less than the sample estimate of σ. It follows in this case that any effort to reduce wage increases in country x is likely to have a weaker effect on capital-labour substitution than such an effort in the countries of the sample. Suppose, on the other hand, that additional information is available, in the form of sufficient time series for country x to enable an estimate of σ to be made. The economic planner would then have two estimates of σ, say, σ_x^t (time series estimate for x), and σ^c (cross-section estimate for the sample referred to above). In general, we would expect that $\sigma^c \geqslant \sigma_x^t$. Both contain bias to an extent that cannot be measured, and are based on hypotheses which may or may not be valid for country x. Thus if our economic planner intends to use information about elasticity it would in general be desirable for additional information to be collected (for example, by surveys of major industries) so as to ascertain which of the estimates is to be preferred.

INDIRECT EMPLOYMENT EFFECTS OF INVESTMENT

3

by J. Krishnamurty [1]

The concept of indirect employment has been used in a variety of ways in writings on economic development, and there are many issues of concept and measurement in that connection which need to be sorted out since they have important implications for technological choice and investment allocation. The different meanings attached to the concept are considered in this chapter, and an attempt is made to relate each of them to a general inter-industry framework. The strengths and weaknesses of the concept and the extent to which it can be applied in the practical context of development programmes of developing countries is also examined.

The implications of indirect employment effects are considered at macro-economic, sectoral and project levels of aggregation. It is often argued nowadays that the emphasis of output and employment planning needs to be shifted from the aggregate growth path to the more crucial question of the output mix and its effects on growth through backward and forward linkages. Such a shift in emphasis has an obvious implication for industrial planning and the choice of particular industries.

For each industry the choice of technology may influence the direct and indirect generation of employment. The direct generation effects are illustrated by several of the case studies in this volume (e.g. Chapter 6 on textiles, Chapter 7 on sugar processing, Chapter 8 on cement blocks). The indirect employment effects of alternative technologies cannot be examined empirically because of the limitations of input-output data. This important point is discussed further below. It is, however, worth emphasising that although most of the discussion here is perforce in terms of inter-industry choice, indirect employment effects can equally well be related to alternative technologies within each industry. It is possible to show that industries and techniques which

---

[1] St. Antony's College, Oxford.

generate a relatively large amount of direct employment might not generate a relatively large amount of total employment, in the same way as activities which generate a relatively large total output might not generate a relatively large amount of total employment. Thus a simple direct employment criterion or a simple output-linkage criterion in the choice of industry and technology could give misleading results. Which industries and techniques yield large direct plus indirect employment effects is therefore a crucial question for employment policy.

In many situations indirect employment effects may not make much difference to the choice of technique. For example, consider the choice between handloom cloth and mill cloth. Since both are cloth and since both are made of mill-spun yarn, indirect effects would be the same. Instead of handloom cloth, however, consider *khadi* (hand-woven cloth made from hand-spun yarn), and the indirect effects are no longer equivalent, because the difference between hand and machine spinning makes for differences in the corresponding indirect effects.

When we consider industries producing a very similar but not identical product, using highly processed but substitutable inputs, indirect effects could be important in determining technical choice. For example, in the furniture and building industries it is possible, at least partially, to substitute bricks and wood for steel and cement. Clearly, the indirect effects of an expansion of cement-intensive and steel-intensive housing would be different from those of an expansion of housing with intensive use of brick and wood. Since construction looms so large in development plans, the choice of brick-intensive and wood-intensive construction would have pervasive indirect employment and output effects.

Even within agriculture there are considerable possibilities of input substitution. The choice between tractors and bullocks or between commercial and natural fertilisers involves taking into account the indirect demand for factors of production and resources in the process of supplying these inputs to agriculture. Again indirect effects are clearly relevant.

Indirect employment effects in technical choice are defined very differently by Pack. [1] In his view, when choosing between two techniques there is "an indirect effect resulting from the saving in initial capital expenditure due to the lower capital-output ratio of the less capital-intensive technique.... The expression for the indirect increase in employment assumes that any capital saved can be used to generate further employment that requires no higher a capital-labour ratio than the labour-intensive textile techniques." Pack also shows that if the Philippine economy of 1970 were to adopt 1950 spinning

---

[1] See Chapter 6, p. 169.

plants and Lancashire looms for their textile industry instead of 1968 plants and Battery-Unifil looms, huge direct and indirect increases in employment would result.

Pack's indirect employment effects do not correspond to those we discuss in this chapter. He is really attempting to provide an employment measure of the saving in capital resulting from a particular technical choice; while we seek to measure the additional employment created as a consequence of the expansion of the input-supplying and output-demanding industries, as a result of the expansion in final demand or output of a particular industry or set of industries.

THE CONCEPT OF INDIRECT EMPLOYMENT

The idea that the expansion of any particular activity involves the expansion of other input-supplying activities is simple and obvious. [1] Economists have always been aware of inter-sectoral relations. In Ricardo, the profits of industry depended on the price of corn; in Walrasian equilibrium, all prices were inter-related and simultaneously determined: and Marshall, for example, clearly was aware of inter-industry transactions in his concept of derived demand. When Kahn introduced the employment multiplier into economics, he clearly distinguished between inter-industry and multiplier effects. According to him—

The increased employment that is required in connection actually with the increased investment will be described as the "primary" employment. It includes the "direct" employment, and also, of course, the "indirect" employment that is set up in the production and transport of the raw materials required for making the new invest-ment. To meet the increased expenditure of wages and profits that is associated with primary employment, the production of consumer goods is increased, and the effect is passed on, though with diminished intensity. . . . The total employment that is set up in this way in the production of consumption goods will be termed "secondary" employment. [2]

Subsequently Keynesian economists emphasised the effect of an autono-mous increase in expenditure on output (and hence employment). The possibi-lity of time lags between rounds of expenditure and inelasticity in the response

---

[1] As Evans and Hoffenberg put it in National Bureau of Economic Research: *Input-output analysis: An appraisal*, Studies in Income and Wealth, Vol. 18 (Princeton University Press; London, Oxford University Press, 1955), p. 55, "the automobile assembler who notes that a finished passenger car requires five wheels, five tyres and twenty-one wheels nuts, and balances his future production and purchasing plans accordingly, is applying a fundamental principle of the input-output approach".

[2] R. F. Kahn: "The relation of home investment to unemployment", in *Economic Journal*, June 1931, reprinted in Kahn: *Selected essays on employment and growth* (Cambridge University Press, 1972), p. 1.

of global or sectoral output, or both, were taken into account. With the incorporation of the accelerator principle, in the models of Harrod and Domar the growth of output in an economy was regarded as being determined by the savings rate and the capital-output ratio. From the point of view of economic planning this meant that once the planned savings rate and capital coefficient were determined, the growth of output would be known, and then the commodity composition of that output could be chosen, subject of course to its being consistent with the assumed savings rate and capital coefficient. Given the composition of output by branch of activity, employment could be estimated using employment norms or coefficients for each activity. [1]

However, output targets are difficult to specify in detail and are often provided only for agriculture and large-scale manufacturing. The food targets usually have some basis and the industrial targets are based on the announced production plans of the government and of private firms. The targets for commercial crops are usually fixed in relation to the likely expansion of manufacturing. In addition, some account is taken of the foreign trade situation and prospects, and a certain increase in services is assumed to be associated with the envisaged expansion of physical output. Quite clearly this approach involves a great deal of guesswork because there are large gaps in the information, and as a result estimates of employment generation are likely to prove treacherous even if based on impeccable employment coefficients.

There also tends to be a certain amount of confusion between final demand and total output. Inter-industry requirements obviously cannot be estimated without an inter-industry table, but what tends to happen is that they are to some extent taken into account in some sectors, and are just ignored in others. We therefore obtain a vector of output for the terminal year of the plan in which some entries are of final demand, some of total output (including inter-industrial uses) and others are a mixture of the two. The resulting employment estimates would clearly be unsatisfactory.

There would be no problem if, for a particular year, we knew the total output (inclusive of inter-industry transactions) and the employment coefficients of each sector. This is often possible for an historical year for which the results of a fairly comprehensive census of production and employment are available. In such a case employment coefficients are derived from the

---

[1] Many refinements are possible here: the marginal rate of saving might be set higher than the average, or it might be interpreted in physical terms, and the number of sectors might be increased somewhat, on analytical, institutional or empirical grounds, but essentially all such approaches aim at determining the level of total output and then obtaining the composition of output in a rather rough and ready way. Using fixed employment-output ratios for each sector estimates of employment are obtained. For a description of this procedure in the Indian context, see Government of India, Planning Commission: *Report of the Committee of Experts on Unemployment Estimates* (New Delhi, 1970), Ch. II.

actual estimates, and the errors, if any, come from the census or survey and are not the result of judgements with little or no factual basis. But even so satisfactory a set of estimates cannot answer two kinds of questions which interest the planner.

The planner might want to know the relationship between final demand and the structure of total output, so that from a given pattern of final demand for some future year he could infer the required composition of output at that date. While a census of production and employment would give him the necessary information for some period in the past, it would not in itself answer his question. Secondly, assuming that the planner has some freedom to choose the composition of final demand, he would need information on the resource implications of alternative choices. He might use sectoral capital-labour and capital-output ratios obtained from historical data, but this would indicate only the factor supplies going directly into the production of output, no account being taken of the indirect demand for factors of production created by an increase in inter-industry requirements. [1]

An input-output framework is of considerable help in tackling such problems. Given its assumptions (on which more later), it gives the pattern of flows of inter-industry purchases so that for the ith industry

$$X_i = x_i + \sum_{j=1}^{n} a_{ij}X_j$$

where X_i is the value of the total output of the ith industry or sector, x_i is the final demand of that sector and a_{ij} is the value of the purchase by the jth sector of the output of the ith sector per rupee [2] of total output of the jth sector. [3]

The inter-industry table does indicate the amount of labour and capital (or wages and profits) directly involved in (or arising from) the production of total output of any sector. In fact the Leontief assumption is made that

---

[1] Ignoring the indirect effects would give highly misleading measures of labour productivity, based on industry measures rather than system measures. We discuss the question in a later section. See W. Gossling and F. Dovring: "Labour productivity measurement: The use of subsystems in the inter-industry approach and some approximating alternatives", in *Journal of Farm Economics* (Lexington, Ky.), Vol. 48, No. 2, May 1966; also S. Gupta and Ian Steedman: "An input-output study of labour productivity in the British economy", in *Bulletin of the Oxford University Institute of Economics and Statistics*, Vol. 33, No. 1, Feb. 1971.

[2] Throughout this chapter, for convenience, "rupee" stands for any measure of the value of output.

[3] Given the a_{ij} which form the coefficients of the inter-industry table, for every vector of x_i, $\begin{bmatrix} x_1 \\ \vdots \\ x_n \end{bmatrix}$, there would exist one and only one vector of X_i, $\begin{bmatrix} X_1 \\ \vdots \\ X_n \end{bmatrix}$. This would in effect satisfy our first requirement that the planner would need to know the link between any particular pattern of final demand and its corresponding pattern of total output. Of course all possible pairs of final demand and total output vectors would not be feasible as resource constraints would restrict choice.

capital-labour and capital-output ratios are fixed for a particular sector and invariant with respect to the scale of production. We are interested in discovering the output and employment implications of the expansion in final demand of a particular sector, and for this purpose we need to know both the factor-output ratios of the different sectors and the increases in output in each of the sectors necessary to sustain the expansion of final demand in our chosen sector. [1]

To maximise employment through the choice of the product mix, the planner should give as high a weight as possible to sectors with high total employment creation per rupee of final demand. While there would be many limits on what the planner could do, he would have some power to control final demand through taxes and subsidies, or directly by changing income distribution and the availabilities of goods and services. Though these constraints might in practice be very complex, the essential logic of the above procedure is unaffected.

The employment potential of a sector, measured as the direct and indirect employment attributable to a unit increase in the final demand of that sector, has in a number of analyses been regarded as the critical ratio for employment policy. [2] Even accepting the adequacy of the static input-output framework, the employment potential measure does not by itself provide a satisfactory criterion because no account is taken of the capital (or other scarce resource) requirements of an increase in final demand. [3] The direct and indirect capital

---

[1] To obtain this, we assume zero final demand for all sectors other than the chosen one, for which we assume unit final demand. We define A_{ij} as an element of the $(I-A)^{-1}$ matrix, l_i as the direct labour coefficient (W_i/X_i), and we find

$$\begin{bmatrix} l_1 & & & \\ & l_2 & & 0 \\ & & \ddots & \\ 0 & & & l_n \end{bmatrix} \begin{bmatrix} A_{11} & \bullet\bullet\bullet\bullet\bullet & A_{1n} \\ \vdots & \ddots & \vdots \\ A_{n1} & \bullet\bullet\bullet\bullet\bullet & A_{nn} \end{bmatrix} \begin{bmatrix} 1 \\ \vdots \\ 0 \\ 0 \end{bmatrix} = \begin{bmatrix} l_1 A_{11} \\ \vdots \\ l_n A_{n1} \end{bmatrix}$$

This means that to increase final demand of sector 1 by 1 rupee, $l_1 A_{11}$ rupees' worth of labour would be required directly and $\sum_{i=2}^{n} l_i A_{i1}$ rupees' worth would be required indirectly. If w_i is the wage rate of the ith sector, the direct and indirect employment created would be: $\sum_{i=1}^{n} l_i A_{i1}/w_i$; out of this $l_1 A_{11}/w_1$ would be direct employment and $\sum_{i=2}^{n} l_i A_{i1}/w_i$ would be the indirect employment created.

[2] See, for example, B. R. Hazari and J. Krishnamurty: "Employment implications of India's industrialization: Analysis in an input-output framework", in *Review of Economics and Statistics*, Vol. LII, No. 2, May 1970.

[3] In our earlier equation we might replace $\begin{bmatrix} l_1 & & \\ & \ddots & \\ & & l_n \end{bmatrix}$ and substitute

requirements of sustaining a unit increase in final demand of the sector would be $\sum_{i=1}^{n} k_i A_{i_1}$. The corresponding labour requirement is $\sum_{i=1}^{n} l_i A_{i_1}/w_i$. Given the scarcity of capital and the abundance of labour, we would seek so to choose the final demand vector that the capital-labour ratio is the lowest.[1] This is obviously an over-simplification, since the over-all capital intensity will be only one of many considerations the planner would keep in mind. But it is still important to note that this approach makes it possible for the planner to be aware of the capital and labour implications of any particular choice he may make. However, the weight he attaches to capital intensity in relation to other considerations is an entirely different matter.

It is often suggested that industries with high forward and backward linkages should have priority in expansion, on the ground that their expansion triggers off growth along a wide front by generating a substantial demand for inputs from other sectors, or by causing substantial increases in the output of industries to which they supply inputs, or in both ways simultaneously. On this basis a key sector is defined as one for which the sum of backward and forward linkages is relatively large. Key sectors could alternatively be defined in terms of their employment, capital or other resource use effects. To do this we need to know sectoral labour (or capital or specific resource) coefficients.

We now seek to relate the factor and output implications of a unit expansion either in the final demand of the jth sector, or in the supply, for inter-industrial use, emanating from the ith sector. In practice it is more convenient to conceive of a unit expansion in the output, but not the final demand, of the jth sector. In this case, the backward linkage in output terms would be $\sum_i A_{ij}$ and the forward linkage would be $\sum_i A_{ji}$ and the total linkage would be $\sum_i (A_{ij} + A_{ji})$. In a similar way the capital and labour implications of total linkages of a unit expansion in the output of the jth sector available for inter-industry use would be $\sum_i k_i (A_{ij} + A_{ji})$ and $\sum_i l_i (A_{ij} + A_{ji})/w_i$. Using this method we can rank sectors and industries in terms of their capital intensity ($\sum_i k_i (A_{ij} + A_{ji})$ /

$$\begin{bmatrix} k_1 & & & \\ & \bullet & & \\ & & \bullet & \\ & & & \bullet \\ & & & k_n \end{bmatrix}$$ which would be the diagonal matrix of capital coefficients.

[1] In other words, out of all feasible final demand vectors that the planner is considering, he would choose that for which the ratio:

$$\frac{\sum\limits_{j}^{n} \sum\limits_{i}^{n} k_i A_{ij}}{\sum\limits_{j}^{n} \sum\limits_{i}^{n} l_i A_{ij}/w_i}, \text{ is the lowest.}$$

$\frac{\Sigma}{i} l_i/w_i(A_{ij} + A_{ji}))$ or in terms of their capital-output ratios $(\frac{\Sigma}{i} k_i (A_{ij} + A_{ji}) / \frac{\Sigma}{i} (A_{ij} + A_{ji}))$ or in terms of their labour-output ratios $(\frac{\Sigma}{i} l_i/w_i (A_{ij} + A_{ji}) / \frac{\Sigma}{i} (A_{ij} + A_{ji}))$.

These ratios need to be carefully distinguished from conventional ratios for which the same terms are used. We are taking into account the indirect capital and labour requirements of the expansion of a sector: these include the capital and labour inputs absorbed by sectors in the process of supplying inputs to our sector, as also the capital and labour requirements of sectors receiving inputs from our sector and, as a consequence, producing additional output.

LIMITATIONS OF INPUT-OUTPUT METHODS

At this point we consider briefly the rather well known defects of input-output analyses of the open static variety. We have to assume that each commodity is supplied by a single sector with one method of production only, and that the sector in question produces no other output. The inputs purchased by a sector are generally assumed to bear a fixed proportion to the output of the purchasing sector. We have also to assume away external economies and diseconomies. As long as we consider an input-output table for a particular year in the past, these limitations are not serious. The estimates of flows of inputs and of final demand in value terms are in a sense the result of a whole series of adjustments of commodity and factor prices. But the moment we use such a system for projection, the assumptions come into play and cannot be ignored. The total output, when projected on the basis of a future final demand vector and historically determined production coefficients, is assumed to be unaffected by changes in relative prices and availabilities of factors and commodities which might take place and which would alter the actual values of the coefficients. [1]

The key assumption mentioned above, of only one method of production, is of course very restrictive, and also contradicts the theme of this volume. In theory it is possible to envisage an input-output table in which alternative technologies might be included in the form of separate vectors of material and factor input coefficients. In practice this data requirement is too comprehensive and the input-output coefficients normally calculated are averages over a spectrum of technologies and products.

---

[1] For excellent discussions of the assumptions and limitations of input-output analysis, see H. B. Chenery and P. G. Clark: *Interindustry economics* (New York, Wiley, 1959), and National Bureau of Economic Research: *Input-output analysis: An appraisal*, op. cit., especially pp. 9-21, 53-124, 137-168 and 169-173.

It is of course possible to take account of many of the other objections to the basic model. Dynamic input-output methods and activity analysis do attempt to do precisely this. But in the context of the rather shaky data base of the less developed countries we face the hard choice between following methods that ignore indirect output and employment (some of which we outlined earlier) or methods that are powerful and useful, even if based on rather restrictive assumptions.

There are two situations where the proportionality assumptions of input-output analysis can be relaxed to some extent. First, we might examine an economy over a period in the past with an inter-industry table each for the initial and terminal year. By making suitable assumptions about the path of changes in coefficients between the two points of time, or by examining the values of parameters at the two points of time, deductions that are relatively independent of the assumption of the fixity of coefficients can be made regarding the behaviour of the economy.[1] However, this is not of much help in projecting inter-industry relationships into the future, and yet that is essential for planning. It could, of course, be assumed that the pattern of change in input coefficients revealed for the past continues in the future, but clearly such extrapolation implies the assumption that relative prices will move in accordance with the projection so that the assumed input coefficients matrix for the terminal year of the projection does in fact come into being. Secondly, the proportionality assumption can be diluted somewhat in respect of employment effects. Chenery and Clark, for example, claim that since employment does not have a feedback into the model, and since we are simply calculating some further economic implication of the indicated pattern of expansion, "employment effects are probably best estimated in this way since they are non-proportional and vary from industry to industry because of such factors as nature of work, productivity trends and variation in hours".[2] This would be more difficult in the case of capital though we could assume, with some loss of realism, that the stock of capital available for a given year is already known and is therefore independent of the production coefficients.

Non-proportionality in employment effect cannot, however, have much play because there is in fact a certain degree of feedback of some of the consequences on the employment front on to the pattern of input-output relations. In terms of our earlier definitions we have $W_i = l_i X_i$ where W_i is the wage bill, l_i the employment coefficient and X_i the total output of the ith sector. We can also write $W_i = w_i h_i L_i$ where w_i is the wage per hour, h_i is the average

---

[1] See for example Howard Pack: *Structural change and economic policy in Israel* (New Haven and London, Yale University Press, 1971), especially p. 49.

[2] H. B. Chenery and P. G. Clark, op. cit., p. 146.

number of hours worked by each labourer and L_i is the number of labourers engaged, in the ith sector. The latter equation could be rewritten as $L_i = W_i/w_i h_i$ so that the numbers directly employed in the ith sector would depend on the wage bill, the hourly wage rate and the average number of hours worked by each labourer. The wage bill would have to be a fixed proportion of the total output of the sector; if it were not, there would be commodity and factor substitution along with output expansion and this would violate the assumption of fixed input coefficients. Similarly, the wage rate per hour of standardised labour must also be deemed fixed to avoid the possibility of commodity and factor substitution. This, of course, means that W_i/w_i, i.e. the number of man-hours put in that industry, is a fixed proportion of total output X_i. The number of workers employed, L_i, could vary widely depending on the number of hours each one works, and in fact this is all the freedom of manoeuvre that can be accommodated within the rather restrictive assumptions of input-output analysis.

Let us see how this could be used in practice. Assume we have an inter-industry table for 1970 and have obtained the inverse. For the vector of final demand (x_i^0) substitute an assumed or planned vector (x_i^1) relating to the year 1975, the terminal year of our planning model. The vector of total output (X_i^1) can easily be obtained using the inverse and the final demand vector. For any sector j we can obtain the number of man-hours worked, W_j/w_j, equal to $l_j X_j/w_j$. We know the values of l_j and w_j which are assumed to remain at their 1970 level and we also know the value of X_j determined for 1975 from the inverse and the assumed final demand vector for 1975. The value of $L_j (= W_j/w_j h_j)$ would depend on our assumption regarding h_j, i.e. the manner in which man-days or man-years are distributed among workers. The number employed in the jth sector in 1975 would be larger (or smaller) depending on whether we assume a lower (or higher) value of h_j which also need not be constant between 1970 and 1975. In fact a whole family of different estimates of the numbers employed in each sector could be obtained by letting h_j take different plausible values. [1]

Suppose that the choice were between cotton textiles and metal products. The number of man-hours of work created by a unit expansion of the final demand for metal products might be much larger but the number of jobs created might be relatively small, being mainly in high intensity sectors like

---

[1] If we are interested in inter-sectoral comparisons of direct and indirect employment creation, our analysis has relevance. As we have already shown, a unit expansion of sector 1 creates $(l_1 A_{11} + \ldots + l_n A_{n1})$ as the wage bill. This would imply $\sum_{i=1}^{n} \dfrac{l_1 A_{i1}}{w_i}$ man-hours of work and $\sum_{i=1}^{n} \dfrac{l_1 A_{i1}}{w_i h_i}$ jobs. We now have two criteria, the man-hours and the number of jobs created, to help us in choosing between alternative final demand vectors.

Table 8. Total employment generated by a final demand of 10 million taka in the *j*th sector in Bangladesh, 1964-65

Sector	Total number of jobs created	
	Adjusted	Unadjusted
Agriculture	1 127	12 886
Industry	7 382	9 809
Construction	5 329	7 198
Electricity and gas	1 701	2 200
Transport	4 556	4 785
Trade	7 972	7 992
Government and services	5 178	5 416

Source: Iftikhar Ahmed: "Sectoral employment response in an input-output framework: The case of Bangladesh", in *Bangladesh Economic Review* (Dacca), Vol. I, No. 3, July 1973, p. 322.

iron mining and steel making. On the other hand, the expansion of final demand for cotton textiles might create many low-intensity jobs in agriculture and rural services. Our analysis makes it possible to take explicit note of the possibility of under-employment and also of different choices between alleviating under-employment (involving a high h_i) or expanding the number of jobs each at low intensity. This is relevant not only for choosing between sectors in final demand but also for deciding on the strategy for a particular sector—on whether to expand employment or reduce under-employment.

Let us try to apply the distinction between man-hours and jobs created to the real world. Consider the Bangladesh economy in 1964-65 for which we know the direct and indirect employment created in each of its seven sectors, assuming in each case that there was a 10 million taka level of final demand in the chosen sector, all others having zero final demand. In table 8 we give the total employment generated by the expansion of final demand in each sector. We present two estimates, one without adjustments and the other assuming that all increases in labour demand in agriculture take the form of fuller employment, not new jobs. It would appear that this assumption substantially alters the ranking of sectors. Of course, we have made a rather extreme assumption, but it must also be remembered that if we had a more detailed sector breakdown, more interesting results might arise.

There is, therefore, a limited degree of flexibility in the choice of labour coefficients. The total number of jobs created per unit increase in final demand of a particular sector could vary depending on the number of man-hours worked per worker in each sector—which would depend on a number of economic, political and other considerations and is not "given" in the sense that the wage bill or the hourly wage rate are. In actual practice even wider latitude is possible. The case for input-output analysis rests not on an assumption that it provides a total representation of reality but on the fact that it provides a reasonable approximation to it. Provided that the input coefficients are not very sensitive to changes in factor and commodity prices, some change in those prices could be tolerated by the system without any serious problems. Some direct ad hoc changes in input coefficients can also be made to take account of new methods of production, or for any other reason, without having to obtain fresh estimates of the entire set of input coefficients. To the extent that this is possible the practical utility of input-output analysis is greatly enhanced.

A simple static input-output model cannot take into account technological change within a sector because the inter-industry and factor-use coefficients are fixed and in theory cannot be varied. In addition, because a sector usually covers a number of products we have to assume that the internal composition of a sector does not change in such a way as to cause changes in the sectoral coefficients. In actual applications, sectors could be demarcated on the basis of known uniformities. For example, traditional and modern branches of manufacturing might be separated if it is possible and desirable to do so. [1] This might be done for cotton textiles; where there is a wide spectrum of techniques in operation, the industry is often large and bifurcation makes sense. In a number of cases the quality of the available information might militate against attempting disaggregation. Some ad hoc adjustments of coefficients may be possible and desirable, provided that they do not upset the system unduly.

For examining the effect of different technical choices on total employment, it would be necessary to have entirely new tables that contain an input vector for each technique. A feasible approach would be to trace the employment implications of a major change in the purchased inputs of different techniques by actual sampling in one industry, or by comparing existing vectors for countries at different stages of technological development. For example, the input vector for weaving in the Federal Republic of Germany and in the Republic of Korea might be incorporated into the Indian input-output table to ascertain the differential employment effect (direct and indirect) of the choice of say, the Korean labour-intensive method versus the German one. [2]

---

[1] The only attempt to separate the modern from the traditional sector in the input-output framework is by Thorbecke. See ILO: *Sharing in development: A programme of employment, equity and growth for the Philippines*, op. cit., Special Paper 19, on inter-sectoral linkages and indirect employment effects in the Philippines. In that study the traditional and modern components have been treated as separate sectors for transport and communications, and consumer goods. While traditional consumer goods ranked fourth, modern consumer goods ranked twenty-third out of the 33 sectors ranked in terms of employment generation (direct and indirect) per unit of final demand.

[2] This point is due to Howard Pack.

It might be better in many cases to examine the problem at the industry or firm level rather than at the economy-wide or inter-sectoral level. In many cases the total factor and resource implications of alternative techniques can be roughly estimated. Whatever the criterion for technical choice—capital intensity, capital-output ratio, surplus per unit of capital, foreign exchange or other scarce resources required per unit of output—the actual values of the ratios would depend on whether indirect effects are taken into account or not.

The input-output system is particularly suited to deal with the choice of the product mix and to indicate the output and factor-use implications of alternative patterns of final demand. Here again, restrictive assumptions of fixed coefficients are involved.

In simple input-output exercises where transactions with the rest of the world are not clearly distinguished, estimated indirect effects may not take place within the country but instead may arise in input-supplying industries located in other countries. While exports and imports directly entering into final demand can be easily handled, a matrix of sectoral import coefficients (m_{ij}) is required to take account of imported inputs. These m_{ij} matrices would indicate the imports of the ith commodity needed to produce a unit of output in the jth sector. The m_{ij} along with the existing a_{ij} make possible estimates of import effects, but involve the rather restrictive assumption that the share of imports in the supply of inputs from the ith sector going to produce a unit of output in the jth sector is constant. This means that although we can calculate the import implications of alternative final demand vectors, we would need to assume a fixity in the share of imports in the total supply of inputs originating in each sector. We might seek to reduce imports by choosing an appropriate final demand vector, but this really implies choosing a composition of total output to achieve that objective without a reduction in any of the import coefficients. However, it is possible for projection purposes to adjust the individual import coefficients revealed in the past to take account of changes expected in the future. It is also necessary to plan for an appropriate change in some or all import prices in order that the adjusted import coefficients will become a reality. This could be achieved through tariffs or quotas, or by altering the exchange rate. While this might work well in practice, there would still be serious problems due to the intrinsic inability of the planner to control economic activity outside his own economy.

Another problem in using input-output analysis for planning is that the contemplated expansion in output may not occur. The possibilities of expanding employment may be restricted by the inelastic supply of wage goods and the fixity of real wages. The contemplated expansion in employment can take place only within these limits, and it is no longer enough to calculate indirect and direct employment effects; we have to verify whether there is an adequate wage goods output to sustain them. [1]

From many aspects a more useful way of proceeding would be to estimate the employment that can be sustained in an economy in a particular year and then examine different final demand vectors consistent with this ceiling and obtain the capital (or scarce resource) implications of each. The longer the period we consider the less serious would be the ceiling set on employment possibilities and the greater the scope for indirect employment effects. The wage goods limitation does not invalidate the concept of indirect employment; it is just one of a number of considerations that militate against a naïve approach under which the maximisation of total employment becomes the over-riding objective of economic policy. From the viewpoint of employment planning, the concept of indirect employment is clearly relevant, but indirect employment has real resource costs which can be measured, and there are over-all limits to employment expansion set by the availability of wage goods. For an economy with a given factor endowment, when we choose a final demand vector we do consider its implications in terms of factor use. However, no set of criteria which ignore indirect employment effects would be complete; at the same time these effects as such do not constitute adequate criteria. The inter-industry framework provides us with a means of linking the output, capital, labour and foreign trade implications of an increase in final demand. We need to know these relations even if precision is not always attainable in their measurement, and even if these relations are not as stable as we should like them to be.

SOME EMPIRICAL TESTS

It might be argued that indirect effects exist but can safely be ignored. In other words, from the point of view of decision making, it may not matter very much if we do not know the indirect employment component. This is clearly a question which has to be decided on the basis of the available empirical evidence.

Gossling and Dovring do show that while theoretically similar indices of labour productivity of agriculture in the United States since 1919 show the same broad pattern, the inclusion or exclusion of indirect effects from the index does make a considerable difference to the resulting estimates. In particular, the contribution of increased productivity in non-agriculture is taken into account in the productivity estimate for agriculture. They show how the number

---

[1] See Amartya Sen: *Employment, technology and development*, op. cit.

of man-years contributed by non-agriculture to agriculture (in producing inputs for it) remained virtually constant since 1919. The use of an index which takes into account indirect effects indicates better the sources of the productivity increase. [1]

Gupta and Steedman emphasise the distinction between industry and system productivity. As they argue—

It is quite conceivable, for example, that over a certain period the industry in question might substitute manufactured inputs for labour in its production process, with the result that its normally measured labour productivity rises rapidly. But when the labour used in the production of these manufactured inputs is taken into account, it might well be found that the total amount of labour used, somewhere in the economy, to produce one unit of final output of the industry in question has fallen only slightly. In other words, the *industry* measure of labour productivity, in which gross output of an industry is related to industry employment, may move quite differently from the *system* measure of labour productivity, in which final output of the industry is related to the *total* amount of labour used in its production . . . [2]

For the British economy they find that for 1954, 1960 and 1966 the rankings of industries by direct and by total labour use were not very different. The rank correlation coefficients were 0.69, 0.69 and 0.68 respectively, and were highly significant. At the same time, while the over-all rankings could be fairly close, for particular industries the ranks might vary widely.

It is interesting to compare the ranks of different sectors on the basis of employment potential per unit of final demand (including indirect employment) on the one hand and direct employment per unit of output on the other. We have performed this exercise for India and the Philippines using the Hazari-Krishnamurty and Thorbecke estimates. [3] For India the Spearman rank correlation coefficient is positive and significant but rather low (0.28). For the Philippines the rankings are very close, the correlation of ranks being 0.9. So in the case of the Philippines there may not be much of a difference caused by using one ratio rather than the other, whereas in India that is not the case. It is also interesting that whereas in India the ranks of total output and of total employment per unit of final demand bore some similarity, the Spearman coefficient being 0.38, in the Philippines the corresponding coefficient was −0.3.

The limited evidence is certainly enough to suggest that indirect employment effects in developing countries are important, and that ignoring them can in some cases lead to serious errors; also a policy of maximising output

---

[1] See W. Gossling and F. Dovring, op. cit.

[2] See S. Gupta and Ian Steedman, op. cit., p. 21.

[3] See B. R. Hazari and J. Krishnamurty, op. cit., and ILO: *Sharing in development: A programme of employment, equity and growth for the Philippines,* op. cit., Special Paper 19.

creation per unit of final demand might conflict with maximisation of total employment per unit of final demand.

SOME FURTHER IMPLICATIONS

While we have discussed indirect employment effects in the context of economy-wide planning and inter-industry analysis, such effects are also capable of being investigated in a much more general manner. For instance, there are a number of analyses of the inter-relationships between agriculture and manufacturing (in some cases, non-agriculture) in the process of economic development. [1] The process of growth can be visualised as a series of inter-sectoral interactions. It is also meaningful to consider indirect employment effects at the level of an industry or a project or a specific scheme. In such cases the estimation of indirect effects may be approximate and will not normally require the construction of an input-output table for the economy as a whole. In many ways such methods predate input-output methods and may be free of some of the restrictive assumptions of the static input-output systems.

Agriculture-industry interactions

The growth of the two sectors generates a number of interactions. The growth of agriculture, which has a large weight in most of the less developed countries, has straightforward final demand linkages. The growing agricultural income is increasingly expended on industrial goods, thus providing a stimulus for industrial expansion. At the same time there are forward linkages in the form of agricultural raw materials supplied by the agricultural sector to the non-agricultural sector. Similarly, the expansion of the latter sector can be viewed as affecting the former in a number of ways. To some extent the growth of non-agricultural income provides additional demand for agricultural (food) produce. At the same time the demand for agricultural raw materials can be viewed as a backward linkage and the supply of tractors, fertilisers, pesticides, and even cement and bricks, as inputs going into agriculture. To the extent that agricultural expansion is contingent on the more intensive use of the area already under the plough, the dependence on the non-agricultural sector as a source of inputs necessary to increase agricultural production will be all the greater. Also, provided that a substantial proportion of the industries established in the course of development process agricultural raw materials, this could give rise to considerable direct and indirect effects.

---

[1] See Bruce F. Johnston's masterly survey: "Agriculture and structural transformation in developing countries: A survey of research", in *Journal of Economic Literature*, Vol. VIII, No. 2, June 1970; also UNIDO: *Industrial development survey*, Vol. IV (New York, United Nations, 1972), Ch. V.

The direct employment effects of the growth of the agricultural sector would depend on the particular technology adopted. New methods of production need not be labour-displacing, but in practice they all too often tend to be. For instance Raj Krishna [1] has shown that on balance the New Agricultural Strategy as it has developed in some parts of India would tend to be labour-displacing, largely owing to the introduction of tractor technology. He considers the indirect effects, both through increased incomes and the multiplier process and through technical interdependence. The coefficients for the agricultural sector are adjusted to take account of the new technology in agriculture, but for non-agriculture the coefficient estimated from the 1964-65 data are retained. While the indirect effect on employment of the expansion of the non-agriculture sector in response to the growth of agriculture and its input demands is positive, Raj Krishna suspects that even if agriculture grows at about 5 per cent per annum, the net effect—taking into account direct and indirect effects—could be negative if the labour force grows at more than 2 per cent per annum, which in the Indian case it probably will. Raj Krishna's estimates of the indirect effects could be over-generous. Given agriculture's weight in the total output and the input requirements of the New Agricultural Strategy, the internal composition of the manufacturing sector would be substantially altered in the process. [2] For example, the share of fertilisers in manufacturing output would substantially increase, and this may not prove employment-intensive in view of the indirect effects. On the other hand it is possible that the expansion of the agricultural sector generates incomes that are spent on goods and services which are labour-intensive if indirect effects are taken into account. The indirect effects are in principle measurable, and a certain kind of agricultural expansion could have a negative employment effect. What is required is to estimate the income elasticities of demand for different commodities as incomes increase, to see the extent to which the commodities for which demand increases with level of income are labour-intensive, directly and indirectly, and to evaluate the employment implications of changes in the composition of the output of the non-agricultural sector. This, of course, would require a multi-sectoral model, and a great deal more information than is currently readily available.

It is clearly necessary to estimate the total employment effect of fertiliser production. Unfortunately, our data for India do not include satisfactory figures

---

[1] Raj Krishna: *Measurement of the direct and indirect employment effects of agricultural growth with technical change*, paper presented at the Ford Foundation seminar on technology and employment in developing countries, New Delhi, March 1973 (mimeographed, 1973).

[2] To put it more generally, the examination of the impact of changes in the coefficients of the agricultural sector, keeping the coefficients of the non-agricultural sector constant, could be misleading; this is the real problem of using a two-sector model and Raj Krishna is fully aware of it.

for this sector. The direct employment effects are demonstrably small, but what of the indirect effects? The backward linkages are likely to be low and the forward linkages, though substantial, are already taken into account in the agricultural sector. Fertiliser production might have important lateral effects: for instance it might stimulate the development of a heavy chemical industry, which though excellent in itself is unlikely to increase employment to any great extent.

The real scope for the expansion of employment opportunities seems to lie in the forward effects of expansion of the agricultural sector. There is no doubt that the expansion of processing industries using raw materials from the agricultural sector would directly and indirectly create considerable employment. It is striking that the top ten sectors in terms of employment potential per unit of final demand in the Indian economy are (1) *gur* and *khandsari* [1], (2) other tobacco products, (3) sugar, (4) oilseeds, (5) cashew nut processing, (6) flour milling, (7) fruits and vegetables, (8) starch, (9) *vanaspati* [2], (10) raw silk. Of these, seven (1, 2, 3, 5, 6, 8 and 9) involve processing the products of agriculture, and the remaining three involve agriculture or activities related to it. For the Philippines the top ten sectors were (1) maize, (2) rice, (3) traditional transport, (4) traditional consumer goods, (5) sugar cane, (6) pineapple, (7) fishing, (8) livestock, (9) traditional services, and (10) other food crops. Of these, the bulk appear to be primary-sector activities, but this is because much of agricultural processing comes under traditional consumer goods, which is a sector with a high rank. It is also significant that traditional transport and traditional services rank high; quite possibly these are also linked with the direct and indirect processing of agricultural produce.

The evidence for India and the Philippines certainly suggests that the expansion of agricultural output would bring in its wake the growth of processing industries affording considerable employment opportunities. The agro-processing industries mentioned above are often relatively labour-intensive, and are run on a small scale.

Indirect employment effects outside the input-output framework

Public works

Public works, as an instrument of employment creation in the event of regional or national disasters or as a means of making amends, albeit partially, for the limited job creation envisaged in the national plan, are clearly exceed-

---

[1] Traditional sweetening agents derived from sugar cane.
[2] Cooking fat of animal origin.

ingly important in developing countries, where even the indirect employment implications can be very significant and yet are not taken into account.

The elements of this approach were known in the 1930s, when in the United States rival agencies of the Government were arguing the impact of their schemes of public spending in creating employment: there was already an awareness of the fact that the indirect effects had to be taken into account, and crude calculations were made of the additional employment which might be created in activities supplying inputs or meeting the increased demand for goods and services generated by the expenditure of incomes created directly by the project. At this time there were no input-output tables, nor were the participants in this dispute aware of the possibilities of the input-output method; yet they tackled the problem using methods approximating to those with which we are now conversant. [1] Much later, in 1958, Strout [2] showed that the employment effects of different schemes of government spending could be measured in an inter-industry framework and, in particular, that ignoring the indirect effects could give highly misleading ranks to the schemes.

It is very significant that indirect effects are ignored in schemes for public works in developing countries. This is particularly surprising since these programmes have the explicit objective of maximising the employment created per rupee of expenditure. It is instructive in this context to look at the approach adopted in evaluating rural manpower projects in India. No attempt was made to obtain the indirect employment effects, and consequently it was claimed that—

employment is low where the proportion of cost on establishment and raw materials is high. Under soil conservation, employment generation in Penikonda (Andhra Pradesh) is very low, because 96 per cent of expenditure is incurred on establishment and materials. The same trend is noticeable in Kotra and Alland. This is also the case for minor irrigation and other schemes. In Kotra (Rajasthan), Mangrulpir (Maharashtra), and Alland (Mysore), the relative employment potential of different works varies according to the extent of costs involved on establishment and materials.[3]

Two sets of considerations arise when the concept of indirect effects is applied to public spending avowedly for employment creation. First, within the traditional alternatives, the choice must take into account the quantum and location of the indirect employment effects. For example, in an area of localised

---

[1] This is described in some detail by Evans and Hoffenberg in National Bureau of Economic Research: *Input-output analysis: An appraisal,* op. cit., pp. 54-55. The dispute between the Works Projects Administration and the Public Works Administration was resolved by the Construction Division of the Bureau of Labor Statistics, which estimated the indirect effects on a sample basis by tracing the pattern of purchases.

[2] See Alan M. Strout: "Primary employment effects of alternative spending programmes", in *Review of Economics and Statistics,* Vol. XL, 1958, pp. 319-328.

[3] Programme Evaluation Organisation, Planning Commission (India): *Report on evaluation of rural manpower projects* (Delhi, 1967), p. 9.

drought, employment created outside the area has no positive value: similarly, other things remaining the same, one would prefer a scheme for which the total effects are more concentrated on the poor than one under which this is not the case. These are considerations which one would normally apply to direct employment effects, and their extension to indirect effects is only logical. The second consideration is more fundamental and wide-ranging: if employment per rupee is to be maximised, should expenditure be directed to construction or to something else? One could reasonably argue that government subsidies to rural labour-intensive processors of agricultural raw materials might in some cases create more employment per rupee of government expenditure and involve a fuller utilisation of rural skills and capital stock. As noted earlier, such industries often have high employment potential—taking into account indirect effects—and do not require labour to come out of the household to obtain wage employment. There would of course be problems of bottlenecks or inadequacies in the supply of agricultural raw materials, and these problems are likely to be severe, especially in a year of drought, flood or other natural calamity. At any rate, schemes aimed directly at employment creation cannot be ranked in a meaningful way without taking indirect effects into account.

Project appraisal

When economists engage in project evaluation, they are attempting to evaluate specific planned investment to produce one or more products; they treat the rest of the economy in a somewhat summary fashion, characterising it in terms of the values for a limited set of variables like the shadow wage rate, the accounting rate of interest and the social rate of time discount, reflecting factor supplies and alternative demands over time. Each evaluator has to decide how far back and how far forward he proposes to examine inflows and outflows. For example, in evaluating a paper and pulp mill, he would examine the employment and wages indirectly created in logging. [1] Similarly, for a water supply project, the evaluator would estimate the employment effect on the farmers who would receive the water from the project. [2] In the first case some backward linkages are examined, in the latter some forward linkages; and some arbitrariness is involved. But what the evaluator is interested in is not so much estimates of indirect employment effects as their valuation

---

[1] This is done by Michael A. Greig in "The regional income and employment multiplier effects of a pulp and paper mill", in *Scottish Journal of Political Economy* (Edinburgh), Vol. XVIII, No. 1, Feb. 1971; see also UNIDO: *Guidelines for project evaluation,* op. cit., Ch. 19.

[2] UNIDO: *Guidelines for project evaluation,* op. cit., Ch. 20.

(positive or negative). The differences therefore relate not to the existence or measurability of indirect employment effects of the project but to their social evaluation.

It is generally agreed that employment effects can be measured, but one view is that they are quantitatively unimportant and can be ignored in project evaluation. This view is difficult to sustain since the available evidence for developing and developed countries shows that a number of activities generate considerable indirect employment through backward linkages, in some cases indirect employment far exceeding direct employment.

A more plausible position is that indirect employment effects can be substantial, but are not very relevant to project evaluation. Four reasons are given in support of this position. First, projects do not "cause" forward linkages since the latter arise from the supply of the commodity in question, and not specifically from its being domestically produced [1]; so they are not relevant to the calculation of the net benefit derived from the project. Secondly, the expansion of the project industry, A, confers a gain to the input-receiving industry, B, in the form of an external economy, i.e. by lowering the price of inputs, but this is accounted for already since the output of A is valued at the pre-external economy (higher) price. [2] Thirdly, it is argued that backward linkages are rarely substantial and that an industry S could come into being supplying inputs to an industry A located in some other country. In other words, setting up A in a particular country is not a necessary condition for the emergence of industry S in that country. [3] So the setting up of S is not really a benefit derived from starting A, and should not be included in the calculation of net benefit. Fourthly, when backward linkages appear large, this is an indication that the coverage of the project needs to be extended and that a whole complex or agglomeration of industries needs to be examined. [4]

Once we have estimated the employment "caused" by a project, how do we evaluate the benefits that arise from the increased employment? One procedure is to choose projects which produce a net benefit over the period under consideration. In that case employment is not a direct objective in itself, but is a benefit to the extent that it is an instrument for improving the distribution

---

[1] See I. M. D. Little and J. A. Mirrlees: *Project appraisal and planning for developing countries* (London, Heinemann, 1974), p. 341. They argue: "In short, the desirability of starting industry A which uses the output of B is no good reason for starting B if B does not pass the test. Industry A can always be started using imported inputs".

[2] Ibid., p. 338.

[3] Ibid., pp. 342-345.

[4] Ibid., pp. 345-347; for a contrary view on this and other objections raised against linkages see Frances Stewart and Paul Streeten: "Little-Mirrlees methods and project appraisal", in *Bulletin of the Oxford University Institute of Economics and Statistics*, Vol. 34, No. 1, Feb. 1972, pp. 78-81.

of income and a cost to the extent that it leads to a shift of income away from investment. We are really taking consumption as our basic measure and treating employment in terms of its effect on consumption. [1] Another way would be to regard employment as implying a commitment to consumption in so far as wage earners tend to consume all or most of their wages. [2] This is good since in this way the benefits are distributed, but it is bad in that the objective of raising the savings rate through the mechanism of project selection may not be attained.

The view that the real constraint on development is savings, not investment, has obvious implications for the question of linkages. Linkage effects do not generally bring into being additional savings. Instead, by expending investment opportunities they increase the pressure on existing savings. Also, since the supply of wage goods tends to be relatively fixed, it sets a limit on employment creation. This means that employment creation through linkages is at the cost of possible employment creation elsewhere in the economy. [3]

Some qualifications need to be made to the above arguments. In the first place, in certain situations linkage effects may lead to increased savings. In this case one benefit of the project A could be that it leads to increased savings in industry S or industry D, which respectively supply inputs to and demand outputs from A. This could happen for small entrepreneurs who might raise their savings rate or increase the utilisation of their existing capital stock when new opportunities appear. This might be the case with rural labour-intensive processing industries and agriculture-based activities in developing countries. Also for many less developed countries both backward and forward linkages with the world economy might be very difficult to forge owing to market and non-market international restrictions on entry. In such situations the setting-up of industry A is necessary for industries S and D to emerge.

Once A, S and D are in operation substantial economies may arise and the industrial structure of the country would be greatly strengthened. But in such clear cases A, S and D should be treated as an agglomeration and analysed accordingly. Also, in most economies where public project evaluation is practised some degree of aggregative planning would also be needed and such inter-relationships could be taken into account.

---

[1] This is basically what is done in UNIDO: *Guidelines for project evaluation,* op. cit., see in particular pp. 97-98.

[2] See for example I. M. D. Little and J. A. Mirrlees: *Project appraisal and planning for developing countries,* op. cit., pp. 169-176.

[3] See Amartya Sen: *Employment, technology and development,* op. cit.

CONCLUSIONS

In conclusion, we stress the theoretical and practical importance of indirect employment effects. At the theoretical level they can be measured and at the practical level they are by no means insignificant. Although they are usually estimates by methods involving the assumptions of fixed coefficients, it is possible to introduce flexibility into the analysis by making some adjustments to the input-output coefficients and alterations in the output-employment ratios.

The objectives of planning can rarely be equated with employment maximisation: while indirect effects have to be taken into consideration in looking at the factor-use implications of a particular pattern of expansion of output, other considerations like the maximisation of the present value of the anticipated stream of consumption may be more crucial. Even outside the framework of economy-wide multi-sectoral models the concept of indirect effects has important applications. It can be used in a two-sector model with adjusted coefficients to determine the effect of changes in technology in a given sector on the over-all employment situation. Even at a more general level, the interactions between agriculture and the non-agricultural sectors cannot be understood without invoking the notion of inter-sectoral flows and indirect employment effects. We have been able to show that in an important area of public expenditure in less developed countries, i.e. on public works, the indirect employment effects are not estimated; yet these are both relevant—since the objective of public works is employment maximisation—and capable of being measured in a rough and ready way. We also show that indirect employment effects can be measured for industrial project evaluation, but do not carry much weight in project choice unless it can be shown that they help to mobilise savings and other resources which would otherwise have been idle. Finally, we show that indirect employment effects can be important in certain situations influencing technical choice. In sum, the concept of indirect employment effects is an important one, and whatever one's economic objectives it needs to be measured.

PART II

EMPIRICAL CASE STUDIES

CHOICE OF TECHNIQUES FOR CAN MAKING IN KENYA, TANZANIA AND THAILAND

4

by C. Cooper, R. Kaplinsky, R. Bell and W. Satyarakwit [1]

There are obvious reasons why questions about the choice of techniques of production come into the discussion of employment policy. To put things crudely, economists think of techniques or production simply as particular combinations of machinery (or investment) and labour that can make a certain product. Then, if there is a range of such techniques available, each defined by a different amount of investment per worker, the economic problem is to specify the optimal technique assuming that we know the cost of capital and the cost of labour. The link to employment policy is simply that if there are unemployed workers, techniques which use relatively little scarce investment capital per worker (that is, labour-intensive ones) may become socially desirable or "optimal" from society's point of view. [2]

There are, as always, lengthy theoretical arguments among economists about the validity of this idea. However, these do not concern us directly. This chapter is about empirical questions, and particularly about whether or not there is a basis in reality for the assumptions that are made in the theoretical arguments about the choice of techniques.

There are two main types of objection about these assumptions. One is that as a matter of fact there is no choice of technique at all, particularly in

[1] Science Policy Research Unit and Institute of Development Studies, University of Sussex. The material in this chapter results from a case study which the authors are preparing for the ILO World Employment Programme project on technology and employment. The authors would like to stress that this chapter contains only preliminary findings of the study and has been written before all the results have been worked out. It is hoped to publish a complete account of research and final results separately in the form of a monograph. The authors would like to thank the management of various can-manufacturing companies in Kenya, Tanzania, Thailand, the United Kingdom and Hawaii for their close co-operation and kind help with this research.

[2] The arguments are more complicated than this suggests. For example, there are problems about choosing a rate of growth of employment as well as choosing the initial level of employment which can be attained with a fixed amount of investment. See A. K. Sen: *Choice of techniques*, op. cit., and UNIDO: *Guidelines for project evaluation*, op. cit.

manufacturing industry. In other words, it is argued that once the product has been specified, it will turn out that there is only one economically efficient [1] way of making it. [2] Two arguments (at least) are used to support this view. The first is that the technological advance of innovation in the advanced countries generally results in new processes which render all older ones inefficient. These new processes are also more capital-intensive than older ones, and so less "appropriate" to the factor endowments of the developing countries (even though, they are, by hypothesis, the optimal ones). In fact, this is an assertion about efficiency in the economic sense irrespective of comparative technological efficiency: innovation may or may not render older processes economically inefficient; whether it does so or not can be decided only by looking at the circumstances of particular cases.

The second line of argument is that most innovation is "product innovation" in which new products displace old ones in the market, and in which very few processes—sometimes only one—are developed, especially to make the new product. Once again it is said that the newer processes associated with newer products are more capital-intensive than the processes used for older products. It is argued that the problem of choice consequently boils down to making a choice of product, there being no choice of techniques for making a particular product. This is also an empirical question, and it can only be examined empirically.

The second type of objection to the way in which the choice of techniques question is handled in economic theory is essentially that the models used are so over-simplified as to be nearly useless. There are many criticisms that run along these lines: a common one is against the assumption that there are only two factors of production (that is, capital and undifferentiated labour), on the grounds that the availability of other inputs like skilled labour or supervision are often critical in real situations. [3] There are other criticisms too, most of them along the lines that it is inadequate to describe how actual choices are made simply in terms of the relative prices of factors of production. In other words, various external conditions like the size of the market, its organisation,

---

[1] Economic efficiency is referred to here in a formal sense: if the authors state that there is only one economically efficient technique, they mean that although there may be other engineering possibilities, they all require more investment and more labour per unit of output and hence will not be optimal at any combination of factor prices.

[2] Frances Stewart in particular has developed this argument. See Frances Stewart: "Choice of technique in developing countries", op. cit.

[3] See, for example, ILO: *Employment, incomes and equality: A strategy for increasing productive employment in Kenya*, op. cit., where it is argued that the scarcity and cost of supervisory labour is an important factor in preventing the use of labour-intensive can-sealing techniques. The labour-intensive techniques use unskilled workers, and because of imperfect sealing, reject rates would tend to be very high without a large amount of supervision.

the organisation of the labour force, expectations about changing wage rates and the like enter into decision making, and the simple static model does not take them adequately into account. Simple models can always be defended on the grounds that they do not set out to mirror reality but only to explore the logical consequences of some of its salient characteristics; there is no doubt that the simple choice-of-techniques model has been illuminating from this point of view. The problem remains, however, that it has also been used to prescribe policy, and that if some of the things it leaves out are in fact common characteristics of investment decisions in the real world, then the policy recommendations that are based on the model might well be misleading.

So we are left with two main questions: first, whether there is really a choice of techniques for making particular products, or only a choice of products: and secondly, if there is a choice of techniques, what are the factors that explain the decisions that are actually made by entrepreneurs in developing countries? We cannot give general answers to these questions, and we do not set out to do so. The rest of this chapter is about a case study of a particular process—for manufacturing tin cans—which throws some light on the broad questions without providing final answers. First we briefly give reasons for the choice of can sealing for study, and describe the can-making process and its various products. This is followed by an examination of the question: are there a number of economically efficient technologies for making cans? The next section is a review of the factors which actually determined the choice of technology, and a final section contains our conclusions.

CAN MANUFACTURE AS AN EXAMPLE

Why did we choose can-manufacturing processes? Part of the answer is: by an accident of history. It happens that a preliminary study of some can-manufacturing sub-processes was carried out in the course of the ILO employment mission to Kenya. [1] Some of the results of that work seemed sufficiently interesting to justify a closer look at the industry. We have already mentioned one of the conclusions of the initial study in Kenya: it seemed to show that the scarcity of supervisory skills was an important constraint on the use of labour-intensive alternatives. This, in particular, was felt to be worth following up.

There are some other reasons too. First of all, this is part of manufacturing industry, and it is widely argued that the limitations on the choice of techniques

---

[1] See ILO: *Employment, incomes and equality: a strategy for increasing productive employment in Kenya,* op. cit., Technical Paper 7, "A case study of choice of techniques in two processes in the manufacture of cans".

are most severe in manufacturing. So the can-manufacturing process seemed to offer us an opportunity of finding out what these constraints were, and how binding they were. Furthermore, there has been a good deal of innovation in can making, particularly through automation and the gradual development of a continuous process, so it seemed as if this industry might give us an opportunity to investigate how innovation, and possibly product changes too, affected the choice of techniques in developing countries.

Can-making processes and products

The can-making process involves the transformation of tin-plated sheet steel into an open-ended can which is ready to be filled. For most types of tin can, manufacture consists of two separate sequences of operations—the production of the ends for the can body and the forming and assembly of the can body itself. All the cases discussed in this chapter are concerned only with the latter sequence of operations.

The process consists of a number of stages. First, the sheet steel has to be slit into blanks from which the can bodies are made. Tin slitting can be done by more or less completely automated machinery, or by more labour-intensive processes. The most labour-intensive method now in use appears to consist of a manually controlled and fed guillotine powered by a foot-operated mechanism. However, more labour-intensive methods have been used. For example, in the nineteenth century in England, blanks were cut by tin snips (large scissors). Secondly, the can body is formed from the blank. Body making consists of a number of sub-processes. [1] In an automated line these processes are all performed continuously in a body-maker or side-seamer system; in labour-intensive systems the processes are separated and can be done on a batch production basis. Thirdly, the end of the can body is flanged in order to take the can end. Again there are various possible degrees of mechanisation. The fourth major process is end-seaming, whereby the end of the can is joined to the body. A range of methods of differing labour intensity are used. The final stages of the process consist of testing (where the can seams are tested under air pressure) and packaging, where once again, there are various possible processes, ranging from a very simple manual one to a fully automated packer. In automated plants the link between the various stages in this process is a conveyor system; in less automated units various combinations of manual transfer are possible.

---

[1] For example, forming, which curves the blank into the can shape; hooking, whereby the sides are hooked into one another to create the seam; bumping, which creates the seam itself; soldering, which fixes the seam; and cooling.

There is of course a very wide variety of can types and specifications. Leaving aside highly specialised products like the cans used for aerosols, cans are usually classified as either "open-top" or "general". "Open-top" cans are typified by the familiar tinned-food can. The end-seamed can body and the top are delivered to the customer, who seals the can after filling. Almost all packed and sealed food cans are subsequently processed at high temperatures; cans for this purpose must therefore meet demanding specifications. This is usually less true for the "general" line of cans, which includes a very wide variety of types. We considered only two types in this study—the shoe-polish tin and the rectangular-section 20-litre kerosene tin used in East Africa. Most of our data are about open-top cans; however, there is a wide variety of such cans. Open-top specifications can vary in terms of size, types of tin plate, lacquering and basic design. As far as basic design is concerned, open-top cans may take all conceivable shapes: there are not only round ones but also oval cans and rectangular meat cans. Also, seaming methods can differ: some cans are "lock-seamed" (the process we described above is the lock-seaming process) whereas others are "lap-seamed". Moreover, whereas most seams are soldered, some are glued. One of the problems we had in studying the choice of techniques in the industry was the problem of holding the product specification constant.

Range of cases covered

The data for the study were collected in three countries—Kenya, Tanzania and Thailand. The intention was not so much to make a formally structured international comparison, but more to cover a number of different economic circumstances that might influence the technical choices that were actually made at the enterprise level. As it turned out, we obtained more detailed data on a particular process (open-top can manufacture) from Thailand than from the East African countries, whereas in the latter countries we were able to collect data on a greater number of different processes, though in less detail. These differences result mainly from differences in time and opportunity. The East African comparisons are specified in table 9, and Thai ones in table 10.

TECHNIQUES, FACTOR PRICES AND PRODUCTS

Our approach to the question of choice of technique is to examine whether those that are actually in use for making various types of can in Kenya, Tanzania and Thailand are economically efficient in relation to one another. Suppose that technique X has a higher labour productivity (i.e. output per

Table 9. Choice of techniques studies in East Africa

Can type	Country	Process	Use of can	Coding
Round open-top	Kenya	Capital-intensive	Processed and non-processed foods	A1
	Kenya	Capital-intensive		A2
	Kenya	Most capital-intensive		A3
	Tanzania	Very labour-intensive		A4
Rectangular open-top	Kenya	Capital-intensive	Processed corned beef	B1
	Tanzania	Labour-intensive		B2
Twenty-litre kerosene	Tanzania	Very capital-intensive	Kerosene or cashew	C1
		Labour-intensive		C2
		Most labour-intensive		C3
Shoe-polish	Kenya	Capital-intensive automatic	Shoe polish	D1
	Kenya	Labour-intensive semi-automatic		D2

unit of labour) than technique Y. In this case, for technique Y to be economically efficient by comparison with X, it must have a higher output per unit of capital than X (if labour and capital are the only factors of production). If Y is efficient in this sense the implication is that there is some combination of price of capital (i.e. rate of interest) and wage rate at which the total costs per unit of output with Y will be less than with X. Since Y has a relatively low output per worker and a relatively high output per unit of capital, it will be favoured in cases in which the interest rate is high and the wage rate comparatively low; technique X will be favoured in cases in which the balance is reversed. On the other hand if Y had been inefficient (i.e. had been associated with a lower output per unit of capital as well as a lower output per unit of labour), there would have been no combination of wage rate and rate of interest at which its cost of production would have been lower than X. Y would never be used, even if social instead of market prices for factors were used in the investment decision. [1]

The first question we ask in this part of the chapter is whether the techniques within each group are efficient in this sense. However, that is not enough: we must keep in mind the possibility that even if a labour-intensive technique, like Y for example, is efficient by comparison with a capital-intensive one, it will nevertheless be preferred only at extremely high interest rates and un-

---

[1] However, "inefficient" techniques may sometimes be used in the context of an extreme type of employment policy. The use of very labour-intensive inefficient technology means a sacrifice of present and future output and future employment in return for higher employment per unit of capital in the short run.

Table 10. Choice of techniques studies in Thailand (open-top cans)

Can use	Process	Coding
Milk, fruit, vegetables	Automated capital-intensive	E1
Milk		E2
Fruit	Semi-automated	G1
		G2
Vegetables, pickles	Manual	M1
		M2

realistically low wage rates. [1] In this case, the existence of an efficient labour-intensive alternative is of mainly academic interest, since it is unlikely ever to be used in practice. The second step in the analysis is therefore to examine total unit costs of production with the various techniques available, and to assess the wage rates below which the more labour-intensive techniques might be used. [2]

In each case we begin by assuming that product differences within each group of technologies listed in tables 9 and 10 are sufficiently small to be ignored in the comparisons; however, in reality this is a very crude approximation, so at a second stage we drop this assumption, and examine how far product differences have affected technical decisions.

Kenya and Tanzania

These studies cover the four technology groups that are listed in table 9. For each of these groups we look at the relative efficiency of the techniques, the unit costs of production and product differences and their implications. The East African comparisons are all based on single-shift working which is more or less the rule in all the factories we visited. The data for these comparisons were collected through factory visits. The quantitative data covers whole production lines: it is not disaggregated to the sub-process level. The main items of quantitative data are:

1. Rates of output for the production lines. These are based as far as possible on the actual operation of the production line, generally obtained from monthly production figures in the factory. Some adjustments have been made to allow for the fact that rates of output vary somewhat with the size of can being produced.

2. Labour input per shift and wage rates. Again these figures were obtained from the factory visits. In general it was possible to break down labour inputs

---

[1] In other words at factor prices which are quite unrealistic even in social terms.
[2] These cost assessments are made in the same way as the assessment in Chapter 5 below.

into skilled and unskilled labour. The cost comparisons are all based on a uniform wage rate—the rate ruling in the Kenyan modern sector—so that the effect of multiple wage structures is not considered here. [1]

3. Capital costs. All capital costs are calculated as if we were making a choice of investing in one or other of the techniques currently used in each group. As far as possible, therefore, we used 1974 replacement costs for machinery and buildings. Annual capital costs were obtained by using an annuity formula, assuming an economic life of 25 years for all equipment. There were some difficulties in obtaining replacement costs in all cases, sometimes because lines had been modified and sometimes because parts of the production lines were obsolescent. In these cases we had to make rough estimates. Building costs were obtained from the factory visits; buildings were assumed to have a life of 50 years. The cost calculations in East Africa are based on a rate of interest of 10 per cent throughout.

4. Power costs. Such costs are the main item of non-labour variable costs in the cost calculations.

Round-section open-top cans

There are four production lines in East Africa manufacturing round open-top cans—three fully mechanised continuous lines in Kenya, and one semi-automatic lap-seam line in Tanzania. Table 11 shows that the output of these lines differs considerably: A2 and A3 produce about 10,000 cans per hour, while A1 produces only 4,200; the semi-automatic lap-seam line A4 produces about 1,100 per hour. Labour input for A1 is eight unskilled operators, one skilled mechanic and one supervisor. For A2 there are ten unskilled operators, and on A3, which has automated packing and slitting operations, only four unskilled operators are required in addition to the mechanic and supervisor.

The productivities of labour and capital are shown in the table.

The techniques in use cover a wide range of capital intensity. The most capital-intensive line is A3, the fully automated Kenyan line. The least capital-intensive is the lap-seam line in Tanzania: this relatively labour-intensive line provides nearly 20 times as much employment per unit of investment, and just over 10 times as much employment per unit of output, as the most capital-intensive line.

Table 11 and figure 2 show that three of the technologies are efficient (i.e. A3, A2 and A4 in descending order of capital intensity). It is interesting that both the most capital-intensive (A3) and the least capital-intensive (A4) are efficient technologies. A1, however, is inefficient under Kenyan operating conditions (we do not know why this relatively new line should be so markedly

---

[1] However, it is considered in the Thai comparisons below.

Table 11. Factor productivity: round open-top can manufacture in Kenya and Tanzania

	Production line and date of installation			
	A1 1972	A2 1956	A3 1970	A4 1964
Output per hour (number of cans)	4 200	11 088	9 450	1 084
Labour				
Skilled workers (including supervisors (in brackets))	2(1)	2(1)	2(1)	1.16(1)
Unskilled workers	8	10	4	6
Capital cost (£EA, 1974)				
Machines	140 000	180 000	191 500	12 500
Buildings	7 200	7 200	7 200	1 844
O/K (cans per £EA)	0.03	0.062	0.049	0.087
O/L (cans per worker)	420	924	1 575	151
L/K (workers per £EA)	0.000071	0.000067	0.000031	0.00057
K/L (£EA per worker)	14 000	15 000	31 917	1 746

inferior to the others but the difference may be due to the fact that it is tooled to produce cans of larger size). We are left with three efficient techniques, one semi-automated and two fully automatic. The most profitable technique from the firms' point of view depends upon the relative prices of capital and labour. In order to simulate a realistic investment decision, we have evaluated all techniques at the current wage rates in Nairobi. This slightly artificial assump-

Figure 2. Investment function: round open-top can manufacture in Kenya and Tanzania

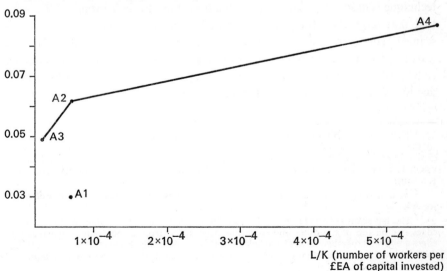

O/K (number of cans produced per
unit period per £EA of capital invested)

L/K (number of workers per
£EA of capital invested)

tion prevents us from comparing actual costs of production in Tanzania with those in Kenya, but it does allow us to conclude what unit costs would be if all the techniques were installed in Kenya and operated as at present.

Unit costs of production are calculated and presented in table 12. At existing factor prices and at capacity utilisation the semi-automated A4 is clearly sub-optimal. So, too, is the more labour-intensive of the fully automated lines (A2), although its unit costs are not far from those of A3. Clearly, however, at current factor prices the optimal technique is A3—the most automated technique. The cost data in table 12 show also that although the labour-intensive A4 technique for making lap-seam open-top cans is efficient by comparison with the capital-intensive technique in terms of use of capital and power, its unit costs at current wage rates in Kenya are very high indeed. [1] Lower wage rates would improve the relative position of this technique, but the wage rate would have to be somewhere round about one-hundredth of its present level in Kenya for it to be preferred. We assume that such a level is well below any realistic estimate of the social opportunity cost of labour in East Africa, and it is therefore hardly conceivable that technique A4 could be preferred even on social grounds. [2]

This therefore appears to be a case in which technological innovation has resulted in new techniques (i.e. A2 and A3) which, although they have not made the old A4 labour-intensive technique inefficient, are so much more productive than A4 that there is no realistic shadow wage rate at which A4 would be preferred.

However, there are also significant differences in product between the A4 technique and the more capital-intensive ones. The labour-intensive technique is used for making lap-seamed open-top cans in Tanzania. [3] We were told that larger tins for packing pineapples in Kenya were previously lap-seamed. However, the manufacturers received numerous complaints about leakages. Head office technicians (the can manufacturer is a foreign subsidiary), concluded that the leaks occurred in the side seam, and might well have been due to the way in which the cans were handled during transport. They recom-

---

[1] They are probably lower in Tanzania, where the technique is actually in use, but even so the variable costs of this technology must be greater than total costs on the more capital-intensive ones at the levels of capacity utilisation implicit in the data. Its continued use in Tanzania may be due to scale factors (the Tanzanian market is much smaller than the Kenyan).

[2] An overwhelming preference for present employment might, however, lead to preference for A4.

[3] The lap seam is formed simply by forming the can cylinder, lapping the longitudinal edges one on top of the other and soldering. This differs from a lock seam, for which the edges are locked together, cleated and soldered. It is worth noting, however, that lock-seaming techniques of much the same capital intensity as the Tanzanian lap-seaming technique are in use in Thailand.

Table 12.   Unit costs of production: round open-top can manufacture in Kenya and
Tanzania (EA shillings per thousand cans)

Unit costs	A1	A2	A3	A4
Capital	0.77	0.37	0.46	0.30
Labour	5.14	2.31	1.48	16.84
Power	0.05	0.01	0.02	0.01
Total	5.96	2.69	1.96	17.15

mended the adoption of more mechanised techniques which produced lock-seamed cans. The main reasons were as follows:

1. It was thought to be too difficult to persuade the handlers to change their methods of storage and transport.

2. The main customer was a multinational enterprise with a considerable international reputation to uphold, and was a large customer not only of the Kenyan subsidiary but also of subsidiaries elsewhere. This customer felt that the cans produced with the more mechanised techniques possessed a stronger side seam, and that internal lacquering could be more effectively applied to them. Therefore it would not be advisable to resist the demands for improvement of product quality.

3. Although it seemed possible to make lap-seam cans of adequate quality, the variability of can quality in an essentially man-controlled operation meant that supervisory costs and rejection rates would be greatly increased.

4. The can producer had to compensate its customer not only for the value of the spoilt cans but also for the contents. As the value of the can is only a small component of final cost there was a considerable inducement to avert risks by mechanisation or at least by supplying a lock-seamed can.

5. In any case, even though the primary impetus to mechanisation was product quality, the lap-seam cans were likely to be cost-competitive only if mechanised lines had to operate at very low capacity utilisation. This was unlikely in the rapidly growing Kenyan market, so there were strong economic grounds for mechanisation.

In the circumstances, the switch to highly mechanised continuous production of lock-seamed cans is explicable. Continuous production methods were apparently the only possibility. It is quite interesting, however, that there was apparently no attempt to seek out a more labour-intensive lock-seaming technology. This is the more interesting in view of the Thai experience, which we discuss later.

Table 13. Factor productivity: rectangular open-top can manufacture in Kenya and Tanzania

	Production line and date of installation	
	B1 (1965)	B2 (1950)
Output per hour (number of cans)	7 200	11 640
Labour		
Skilled workers (including supervisors (in brackets))	2(1)	2.5(1.5)
Unskilled workers	10	15
Capital cost (£EA, 1974)		
Machines	150 000	150 000
Buildings	7 200	7 180
O/K (cans per £EA)	0.048	0.078
O/L (cans per worker)	600	665
L/K (workers per £EA)	0.00008	0.0001
K/L (£EA per worker)	12 500	8 571

Table 14. Unit costs of production: rectangular open-top can manufacture in Kenya and Tanzania (EA shillings per thousand cans)

Unit costs	Production line	
	B2	B1
Capital	0.30	0.50
Labour	2.73	2.75
Power	0.02	0.03
Total	3.05	3.28

Rectangular-section open-top cans

We examined two techniques for making rectangular open-top 12 oz. cans, one in Tanzania and one in Kenya. Table 13 shows that the output of technique B2 is 11,640 cans per hour and the output of B1 7,200 cans per hour. Whereas B2 employs 15 unskilled labourers and 2.5 skilled labourers, B1 employs 10 unskilled and 2 skilled labourers. The table also shows the factor productivities of the two techniques.

Table 13 shows that B1 is an inefficient technique. This is almost certainly due to its low rate of output. We suspect that this is the consequence of a very low rate of capacity utilisation, so that further comparison of the techniques will probably not yield much. As one might expect, there is no set of factor prices which would lead to a preference for B1. The unit costs of outputf rom the two techniques are shown in table 14 and are derived from wage rates currently prevailing in Nairobi. The marginally greater labour intensity of the B2 technique is more apparent than real: it is probably the result of a higher rate of capacity utilisation.

Twenty-litre kerosene tins

We have a sample of three techniques for the production of kerosene tins. In two of these (in Kenya), the tin body is made in two sections and then locked and seamed together. In the more modern, automated Tanzanian technique the tin is formed from a single piece in one operation. The output of the C1 line is 936 per hour, while that of the C2 line is 1,291 per hour and that of the C3 line is 310 per hour. Labour input is 28 unskilled and 2 skilled on the C1 line, 44 unskilled and 1.25 skilled on the C2 line and 16 unskilled and 1.3 skilled on the C3 line. These figures and the resulting factor productivities are shown in table 15. That shows that the most labour-intensive C3 technique is inefficient, and we would expect it to be scrapped in favour of a new line if the market were sufficiently large.

However, C1 and C2 are efficient. Table 16 shows unit costs for the three techniques. There must be some doubt about these data, since the capital cost of the very old C2 technique is hard to estimate with any certainty. Nevertheless, the position is fairly clear. Unit costs are more or less the same with the old C2 technique, which is labour-intensive, as with the new capital-intensive C1 technique. This means that in social terms the more labour-intensive C2 technique is almost certainly to be preferred. The social opportunity cost of labour in East Africa is well below the wage rate in the modern sector in Nairobi, so that even if there is some doubt about the replacement cost of the C2 technique, it is most probably the desirable one from a social point of view (if not from a private point of view also). The C2 technique, however, is only very slightly more labour-intensive than the new one (it allows 20 per cent more employment per unit of investment).

On the face of it, the technological change in moving from C2 to C1 has not had much effect on factor productivities, and has not undermined the relative efficiency of the older technique. It might, however, have reduced materials intensity of production (the new technique makes a can with one side seam whereas the older one makes a can with two), though we would expect this effect to be rather small.

There remains the question whether the single-seamed can is in some real sense an improved product (i.e. whether there has been a product innovation). The answer is probably not: the can with two side seams appears to serve as well as the can with a single seam for sale of kerosene in predominantly rural markets.

Table 15. Factor productivity: manufacture of 20-litre kerosene tins in Kenya and Tanzania

	Production line and date of installation		
	C1 1970	C2 1935 onward	C3 1930 onward
Output per hour (number of tins)	936	1 291	310
Labour			
Skilled workers (including supervisors (in brackets))	2(1)	1.25(1.25)	1.3(0.67)
Unskilled workers	28	44	16
Capital cost (£EA, 1974)			
Machines	115 000	125 000	32 050
Buildings	11 500	4 000	7 488
O/K (tins per £EA)	0.0081	0.0103	0.0097
O/L (tins per worker)	31.2	28.5	17.9
L/K (worker per £EA)	0.00026	0.00036	0.00054
K/L (£EA per worker)	3 833	2 762	1 853

Table 16. Unit costs of production: manufacture of 20-litre kerosene tins in Kenya and Tanzania (EA shillings per thousand cans)

Unit costs	Production line		
	C1	C2	C3
Capital	0.30	0.22	0.28
Labour	7.11	7.15	11.79
Power	0.01	0.04	0.07
Total	7.42	7.41	12.14

Shoe-polish tins

Lastly, we examine the production of shoe-polish tins. We studied two techniques, one automated and one semi-automated; both are used in the same plant in Nairobi. Whereas the D1 automated technique produced 6,000 tins an hour, the less automated D2 technique produces only 1,800. The D1 technique employs 6 unskilled and 0.2 skilled operatives, and the D2 technique employs 5 unskilled and 0.4 skilled. Factor productivities are shown in table 17. The D2 technique is clearly inefficient. Unit costs of production are shown in table 18: the unit variable costs of production of the D2 technique are greater than the total unit costs of production of the new D1 technique, and therefore we might expect it to be scrapped.

In this case, therefore, the result of innovation has clearly been that an older labour-intensive technique has been superseded entirely by a superior,

Table 17. Factor productivity: manufacture of shoe-polish tins in Kenya

	Production line and date of installation	
	D2 (1960)	D1 (1974)
Output per hour (number of tins)	1 800	6 000
Labour		
Skilled workers (all supervisory)	0.4	0.2
Unskilled workers	5	6
Capital cost (£EA, 1974)		
Machines	10 500	16 550
Buildings	1 040	2 240
O/K (tins per £EA)	0.171	0.363
O/L (tins per worker)	333	968
L/K (worker per £EA)	0.0005	0.00037
K/L (£EA per worker)	1 944	2 669

Table 18. Unit costs of production: manufacture of shoe-polish tins in Kenya (EA shillings per thousand tins)

Unit costs	Production line	
	D2	D1
Capital	0.14	0.07
Labour	6.38	2.12
Power	0.01	0.01
Total	6.53	2.20

more capital-intensive one, which, as it happens, also gives a better product. However, there are two interesting complications. The first is that the D1 technique produces tins of higher quality than the D2. In fact, the can-producing company introduced the D1 production line because a new customer company (from outside Kenya) entered the market and made strict demands for very high-quality tins. The decision to invest in the D1 technique was prompted first by these strict specifications and secondly by the assumed expansion of the market for shoe-polish tins because of the arrival of a new polish producer. It is noticeable that the D1 technique would only just be competitive with D2 at the D2 scale of output [1], so that the large increase in the size of the market was probably quite an important factor in the decision to install the automated technology. The second complication is that the

[1] Assuming that there would not be much reduction of labour input at a lower scale of output on D1. This is quite a reasonable assumption for a highly automated process.

high degree of automation of the D1 line is apparently a direct outcome of the customer requirement for a very high-quality tin. There may well be less automated methods available for producing tins, though of somewhat lower quality.

Thailand

Whereas the East African survey covered a number of different types of can, with very different end uses, the Thailand study is much more restricted in product scope: it deals only with round open-top cans used for food preservation. [1]

The Thai situation differs from the East African in some important ways. Can manufacture in East Africa is dominated by a single multinational enterprise, and all our data for the East African comparisons were obtained from subsidiaries of this firm, which is the only supplier of cans, apart from one canning company that also produces cans, but only for its own use. In Thailand the manufacture of cans is divided between enterprises in the modern sector (producing daily products, for example) which make their own cans for their own use, and specialised can producers. Furthermore, the specialised can producers fall into two groups: in the high-wage sector there is a subsidiary of a multinational firm which specialises in can production [2], and there are a number of local firms paying lower wages which also produce cans. The foreign subsidiary sells its cans mainly to firms which export canned goods and to firms supplying high-quality products such as dairy products to the domestic market (these customers are in many cases themselves subsidiaries of foreign firms); while the local firms supply locally owned companies selling their canned goods in Thailand.

The main reasons for this difference in organisation are first that the multinational specialist can producer entered the Thai market at a much later stage than in East Africa; consequently a substantial part of its potential market had already been mopped up by firms producing cans for their own use. Secondly, locally owned Thai industry is more developed than local industry in East Africa; there was therefore a considerable demand from local canners, which provided an opportunity for the development of local can makers. Also, the entrepreneurial and technical abilities for setting up a local can-making industry are probably more developed in Thailand than in East Africa. The

---

[1] Partly as a consequence of this restriction, there is more detailed data for the Thailand study, although for the purposes of this preliminary account we shall not take full advantage of the possibilities for more detailed analysis that the Thailand data present.

[2] It is of some interest and significance that this multinational firm is the same from whose subsidiaries the data for the East African cases were obtained.

result is that there is a range of different types of firms producing cans in Thailand. By and large they produce for different markets and face different factor prices. This in itself suggests that there might be a wider range of techniques in use *in* Thailand than in East Africa, and was one of the reasons why we wished to include a study of Thailand.

For the purposes of this preliminary account we have used the Thai data in the same way as the East African. We shall examine whether other technologies are available (this time for round open-top food cans only), and how product differences might affect the choice between them.

The comparison is based on data from six production lines, listed in table 20, for open-top cans. But first a little more must be said about the techniques and the enterprises which have installed them. The E1 technique is used by the foreign subsidiary which specialises in can manufacture. It consists of a relatively high-speed continuous line, which because of the fragmentation of the firm's market (itself a result of relatively late entry) is used for a variety of cans of different sizes. The E2 technique is used by a firm which sells canned dairy products in Thailand and abroad, and which makes its own cans. The production line is used for one size of can only. It is similar to the E1 technique except that, first, E1 is more flexible and probably embodies some diseconomies; being a multi-product system, as compared to E2; secondly, E2 does not include an automatic tester (testing is done off line), nor an automatic packer (the cans are passed directly to the milk canner), both of which are fitted to the E1 line. Techniques G1 and G2 are rather similar, both introduced by locally owned firms operating with relatively low-wage labour and producing for locally owned canneries serving the Thai market. Most of the cans are used for fruit, vegetables or pickles. The production lines are slower than the E1 and E2: they may be described as semi-automated continuous lines. Whereas the E1 and E2 lines are fitted with equipment from Europe or the United States, the G1 line is Japanese and the G2 is from a developing Asian country. H1 and H2 are labour-intensive lines. They are discontinuous; the workers operate around individual pieces of machinery, and inter-machine transfer is manual. Most of the equipment is made in Thailand. Speeds are low. Lines of this type are used to produce a wide variety of products for the local market. The two H lines are used to produce cans for locally owned companies marketing canned vegetables and pickles in Thailand.

The range of techniques used in Thailand is therefore considerable, and wider than the range in East Africa. We would expect some differences in product quality between these production lines. However, as a first step in the analysis, we leave this question aside; we shall come back to it later on.

The basic data on factor productivities are given in table 19. There are some qualifications to these data. First, the rate of output for E1 is given as

21,600 cans an hour. This is the "expected rate" of output in the factory. The actual long-term average (over more than four years) is considerably lower—about half of this. The loss of output seems to be mainly the result of loss of time through inefficiency in dealing with the problems of changing the line from one can type to another. Both the problems and the loss of output they lead to are in large part a function of the technology. This qualification means that the data on units costs are favourably biased for E1. Secondly, capital costs for E1 include costs of machines for packing and testing, and the labour inputs needed for those machines are also included. The other techniques do not include packaging and testing. This biases the results against E1, but changes none of our conclusions; we make some adjustments later on where necessary. Thirdly, there is another respect in which lines E1 and E2 are not strictly comparable. The first is a flexible line producing many types of can; the second produces only one type at or near the "expected" rates. This difference in the homogeneity of the product probably affects both capital and labour costs. For capital costs the rate of discount is assumed to be 10 per cent, and the life of the equipment ten years.

The main conclusion to be drawn from table 19 is that there is a broad range of efficient techniques: it is true that E1 is inferior to E2, but as we have seen this difference can be ignored because they are not strictly comparable; also H2 is just inferior to H1, but otherwise we have a range of techniques which are efficient, from the most to the least capital-intensive. The least capital-intensive (H1) employs nearly 300 times as much labour per unit of capital as the most capital-intensive (E1); the intermediate types of technique (G1 and G2) employ up to about ten times as much labour per unit of capital as the capital-intensive ones.

The unit costs of production with these techniques are given in table 20. The first line of the table shows unit costs at actual wage rates facing the various firms. Since the wage rates on the G and H production lines are about half of those on the E lines, this favours the intermediate and labour-intensive techniques. It turns out that G1 has the lowest unit costs, followed closely by G2. In the second line of the table costs are adjusted to a standard wage rate, namely the rate payable on lines using the E techniques in the modern sector. At this rate G1 and G2 are still the cheaper techniques, though they are not far off the cost levels of E2, the single-product automated production line.

There is a further adjustment in the third and fourth lines of the table. The fourth line shows unit costs excluding the costs of packaging and testing from technique E1 (which is the only production line that includes these components). The results are broadly unchanged. The intermediate techniques (G1 and G2) have the lowest unit costs, and are followed by the automated ones. The very labour-intensive production lines have higher unit

Table 19. Factor productivity: manufacture of round open-top cans in Thailand

	Production line					
	E1	E2	G1	G2	H1	H2
Output per hour (number of cans)	21 600	19 200	8 760	10 600	720	850
Labour						
Skilled and supervisory personnel	4	4	1	3.5	0.5	0.5
Unskilled and semi-skilled workers	7	2	9	11	9	12
Capital cost (thousands of bahts)	9 019	6 331	1 535	1 230	32.5	56
O/K (cans per thousand bahts)	2.39	3.03	5.71	8.62	22.15	15.18
O/L (cans per worker)	1 963	3 200	876	731	76	68
L/K (workers per million bahts)	1.22	0.95	6.51	11.79	292.3	223.2
K/L (thousands of bahts per worker)	819.90	1 055.16	153.50	84.82	3.42	4.48

Table 20. Unit costs of production: manufacture of round open-top cans in Thailand (bahts per thousand cans)

Item	Continuous automatic		Semi-automatic [1]		Manual	
	E1	E2	G1	G2	H1	H2
Unit costs						
Total						
At going wage rates	35.5	25.2	11.4	17.6	35.9	62.1
At rates of lines E1 and E2	35.5	25.2	17.8	19.4	67.9	98.1
Excluding packaging and testing						
At going wage rates	30.6	25.2	11.4	14.7	35.9	62.1
At rates of lines E1 and E2	30.6	25.2	17.8	19.4	67.9	98.1
Labour contribution to unit costs at going wage rates (percentages)	19.3	12.9	22.5	40.7	87.4	75.9

[1] G1 and G2 were not technically identical; nor were the wage rates observed in the two cases the same (G2 being higher than G1). This explains why the same adjustments have proportionately different effects on unit costs for the two techniques which in the text are discussed as being broadly similar.

costs than the others, even in the first line of the table, where they have the advantage of a very low wage rate. For the labour-intensive production lines to be preferred to the others, the shadow wage rate would have to be very nearly zero—which is unrealistically low, as in the case of East Africa.

There are, of course, other reasons which might favour the H1 and H2 techniques. For example, they use locally made machinery, thus saving foreign exchange and possibly creating external economies in the machine-making sector. Even so, these advantages would have to be given massive weight to offset the low productivity of these particular techniques.

Apparently, then, the process of innovation in this case has made the very labour-intensive techniques uncompetitive. Much the same thing happened in the East African case of round open-top can production, although in East Africa the labour-intensive alternative is not as labour-intensive as in Thailand. A very interesting difference in Thailand, however, is that there are two intermediate types of technique, both developed in Asian countries with what are by world standards low-wage economies (one country being Japan). More interesting still, these techniques have lower unit costs of production than the automated ones, even at the wage rates paid in the modern sector, and even though the automated techniques are both being operated by multinational firms with sophisticated and efficient management. This therefore appears to be a case in which the spectrum of techniques had been extended by the development of some further possibilities. [1] This suggests that it has been possible to offset the effects of labour-saving bias, at least to some extent.

But what about products? A key question is: are the cans produced by techniques G1 and G2 sufficiently good to meet the quality standards in the markets which are at present being served by E1? [2] At first sight the answer is no. We took a sample of cans from G2 (which appeared to have lower quality standards than G1), and had them quality-tested in the laboratories of the specialist can producer (E1). The cans had a number of defects which made them unfit for use in export markets. However, a closer inspection of the kinds of defects in these cans suggested strongly to us that the problem was not inherent in the G2 technique itself but resulted from the way in which the technique was being used [3]: it seemed perfectly possible that with appropriate operating procedures and quality control facilities, the same technique could be used to produce cans of export quality. The only way we could find of testing this possibility was to discover whether there were any Asian exporters of canned food who used cans made with this technology. We made a limited questionnaire survey of exporters of canned goods from one Asian country in which there is a wholly foreign-owned food-canning subsidiary of a foreign

---

[1] See the discussion on pages 105-109 of modern adaptations of techniques that have in themselves fallen into disuse.

[2] It should be noted that the cans produced on the labour-intensive lines H1 and H2 are of very poor quality. When in this chapter we refer to "poor quality", we refer to both the optimum level and the variability of quality achieved. For techniques in which the operations are largely controlled directly by men rather than by machines (e.g. A4 as well as H1 and H2) it may be the variability rather than the attainable level of quality which causes problems.

[3] The final report of our findings contains a detailed analysis of the results of the quality study.

firm, exporting 99 per cent of its output. This subsidiary makes all its own cans, using precisely the G2 technique of our case study; moreover, it exports to markets in which both customer and public health requirements for can quality are very high, namely Japan, the United States, and the Federal Republic of Germany and other parts of Western Europe. We concluded from this survey that there is no intrinsic problem of quality with the G2 technique; the problem is that the production line is not being operated properly in Thailand.

It is true, of course, that unit costs would rise if the Thai firm had to meet international quality standards. To explore the cost of such a requirement we made an analysis of what unit costs would be with the G2 technique if it were used in the institutional context of the E1 technique. We allowed for a higher wage rate, E1-type supervision, testing, quality control and automatic packing. With all these adjustments, there is still a considerable saving in private costs with the G2 technique.[1] Our calculations show that over a ten-year period under these stringent conditions of operation, the G2 technique would give cost savings, by comparison with the E1 technique, of a present discounted value that could be at least as much as 10 million bahts (about £200,000) on certain very plausible assumptions. In social terms savings would be even greater: employment per unit of both output and investment is considerably larger; the foreign exchange costs of investment are much lower per unit of output; there are greater possibilities of building up local capability to copy and manufacture the machinery than with technique E1; and there are greater chances for Thai as opposed to foreign firms to supply the export market and thus to increase the country's foreign exchange earnings.

In short, the problem of meeting quality requirements with the intermediate technology seems to be more apparent than real. The intermediate technique has not been superseded by the automated one because of product quality requirements, although the very labour-intensive techniques certainly have been.

The concept of availability of techniques

On the basis of the case studies, what can we say about the availability of different techniques as far as this industry is concerned?

The development of new can-making techniques has made some existing ones inefficient—the labour-intensive technique for making 20-litre kerosene tins, the semi-automatic technique for shoe-polish tins and one of the very labour-intensive methods for making round open-top cans in Thailand.

---

[1] It also has the advantage that it is used on a smaller-scale line, so that problems of flexible, multi-product production would be easier to deal with than with the E1 technique, as also would be the problem of matching capacity to increases in demand.

105

However, it does not seem to be generally true in this industry that innovation produces a technology that supersedes all the others existing in the sense of making them inefficient. Even in the cases in which a labour-intensive technique is superseded by innovation—for example in the case of kerosene tins in East Africa—there is usually more than one technique left to choose from. It is only in the case of shoe-polish tins that this is not true; and obviously there may be other possible techniques that are not included in our study. It is quite striking that in the case of round open-top cans in both East Africa and Thailand even the very old, labour-intensive techniques are still efficient by comparison with the new ones. The problem with those very labour-intensive techniques is not that they are inefficient in a formal sense but that they have such low productivities that they are sub-optimal for all realistic factor price combinations if the possibility of a zero shadow wage is excluded. This applies to the production of round open-top tins in East Africa and Thailand.

Changes in product have occurred during innovation: the very labour-intensively produced open-top cans in Thailand (and also in East Africa) are of poor quality by comparison with those that are made on automated lines. The same is true in the case of shoe-polish tins, though not for kerosene tins. However, in this industry at least, product changes are not always as decisive as may seem to be the case. In the case of lines developed in Japan and another Asian country and used in Thailand, poor quality is more the result of inadequate operation of the production line than of inherent defects in the line itself. In this important case, a change in product is not unavoidable.

The Japanese techniques are important because they show that efficient techniques can be found which are more labour-intensive than newer capital-intensive ones, but which can produce at much lower private as well as social costs. These techniques are interesting also because they seem to be up-to-date variations of techniques which were used in Europe and the United States in the 1920s or before. These modern techniques derived from the older technologies can be used to make high-quality products. They have been developed in countries where wages are low by comparison with those now paid in the United States and Europe.

In this industry at least, therefore, there are alternative techniques in a real sense, at least in some important cases, and there is no need for a product change that would preclude the use of more labour-intensive techniques. Moreover, the range of techniques available is a good deal wider than the studies suggest. [1] We simply looked at some of the techniques that are actually

---

[1] In general our sample is biased towards the more labour-intensive end of the spectrum of techniques in use. Even the most capital-intensive of the techniques covered in this study are far less capital-intensive than a number of techniques in use in the more industrialised countries.

in use in East Africa and Thailand. There are other techniques in use in Thailand, and yet others in use in Europe and the United States. It is quite conceivable that there may be intermediate variations which have not been covered by the study. One particularly interesting development is likely to extend the available range considerably during the next few years: there has recently been a major innovation in can making (the introduction of the two-piece can). The technique for this is appropriate for very high volumes of output. This technological leap is making available a range of second-hand high-speed lines in Europe and the United States. The price of this type of second-hand machinery is falling quite rapidly, and smaller-scale European and United States can producers are purchasing this equipment. In turn they are disposing of their own range of second-hand, lower-speed equipment at low prices. It is more than likely that this innovation-induced change [1] in the second-hand machinery market will make available a number of techniques that are likely to be both labour-intensive and inexpensive.

Another limitation of our survey is that we have not looked at alternatives at the sub-process level. In fact, there are a very large number of conceivable combinations of sub-processes to make up a production line. If we had taken these into account, it would plainly have increased the number of different possibilities.

The study also shows that the concept of availability of different techniques needs to be more clearly formulated. What does "availability" mean—in particular, what does it mean as far as the private decision maker is concerned? In the first place, it seems clear that there are costs involved in establishing what is available. The curious notion that what is available is in some way immediately known to everyone in the industry, or can be discovered at no cost, is not valid. Our information suggests that the only reason for which the specialised modern can producer in Thailand (using the E1 technique) did not consider the techniques evolved in other Asian countries was that the engineers did not know about them [2]; yet this firm is a large, highly organised multinational enterprise. Indeed, one reason why the firm did not consider these techniques may well be that it is too organised, in the sense that most technological decisions are made by head office personnel, who are much more aware of technology and conditions in advanced countries than in developing countries. However, we spoke to managers of another multinational enterprise (not included in the case studies), which has incurred considerable

---

[1] See Chapter 5 below for a discussion of the way in which innovation-induced availability of second-hand machinery is likely to ensure both low initial cost and low operating cost for entrepreneurs.

[2] See p. 112.

costs in seeking out the techniques which were available and was actually considering the Asian techniques in question for one of its operations.

The range of techniques available can also be widened through engineering adaptation by individual firms. We found one firm in an advanced country that was considering the possibilities of re-tooling its operations in a developing country (to improve quality of output), and another was looking at the possibility of producing new designs based on Japanese technology, specially for its operations in developing countries.

Our observations in Thailand indicated that local manufacture of equipment may be another way in which the range of techniques available can be expanded. This method of expanding the range is distinct from any adaptive engineering or re-design efforts. By this method a particular technique in itself remains unchanged but its capital cost may be lowered because local machine-building costs may be lower than those in the country of origin of the technique. In economic terms, this then becomes an alternative, less capital-intensive technique. This method of securing a more appropriate technique was observed on a number of occasions. One firm copied an imported packing machine at a cost of between 50 and 60 per cent of the cost of the imported machine. Another firm copied Japanese end-making machinery, at a cost again of about 50 per cent of that of the imported equipment. The same type of action was observed in other sections of industry in Thailand, but it is by no means clear how widespread this method is, nor what the possibilities are of using it to increase the range of available techniques. Clearly there are some limits: for one thing the costs of making machinery locally are not lower for all types of machine.

Another possibility lies in recourse to techniques that have been used in the past but are now obsolete. As such, it is perhaps seldom likely to be very useful; but when such a resurrection is combined with some simple engineering, the possibility of generating a wider range of technological possibilities may be quite large. In the case of can-making techniques, it is interesting to note that the H techniques in use in Thailand are in fact relics of techniques used in Europe and North America in the second half of the nineteenth century. The G techniques are those brought into use in the 1920s and 1930s. Between the middle of the nineteenth century and the 1920s the technology of can making evolved, producing a number of techniques intermediate in productivity between the G and H techniques. A superficial examination of these intermediate techniques suggests that they could provide the basis, almost certainly at the sub-process level, for a number of probably efficient modern techniques.

A variation on the resurrection of old techniques seems to be responsible for the availability of the G type of technique found to be optimal in Thailand. As mentioned above, these techniques were developed in the 1920s and 1930s, and now seem to have passed out of use in Europe and the United States. However, the techniques have been preserved in two Asian countries where they are used. It is not purely by preservation that this type of technique has remained available as a competitive alternative: in a number of respects (materials used, accuracy of parts, power units incorporated) they are different from the technique used in the past. These adjustments have almost certainly contributed to the maintenance of the competitive position of this type of technique, particularly with respect to their ability (when operated correctly) to match the quality specifications of cans produced by more "modern" techniques.

The implication is that there are a number of measures that can widen the range of techniques available; therefore what is available is not fixed and immediately known to all decision makers. These measures may range from simple [1] search, through local manufacture or engineering adaptation, to research and development proper. They all involve costs, which are, however, probably quite small by comparison with potential private and social gains. A case in point is provided by the gains that could be made in Thailand by using the Asian technology instead of that evolved in countries that are more advanced in this respect.

OTHER FACTORS AFFECTING CHOICE

For the most part we have looked at the question of choice of techniques as a matter of minimising costs of production given factor prices, more or less assuming a two-factor world (i.e. where factor inputs consist of capital and homogeneous labour). This concept is open to criticism on the grounds that it is an extreme simplification of how a decision about technology is actually made. Other factors may lead to choices that are different from what one would predict using the simple model. What is more important, however, is that such factors may also be serious obstacles to the success of some of the policies that are proposed on the basis of simple models—factor price policies for example. In the course of our research we collected information on factors that influence and sometimes determine the technological decision, but that lie outside the parameters of the simple model. We have already discussed some of these factors—the influence of changes in product, for example—but a number of others remain.

---

[1] Search may, in fact, be far from simple, particularly if it is intended to search for available techniques in the second-hand market.

Organisation and growth of markets

All the unit cost comparisons between techniques have been made on the assumption that the volume of demand will be adequate to allow operation of each technique at, or close to, its normal full-capacity rate of output. This is so even where these rates differ markedly between the cases compared. [1] This, of course, may not be the real decision situation; and since costs are obviously sensitive to the rates of output and hence to the volume of demand, the appropriate and optimal techniques may not be those suggested in the analysis above. We noted that the scale of the Tanzanian market might justify the use of the labour-intensive A4 lap-canning technique for open-top cans which could not be justified in the Kenyan market. Similarly, the D2 shoe-polish tin technique might be preferred to D1 if the scale of output had been less than was assumed in the cost analysis for D2. As it was, the expected rapid growth of the Kenyan market seems to have had some influence on the choice of the D1 technique. Again, the H techniques were probably a sensible choice in Thailand, given their inherent flexibility and the very small particular product markets that they supplied.

It is unlikely that in the particular cases covered in this study the scale of output would change any of the conclusions reached about the inappropriateness of some of the more labour-intensive techniques. However, the general point is worth stressing that two types of situation may be general enough to increase the probability that preference will be given to smaller-scale, more labour-intensive techniques.

The first situation is one in which a market is building up to the level allowing full-capacity operation of more capital-intensive techniques. Preliminary analysis of the Thai data indicates that the G techniques would still be preferred to the E1 technique, even if the unit costs of the former incorporated the penalty of very rapid accounting depreciation over only five years, while the latter was depreciated over 10 or even 25 years. Thus, even if the growth of a market is expected to reach a level justifying the use of a more capital-intensive technique, there may be a period during which the more labour-intensive technique is to be preferred. This situation might even arise repeatedly when the market grew above the level of demand that could be met by the first unit of productive equipment for the capital-intensive technique but before it grew to the level justifying investment in a second unit.

Secondly, it may be an aim of general social and economic policy to seek a wide regional dispersion of industry. Such a policy might be quite reasonable in connection with food production and canning and with can making. The

---

[1] See for example table 19. The H techniques operate at about one-tenth of the rate of output of the G techniques and at less than one-twentieth of that of the E techniques.

scale of production of dispersed units would almost certainly have to be lower than that of the more centralised plants covered in this study. Weighting of the costs to express a preference for a particular location as well as for a socially desirable greater volume of employment would of course favour the techniques with lower full-capacity rates of production, i.e. the more labour-intensive techniques. However, in the particular cases dealt with in this study it has been shown that an unacceptably big employment premium is necessary to bring some of the most labour-intensive techniques into parity with their more capital-intensive counterparts. Adding locational considerations is unlikely to make these techniques acceptable.

On the other hand, this brings us to the question of product specifications. Take for example the case of shoe polish. It is not inconceivable that a type of shoe polish other than "multinational shoe polish" could be manufactured in smaller, regionally dispersed units. It might well be marketable locally in cans of lower quality specification than "multinational cans". The "local cans" in turn could be manufactured with smaller-scale and more labour-intensive techniques (e.g. D2). An evaluation of the whole situation in terms of social costs might well indicate that the smaller-scale, more labour-intensive can-making technique was preferable.

Supervision and skill requirements

In an early case study of can processes carried out in Kenya [1], it was suggested that supervision costs might be a major factor militating against the use of labour-intensive technology. Some of the Kenyan data apparently support the notion that supervision costs increase as a proportion of total unit costs as labour intensity increases. In the case of round open-top cans, supervision costs were 37 per cent of total unit costs on the very labour-intensive lap-seaming line. They were proportionately lower on the more capital-intensive A1 and A2 lines (16 per cent and 13 per cent respectively), but they rose sharply with the most capital-intensive line (A3) where they came to 21 per cent of total unit costs. [2] The Kenyan data, therefore, are not very conclusive: if anything they suggest that supervision costs per unit are lowest with more "intermediate" technologies, but rise with greater labour and capital intensity.

Nor did the Thai study throw much light on the problem. In this case there was no discernible relationship between labour intensity and supervision

---

[1] ILO: *Employment, incomes and equality: a strategy for increasing productive employment in Kenya,* op. cit., Technical Paper 7.

[2] This may not represent the "real" requirement of this line for supervision inputs, but rather the indivisibility of the single supervisor on the line in relation to the lower number of line-operating workers on this more automated system.

costs as a proportion of total unit costs. Supervision costs were generally about 15-20 per cent of total unit costs, but there was no particular pattern to be observed. However, the Thai study does suggest an interesting qualitative difference between supervision on capital-intensive lines and on labour-intensive ones. The supervisor wage rates on automated lines tend to be very high; they are much less on the normal lines. Partly no doubt this is because the automated lines are in the high-wage sector anyway. But there are also some obvious differences between the type of supervision needed on automated lines—generally people with a high level of technical ability to manage machines—and the type needed to organise a large number of workers on a labour-intensive line. Probably the skills needed for organising workers (as opposed to supervising machines) are more readily available in developing countries than those needed for automated lines. If this is the case, the supervision problem may not be quite as general a problem in manufacturing as it is sometimes thought to be.

Imperfect knowledge

There is a good deal of evidence of imperfections in the flow of information about the range of techniques that are actually available. We have already pointed out that even a multinational specialised can-producing firm, with all its advantages of technical skill, organisation, management and other resources, may not be adequately informed of the possibilities. The firm which uses the E1 open-top technique in Thailand is a multinational and seems to have had little if any information about the technologies developed in other Asian countries (G1 and G2)—and this only after twice choosing the E1 technique. Imperfect knowledge can also lead to an exaggerated assessment of risk, which can have a determining effect on decisions. For example, the Thai firm using Japanese equipment bought similar but cheaper equipment from another Asian country for a later investment. Its reasons for making the more costly first purchase were largely that it was uncertain about the reliability of the equipment from the other country. In this sort of situation, one may even encounter the notion that some equipment is more reliable than equipment from some other source simply because it is more expensive, though very similar in design.

Multinational producers

Plant standardisation

The choices made by multinational enterprises can be strongly influenced by the advantages of using more or less standardised plant everywhere. This can cut operating costs very significantly by simplifying repair and speeding up diagnosis, particularly when diagnosis involves communication with the head office; this argument was put to us by a senior engineer from a large Hawaiian enterprise which in all other respects was apparently acutely conscious of factor prices. We have encountered the practice of standardisation by multinational producers in other industries too—notably in tyre manufacture, where it plainly leads to a commitment to very capital-intensive, automated plant. It was probably a consideration in the choice of the E1 technology in Thailand, and possibly also in failures to seek out more labour-intensive variations (as in the case of round open-top cans in Kenya, for example).

Vertical integration

In the cases covered in this study, choices were made by a number of different types of firm, and in general we would expect the nature of individual firms to influence their selection of techniques. However, the evidence from the cases on what the influence might actually have been is somewhat mixed and inconclusive. In the detailed cost analysis of the Thai data (not included in this chapter) we have allowed for some of the more obvious institutional differences between multinational and local firms. Although unit costs are sensitive to these adjustments, the optimal selection is much less so.

In broad terms the multinational firms themselves are similar to each other, but there are none the less some differences among them. For example, technique E1 was chosen by a vertically integrated firm which not only manufactures cans but also generates technology relevant to can making, develops techniques and manufactures much of the relevant equipment. To a considerable extent, then, its decisions with regard to the choice of techniques for can making in a subsidiary in a developing country are likely to be affected by its own involvement in this whole spectrum of activities. The firm may, for example, wish to spread research and development costs as widely as possible, to recoup them as quickly as possible, to maintain a certain level of activity in its machine-building facilities, or to sell refurbished equipment from its plants in developed countries [1], perhaps in response to innovation in the latter. These and other intra-firm considerations may well affect choice at the level of subsidiaries. Whether they actually do (or did) or not depends on details of the decision-making process about which our information does not allow us to come to any conclusion.

Another multinational enterprise operating in Thailand (not included in the six basic cases) seemed much more flexible with respect to the choices it made. For example, one of its first overseas plants, set up in the Philippines,

---

[1] This probably accounted for the kind of equipment used in the East African subsidiaries of the firm.

used techniques that were quite different from those in its home plant. Its Thai plant used techniques which were different from those in the Philippines plant; and although here it used some equipment displaced from its home plant, it also incorporated elements of the G2 Asian technique. Unlike the firm that had chosen the E1 technique, this multinational firm did not have such a great institutionally organised commitment to a particular type of can-making technology or to a particular set of techniques. It specialised in the canning and marketing of food products, and can making was merely a necessary complement to these activities. The firm was probably therefore more easily able to search for and use other can-making lines.

However, it appears that other factors also came into play. The multinational firm operating the E2 technique was similar to this less technologically committed firm: it was also a firm specialising in food processing and marketing, with can making merely a necessary complement, but it showed very limited flexibility and appeared to have made almost no search for alternatives. Although the E2 technique had both a lower initial capital cost and lower unit costs than E1, the different choice is, as we have suggested above, probably to be explained in terms of differences in product rather than by differences in the type of firm. However, as table 20 suggests, techniques G1 and G2 were probably to be preferred to E2; yet neither of the former were considered by the E2-using firm, even when in 1973 it came to consider the installation of a second line. Even after having a subsidiary operating in Thailand for a number of years, this multinational enterprise did not seem to know of the existence of these more labour-intensive and probably lower-cost techniques. Certainly it did not consider them as alternatives to another installation of the same technique as before.

Fear of labour problems

We came across the argument that machines are easier to manage than workers, and that machines do not go on strike. This appeared to have influenced technical decisions towards higher levels of capital intensity—at least in the sense that it is a disincentive to seek out labour-intensive techniques. Managers are also inclined to favour more capital-intensive technologies if it is very hard to dismiss workers or where, as in Kenya, the policy is to encourage firms to increase employment by some fixed proportion of their existing labour force.

The case studies did not yield very much useful information about the extent to which decision makers regarded personnel management as a constraint on the use of more labour-intensive techniques. However, one of the Thai cases (E1) provides some insights into aspects of the reality behind whatever impressions may exist. This firm, using the high-speed automated technique, encountered a number of problems concerned with personnel management. For example, there was a constant problem of inadequate supervision. The firm had to develop and train its supervisors from among workers initially operating the production line. At the same time there were apparently some signs of resentment on the part of the labour force at the limited promotion possibilities to supervisory positions within the plant. On top of this, the difficulties associated with running a high-speed line to produce a wide range of products led to an atmosphere of constant crisis on the factory floor. There was quite a high propensity to strike, and machine down-time was very large. Everybody from the plant manager downwards seemed to be under almost persistent pressure owing to the need to cope with these emergencies. The situation was one of potential and actual labour-management conflict.

Our preliminary analysis in this case suggests that in fact a wide variety of labour-management problems would have been eased considerably if the more labour-intensive G techniques had been used in place of the high-speed E technique: the larger line-operating labour force would have increased the probability of finding a larger absolute number of potential supervisors; the larger number of supervisory posts (though not necessarily corresponding to a larger proportion of the labour force) might have reduced the resentment about limited promotion possibilities; the simpler operating routine of very low change-over frequency resulting from using a number of lines of the G technique would have allowed greater attention to be paid to training in the required supervisory skills, and would also have sharply reduced the number of emergencies and the resulting situations of potential conflict. This is all a little hypothetical [1], but it suggests that one might at least question the often repeated hypothesis that increased labour intensity of production is associated with increased problems of labour management.

SOME GENERAL IMPLICATIONS

The results of the East African and Thai case studies suggest that capital-intensive techniques were used not merely because other more appropriate techniques were not available: a policy of creating more appropriate techniques by research and development or of developing intermediate techniques by

---

[1] We should also stress that the comments are based solely on a single case, possibly peculiar, in Thailand. They should be set in the context of the feeling of some managers in East Africa that increased labour intensity would in various ways generate problems of personnel management.

adaptive engineering probably would not on its own have made much difference to employment in the can-making industry; more appropriate techniques did already exist, but they were simply not used in situations in which it appeared that their use would have been justified.

It seems that it was the selection process, not the nature of the available techniques, that was somehow inappropriate. The decisions were probably not influenced by questions of products and markets; at least such questions did not prevent the use of G techniques in Thailand [1]; the selection of these more labour-intensive techniques did not require any shift in the patterns of consumption and income distribution.

Comparative irrelevance of some selection criteria in practice

It seems likely that policies designed to adjust distorted factor prices would have been equally ineffective in changing the choices made. On the one hand, some techniques (the most labour-intensive among those used in the cases studied) could not have been preferred unless relative factor prices had been changed to a degree that does not seem at all acceptable. On the other hand, other relatively labour-intensive techniques (the intermediate G techniques in Thailand) were already the most economical techniques at existing wage rates—even at the wage rates paid in the "modern" sector of the economy. Given the wide margin of existing cost advantage for these techniques, there is no reason to believe that efforts to make it any wider by manipulating relative factor prices would have achieved very much.

It is perhaps an unusual accident both of our sampling and of the particular history of can-making technology that the optimal selection from among the particular techniques in this study is not more sensitive to marginal shifts of relative prices. This very limited sensitivity is also perhaps due to the fact that we, like most of the companies making the decisions, have considered production lines as complete units. We have described above aspects of the behaviour of one multinational firm in Thailand which seemed to be quite flexible in its technical decision making: this company seemed to be unusually responsive to prices and costs, and also to be more prepared than most of the other firms to consider making up a production line with sub-processes drawn from a variety of different sources; if faced with different relative prices this company might have adjusted the average labour intensity of the whole line by adjusting the mix of techniques within the line.

---

[1] In fact, the management of the firm using the E1 technique did perceive product quality requirements as a probable constraint on the use of the G techniques. This, however, was after they had already installed two units of the E1 technique.

The cases suggest, then, that choices of technique are not highly sensitive to considerations of factor and unit cost. In the previous section we discussed some factors which may dilute the influence of cost considerations on decision making. Some of the factors do not seem actually to have been operative in these cases. For example, we cannot use considerations relating to the scale of the market as even part of an explanation of the inappropriate choices in Thailand; similarly the choice of more labour-intensive G techniques instead of the E1 technique may in fact have simplified personnel management and made it easier rather than more difficult. In fact, these case studies suggest conclusions similar to those of a study carried out in the footwear and sugar industries, namely that some decision makers appear to have been "imprudent". [1] That study suggests that the explanation of this type of imprudence is to be found in the failure of engineers, with their limited vision, and to a lesser extent of economists, to exercise a more decisive influence on investment decisions. The can-making case studies confirm the existence of this kind of limitation, but not in a manner systematic enough to allow any particular categories of decision maker to be identified as tending to be more "imprudent" than others. We must therefore attempt to find the origins of apparently imprudent decision making among the influences that cushion the influence of prevailing prices and costs on the selection process, in particular limited knowledge, risk avoidance and the desire for system compatibility within multi-plant firms.

We have described above some aspects of the differing behaviour of firms which appear to be broadly similar. We should stress the point that it does not seem possible to generalise about the types of decision maker who tend more often than others to make imprudent decisions. For example, a multinational enterprise using its engineers from a developed country was twice responsible for the inappropriate choice of the E1 technique in Thailand. This may give some support to general hypotheses about the malign influence of engineers. However, other information we collected suggests that the influence of engineers is not always malign. The production engineer of one firm, another multi-national enterprise, deliberately sought out information about the availability of a wide range of techniques for its overseas plants. The firm's choice was limited by a desire to maintain compatibility between the systems in its Thai and headquarters plants. However, the costs of compatible techniques were carefully balanced against the costs of other techniques. The firm incorporated selected elements of the G2 Asian technique with elements of other more

---

[1] See J. Pickett, D. J. C. Forsyth and N. S. McBain: "The choice of technology, economic efficiency and employment in developing countries", in *World Development*, Vol. 2, No. 3, Mar. 1974.

familiar techniques in its Thai plant. The firm was quite clearly seeking a least-cost solution to the choice problem at prevailing local factor prices.

We thus find similar firms [1] which behaved differently and made different decisions. In another example drawn from Thailand, two decisions with different resulting technical choices were made by a single firm. The firm using the G1 technique also selected the slightly less capital-intensive G2 technique This was only after the experience of using the G1 technique had increased its ability to cope with the risk perceived to be associated with the cheaper and therefore apparently less reliable G2 line.

Scope for national demonstration centres

It seems fairly clear from our observations in Thailand that inept decisions are apparently quite widespread. (Although we have not directly compared the alternatives found in Thailand with those used in East Africa, the implication is that the situation is similar there.) We have until now described only a few of the more inappropriate decisions made by the firms in the cases studied. A few examples may illustrate the wider range of inappropriate choice of techniques. The firm using technique G2 selected a line including two units of machinery which have never been used and probably will never need to be used for the types of product made. These two items cost 40,500 bahts—about 38 per cent of the total investment made. A number of the fruit-canning firms chose to set up their own can-making lines when even an optimistic assessment of their level of output would have indicated that purchase of cans from a specialist can maker would probably have been more appropriate. Having made that decision some of these firms then chose lines of equipment which incorporated unnecessary items or which called for unnecessarily high-speed techniques. Both of these features of the equipment made the capital costs of production higher than they needed to be.

The East African and Thai case studies suggested, therefore, that steps should be taken to reduce or eliminate various randomly distributed forms of imprudent selection of techniques and to reduce the influence of the factors that insulate decision making from local prices and costs.

A first step would be to assemble detailed information about the available techniques for the manufacture of the particular product—tin cans in this case. However, the case studies suggest that we should be very cautious about proposals to set up technology information centres, particularly their international variants. At least three types of consideration seem to be important

---

[1] As described above, these two firms are similar in only some respects. In others they differ: for example in the degree to which they are committed to can making and to particular types of can-making technology and can-making techniques.

in this respect. The assembled information should actually be brought to bear on the decision-making process. This requires some system that can influence both the entrepreneur in the informal sector in Bangkok and the decision makers of the multinational enterprise in Europe or the United States. It seems unrealistic to expect that the decisions of either would be much influenced if the system consisted only of some international information centre. Without a component at the national level at least, it seems likely that any international system will be ignored.

The second point stems from the fact that the information available to the decision maker probably needs to be of such a quality that it reduces risk aversion. The Thai entrepreneur in case G1 knew of the existence, specifications, costs and source of the less capital-intensive G2 technique, but he did not trust it. The local management of the multinational firm of case E1 similarly distrusted the G2 technique when they became aware of its existence. Interestingly enough, the first steps they took to measure the risk were to visit a local plant using the technique, to see it in operation and to take samples for analysis in their own testing laboratory. It seems likely that to reduce the perception of risk associated with alternative techniques the accumulated information will not only have to be locally available but will also have to be tangible and highly convincing. This seems to point to the need for demonstration plants capable of operating on a production basis, perhaps in a government-controlled production centre.

We noted earlier that the efficiency of a technique (in this case the G techniques in Thailand) was highly sensitive to the way in which the technique was operated. For example, inadequate operating capability ruled out the possibility that the firms actually using the G technique could supply cans to certain quality-demanding segments of the market. Limited operating capability may also severely restrict the average rate of output. Thus it may well result in higher-cost, lower-quality output. This in itself is undesirable, but it will also affect decisions taken by other producers: the observed experience of high-cost and low-quality output with a particular technique is likely to be a strong influence on subsequent choices of technique by the same firm or by others.

In addition, the existence of a kind of operational knowledge about technology might be a considerable help in monitoring and controlling the choice of techniques by multinational enterprises. For example, our information suggests that there would have been considerable social as well as private gains if the multinational can producer had used G techniques instead of E1. In fact private gains from using G techniques are apparently so big that they would almost certainly more than compensate for the private costs of losing inter-plant technical compatibility. Whether they would make up for the loss of the full range of intra-firm benefits is a question that we cannot answer. Obviously, in order to monitor effectively the choice of techniques by multinational enterprises, one would need to be able to make this kind of point. This again is an argument for accumulating detailed working information about the operating costs and performances of different techniques.

It is true that in the first instance most of the gains from lower-cost production through use of an intermediate technique would be appropriated by multinational enterprises. This obviously raises questions of how the gains will be distributed between the national economy and the firm (which may simply increase the payments of profits abroad). The provision of operational knowledge would therefore have to be accompanied by adequate control over the activities of foreign firms. Even in the absence of such control there would, however, be some social gains, particularly in terms of more employment per unit of output.

In general, then, we suggest that any attempt to influence the choice of production techniques must be very firmly grounded in, and very closely related to, local capability to operate them. The institutional instrument of such a policy must be a centre of excellence in existing technology.

Such a policy is not necessarily an alternative to the manipulation of relative factor prices; indeed, it is probably a necessary complement to it. However, if it is the case, as we suggest on the basis of these case studies, that the decision-making system is relatively unresponsive to relative prices, then efforts to manipulate prices so as to change the decisions made are unlikely to succeed unless the manipulation is massive—so massive that it would almost certainly generate other undesirable economic distortions.

Similarly, the use of an institution dealing with existing technology to influence decision making is not simply an alternative to policies designed to expand the range of techniques that are available for choice. Again the two approaches are complementary. If, as the case studies suggest, the more labour-intensive techniques available are not chosen despite lower costs at prevailing factor prices, then simply making available yet more possible techniques is unlikely to achieve a great deal. Excellence in operating and making decisions about available techniques will be both a stepping-stone towards and a necessary complement to design and production engineering for local manufacture with the technique; to engineering modification of techniques; and to any strategy of advance, through research and development, towards techniques more appropriate in developing countries than those evolved in the developed countries.

Clearly, a policy of the type we outline requires the use of resources. Can this resource use be justified? The analysis of the Thai case studies can be used to suggest a very approximate answer. We have already indicated, with reference to a single decision in Thailand, that the discounted present value of a more appropriate decision than that which was made might be as high as 10 million bahts (£200,000). To this must be added the value of the various social benefits which we noted would have resulted from such a decision. We must then include a valuation of the social and private costs of all the other "imprudent" decisions which we indicated had been made in the can-making sector in Thailand. Where one complete unit for the G type technique would have cost about 1.2 million to 1.5 million bahts, and where the total costs of employing an experienced, expatriate can-line engineer might be almost about 500,000 bahts a year, it is likely that the rate of return on the use of resources to improve decision making could be quite substantial.

Having attempted to indicate roughly the possible returns of a change of policy in this particular case, we should stress that in making any cost-benefit evaluation of this type of use of resources to improve decision making one must, of course, carefully consider the true opportunity costs of the particular use of what may be, in some countries, very scarce engineering and technical resources. However, the evaluation must also take into consideration the fact that there may be significant external economies to this type of resource allocation. The necessary types of decision making and operating capability must to a large extent be developed from more general technical skills through a process of learning by doing.

SECOND-HAND EQUIPMENT IN DEVELOPING COUNTRIES: JUTE PROCESSING MACHINERY IN KENYA

5

by C. Cooper and R. Kaplinsky [1],
with the collaboration of R. Turner

It is hardly surprising that the recent preoccupation with problems of unemployment should have revived interest in the use of second-hand machinery in developing countries. It is often argued that second-hand machinery is particularly appropriate to countries where investment capital is scarce and labour plentiful, because compared to new machinery it saves investment costs per worker. So if the unemployment problem is primarily the result of a shortage of investment, it seems sensible to look at the possibility of using second-hand machinery as a labour-intensive alternative, particularly in manufacturing industry where the choice of new technologies is rather limited.

There are many counter-arguments. For example, some people argue that second-hand machinery is nearly always inefficient because while it may save investment costs per worker compared to new equipment, it nearly always entails both higher investment costs per unit of output and lower labour productivity. Also, they argue that spare parts are an insuperable problem, maintenance costs are prohibitive and installation bedevilled by special difficulties that enterprises in developing countries cannot overcome.

There is surprisingly little factual evidence on either side of the argument, even though the question of whether second-hand machinery is "useful" or not is basically an empirical one. Consequently, we thought it would be illuminating to make an empirical case study of the economics of second-hand machinery. This chapter gives the main results of the case study. It starts with a brief review of the theory of second-hand machinery. It continues with an examination of the market for second-hand jute processing machinery in Dundee

---

[1] Science Policy Research Unit and Institute of Development Studies, University of Sussex. This chapter is a summary of the findings of a case study the results of which are discussed in more detail in C. Cooper and R. Kaplinsky: *Second-hand equipment in a developing country: A study of jute processing in Kenya* (Geneva, ILO, 1974).

and then with a comparison of the performance of second-hand and new machinery of similar types in a Kenyan factory. The last section reproduces some conclusions of the case study.

THEORY OF SECOND-HAND MACHINERY

The first step towards understanding what Amartya Sen has called "the usefulness of used machines" [1] in developing countries is to work out how an owner might value machines that are already installed in his factory, and to compare this to the value that they might have in a country where the rate of interest or profit is higher.

What is the value of a machine that is already installed?

Once the machine is installed, its value no longer depends on what was paid for it but on what it will earn in the way of rent in the future. The rent it can earn depends on the difference between the unit variable costs of operating the machine and the price of the final output from it. In a competitive market, the price of the output equals the total unit costs of making it with the most efficient new machine (i.e. the "best practice machine") that is available on the capital goods market. So, the rent per unit of output from the installed machine is equal to the difference between total unit costs on new machines and unit variable costs on the installed machine. This rent must cover depreciation on the installed machine, as well as giving the owner some net return. The value of the machine to him can be imputed from this net return if we know the rate of interest or profit.

The theory is that the owner will go on using his machine so long as its operating value is greater than zero [2], or if it has some value as scrap, so long as its operating value is greater than scrap value. However, if someone offers the owner a price for the machine greater than its value in use, he would be wise to part with it. So if there is a market for second-hand machinery, the prices ruling in the market may determine when the owner sells his old machines and buys new ones.

For the moment, however, the second-hand market will be left out of account, and a little more attention will be devoted to the factors that determine the value which the owner puts on installed machinery when there is no demand for used equipment.

---

[1] Amartya Sen: "On the usefulness of used machines", in *Review of Economics and Statistics*, Vol. XLIV, No. 3, Aug. 1962.

[2] The value of the machine drops to zero when total unit costs on the best available new machine falls below the unit variable costs on the installed one. For example, an innovation may result in a sharp reduction in total unit costs on new machines, or the unit variable costs of the old machines may increase with age.

Some of these factors are quite easily measurable. For example, the value of installed machines depends, inter alia, on the price, rate of output, and labour productivity of new machines for the same output. If new machines appear on the market at a lower price than before with a similar or greater rate of output and labour productivity, the value of installed machinery will fall, because the improvements in new machinery reduce the total unit costs of production, and in consequence reduce the market price of the output (under competitive conditions). The result is that the rents that can be earned from installed machines will fall, and so also their value. In the same way, an increase in unit variable costs on old machines—due to increased maintenance costs perhaps, or to a fall in their rate of output—will reduce their value, even if the prices and performance of the machines are unchanged.

However, the value of installed machines also depends on expectations, which are subjective and hard to quantify. Obviously, the rents that can be got from old machines depend on factor costs such as wage costs. If machine owners expect the wage rate to increase, and if new machines have higher labour productivity than old, owners will put a lower value on installed machines. Also, owners who think that the prices of new machines will fall in future, or that innovations will appear with higher rates of output or labour productivity than those that are currently on the market, will place less value on installed machines. In general, any expectations of falling product prices—whether due to innovation or, for example, to an anticipated weakening in demand—will pull down the value of installed machinery [1], perhaps below scrap value.

The value of installed machines may be thought of as the lowest price at which the machine owner is likely to part with them. One consequence of this is that it might be very difficult to determine what the price is in general terms: it will be different for different machine owners because it depends ultimately on their expectations of the future.

There are some conclusions to be drawn from the foregoing discussion. Anything that reduces the value of installed machines is likely to increase the supply of old machinery on the second-hand market at any given price. [2] Broadly speaking, there are three reasons why old machinery becomes available:

(a) because the variable costs of the machinery have increased with age, and perhaps because wage costs for operators and maintenance in advanced countries have risen, while labour productivity on old machines has fallen, compared with that on new ones:

---

[1] Such expectations will also reduce the equilibrium price of new machines in the capital goods market.

[2] This idea of a price level for second-hand machinery has to be reconsidered for the case of jute machinery, as will be seen later.

(b) because technically more advanced equipment has reduced the value of installed equipment; or

(c) because there is a decline in demand for the products.

These reasons are not independent of each other. In case *(a)*, for example, the important factor is the relationship between variable costs on old equipment and the price level. The price level, however, is a function of technology, since it depends on total unit costs on new machines. In both cases *(b)* and *(c)* the important change is the fall in product prices under competitive conditions. [1]

However, for practical purposes the distinctions are important. Machines which become available for reason *(a)* are likely to have very high running costs and low labour productivities. Machines that are available for reason *(b)* may be obsolete in both developing and advanced countries, and should be treated with caution. On the other hand, it is possible that machines which become available because of a downturn in demand will be technically up-to-date and quite new.

So far, attention has been concentrated on the value of installed machines to their present owners. The next question is, how do potential buyers of used equipment (particularly buyers from developing countries) value it?

Let it be assumed that firms in the developing countries choose between new machines (models that are also available to firms in advanced countries) and second-hand machines from firms in advanced countries. In order to maximise their profits, they should choose the equipment which gives the lowest total unit costs at their factor prices. Let it also be assumed that both new and second-hand machines perform in the same way in an advanced and in an underdeveloped country (so that rates of output, machine productivity and labour productivity are the same in the two countries). Assume also that there are no transport costs, installation costs or differences in maintenance costs between the two countries. The only significant differences are that the rate of interest in the developing country is higher than in the advanced and the wage rate lower. If we make these assumptions it turns out that the choice between new and second-hand machines depends only on the relative prices of the machines. [2]

Usually the price of new machines can be taken as given, i.e. as a world market price. In that case there is some price for second-hand machines at which the firm will be indifferent between them and the new machines. This

---

[1] However, the analysis does not depend on the assumption of perfect competition. It will work equally well for oligopoly, provided we assume profit maximisation. See W. E. G. Salter: *Productivity and technical change* (Cambridge University Press, 2nd ed., 1966), pp. 90-93.

[2] Strictly speaking, we must also assume that businessmen who wish to buy the old machines have the same expectations about future innovations as the machine owners, so that they make similar assumptions about machine life.

price is the maximum price which the businessman in the developing country can pay for second-hand machinery. If the price is higher he will prefer new machines, because they will give lower total unit costs. The important point about this maximum price is that on the assumptions we have made, it is always higher than the minimum price at which the machine owner in the advanced country is prepared to part with the machine. [1]

This is an important result because it is really the basis of arguments in favour of using second-hand machines in developing countries. It implies that if firms in the developing countries can buy second-hand machinery at or near the price which reflects their value in use to the owners in the advanced countries, it will always be more profitable—from the private point of view—to use the second-hand equipment. This is the first step in the argument. The second is that second-hand machines must also have the advantage from a social point of view: they save on foreign exchange; they will benefit more than new machines if the shadow wage for labour is below the market wage, and so on. Thus, the differential advantage of second-hand machines over new ones is likely to be increased if social, as opposed to private, accounting is used. Finally, if policy makers put special weight on employment generated per unit of investment (a "pure employment objective") [2], second-hand machinery will in most cases be even more favoured over new equipment.

However, the argument rests on some challenging assumptions, and much of the debate about second-hand machinery centres on those assumptions

---

[1] Amartya Sen, "On the usefulness of used machines", op. cit. See also the monograph on which this chapter is based (C. Cooper and R. Kaplinsky, op. cit.). This result holds even if wage rates are the same in the two countries; all we require is that the rate of interest and profit in the developing country should be higher than in the advanced.

Suppose that the second-hand machines are priced at their value in operation to the firm that owns them. This is the price that equalises the total unit costs on new and old equipment at the rate of interest prevailing in the advanced country (and also at the wage rate in that country—but we will leave wage rate differences out of account). At this price for second-hand machines, their unit capital costs will be a smaller proportion of total unit costs than with new machines. If now the rate of interest is increased, while maintaining the prices of old and new machines at their original level, total unit costs on both old and new machines will increase, because of increasing unit capital costs. The proportionate increase in unit capital costs will also be the same on both old and new machines. But because unit capital costs are absolutely smaller with old machines, the absolute increase in unit capital costs, and hence in total unit costs, will be less with old machines than new. Then, since at the original level of the rate of interest total unit costs were the same for both types, it follows that at higher rates of interest total unit costs will be lower on the second-hand machines than on the new. This is the argument which Sen develops. A corollary is that at the higher (developing country) rate of interest, the price for second-hand machines at which their total unit costs are equal to those on new machines will be greater than at the lower (advanced country) rate of interest. This profit-equalising price in the developing country is the maximum price which a firm in the developing country can afford to pay for the machines, and it is necessarily higher than the minimum price at which the owner in an advanced country will be prepared to part with them.

[2] On the question of the pure employment objective see UNIDO: *Guidelines for project evaluation*, op. cit.

rather than on the logic of the argument itself. Before we look at the empirical results, we shall list some of the assumptions and discuss the consequences of dropping them.

1. There is the assumption that the second-hand machinery can be purchased at or about the minimum price which is acceptable to its owner, or at least well below the maximum price which the firm in the developing country can afford to pay for it. This is really an assumption about the organisation and structure of the market for second-hand machinery. In his analysis, Sen justifies this assumption on the grounds that ". . . the underdeveloped country [provides] only a very small part of the total demand for machinery of any age group". However, as we shall see in the case study, a thin market of this kind can have quite different implications for price formation. There is also the possibility that a substantial demand for second-hand machinery might push prices up, to the point where it becomes privately unprofitable in at least some developing countries (i.e. those where interest rates are not so high and wage rates not so low compared with those of advanced countries).

2. The argument does not cover the possibility that the old machinery may be available because of an innovation. A radical innovation could make the machinery obsolete in the advanced country, but equally, it could mean that the maximum price that firms in the developing countries can afford to pay for it is less than its price as scrap. In this case, there is no basis for a transaction and there will be none, unless someone makes a mistake (which can happen). In some rather muddled arguments against second-hand machinery, it is asserted that since second-hand machines generally become available because of innovation in the advanced countries, they are always sub-optimal —even in developing countries. This is not a very sound argument. First, as already pointed out, the supply of second-hand machinery does not depend only on there being an innovation in the advanced countries. Secondly, even if there is an innovation, it does not follow *a priori* that the second-hand machines will be sub-optimal in the developing countries. The point is only that they may be—whether they are or not is a matter of fact, not logic. Thirdly, if the machines are sub-optimal, presumably no one will wish to buy them anyway.

3. Transport, building and installation costs are left out of the argument. Generally speaking, these costs are the same for second-hand as for new machines. If they are large they may substantially reduce the maximum price that the firms in the developing country can afford to pay for the second-hand machines themselves. It is possible that this maximum price could fall below the minimum price the machine owners are prepared to accept.

4. The argument assumes that rates of output on new and second-hand equipment are the same in the advanced as in the developing countries. Expe-

rience in many industries suggests that this is extremely unlikely. Of course, new equipment may be affected just as much as second-hand by sub-optimal working, but there may be cases in which second-hand machines are more seriously affected. For example, new machines are often installed by the machine makers, whereas second-hand ones may be installed by the purchaser himself. Faulty installation could be one reason why the loss of output on second-hand machines is greater than on new.

5. The argument does not take account of problems of maintenance and spare parts, which may increase the variable costs of production on second-hand machines more than on new machines.

6. Finally, the argument leaves out the role of expectations and risks in the investment decision. Both the maximum buying price for second-hand machinery in developing countries and the minimum selling price in advanced countries depend on business expectations as well as on more easily measurable things like machine performance. For example, potential buyers of second-hand machinery may, for one reason or another, have stronger expectations of a future innovation than the sellers. This could drive the maximum price they are prepared to pay below the owner's minimum selling price, and destroy the possibility of a transaction. Alternatively, buyers may be ignorant of forthcoming innovations that sellers already know about, so that they take an excessively optimistic view of the value of second-hand machinery. An important point which is not considered is whether there are special risks in buying second-hand equipment, greater than with new. The case study suggests that there may be.

When we look more closely at the assumptions underlying the argument, the rather optimistic view about the usefulness of used equipment which comes out of the theoretical argument is at least attenuated. Precisely how much importance attaches to these matters is a question that can be answered only by looking at the facts of particular cases. The case study which follows is an attempt to do this. Obviously, it cannot give a basis for broad generalisations, but it can illuminate some of the problems that arise with the use of second-hand machinery in reality.

THE DUNDEE MARKET FOR SECOND-HAND JUTE PROCESSING EQUIPMENT,

This section deals with the market for second-hand jute processing equipment in Dundee, Scotland; Dundee is or has been the most important centre for the jute processing industry in the advanced countries. This section is more concerned with the organisation of the market than with its quantitative

aspects. Before the results of the study in Dundee are examined a brief description of jute processing needs to be given.

Jute processing is not very complicated in principle, although it is technically sophisticated in practice. Jute is delivered to the processing plant in
bales of fibre. The bales are opened, and the tangled fibres are spread out
(there are four processes involved here—bale opening, batching, breaking
and finishing). Next, the tangled fibres are combed out in a number of stages
so that they lie parallel to one another. This is called "drawing".

After drawing, the jute is in the form of a long flat strip of parallel jute
fibres about 1.5-2 inches wide. The strip is crimped to give it greater tensile
strength. These strips of aligned fibres are called "slivers".

At the next stage the sliver is spun into yarn on a spinning frame. It is
then prepared for weaving by treatment and by winding on the appropriate
delivery spools. The yarn is then woven on various types of loom to produce
jute cloth.

The detailed study in the Kenyan factory, which is described in the next
section, is concerned with only two types of jute equipment. These are circular
looms, which are used for weaving a tube of cloth for making maize bags,
and spinning frames. However, in looking at the market for second-hand
equipment we took account of other types of machinery as well.

Supply

A first conclusion is that the supply of second-hand jute machinery has
been affected by all three of the conditions previously mentioned, viz. ageing
of installed equipment (without noticeable innovation in new equipment);
innovations; and declining demand for products from the United Kingdom
jute processing industry. We can designate three periods according to which
of these factors was dominant.

First there is the period up to the Second World War. In this period
there was not much innovation in jute processing. The supply of second-hand
equipment was mainly determined by the mechanical deterioration and the
consequent increase in unit variable costs on installed machines. This kind
of deterioration is slow on many kinds of jute machinery (which tends to be
mechanically simple and very robust), so a lot of the equipment that became
available in this period was very old—some of it had been in use for between
80 and 100 years.

The situation changed after the war. From 1945 until about 1960 there
was substantial demand for the products of the United Kingdom processing
industry, and also a considerable amount of innovation in machinery. Some
of the innovations led to large-scale replacement of installed capacity. The

replacement of the old rove spinning capacity by so-called "slip-draft" spinning immediately after the war was an important case in point. Other innovations, however, actually reduced unit variable costs on installed machines and had the effect of slowing down replacement. A particularly important one was the automatic loader. This made it possible to replace shuttles without stopping the loom; it increased labour productivity in weaving by a factor of 4, for a rather small investment outlay per loom.

Since 1960 the supply of old machinery has been affected partly by innovations, but mainly by the declining market for the products of the United Kingdom jute processing industry. Demand fell off for a number of reasons. For one thing, the Government of the United Kingdom reduced tariffs and did away with quotas on imported jute cloth and bags, so that Indian and Pakistani manufacturers ran off with a large part of the bag market. Secondly, substitute materials came on to the market—particularly paper and polypropylene for making bags. And thirdly, agricultural demand for jute bags fell off because of bulk handling. The United Kingdom jute processing industry diversified. Most firms withdrew from the bag market and concentrated on high-quality jute cloth, particularly broadcloth for making tufted carpet.

The main result of this was to increase the supply of used narrow looms, which had been used mainly for bag cloths. There was a substantial market for these in Africa, and later on in Thailand. Also, used circular looms (which are entirely specialised to bag making) became available. In the initial part of this period the supply of spinning and drawing equipment was not increased as much as the supply of weaving machinery. Diversification into broadcloths did not require any major changes in spinning capacity.

Later on, however, after about 1965, the supply of used spinning frames increased, partly because of innovation. A new type of spinning frame, called the "apron-draft" frame, displaced some of the slip-draft frames that had been installed just after the war. But the decline in the broadcloth market (which was taken over in its turn by substitute materials, particularly polypropylene) meant that there was substantial excess capacity in both spinning and weaving equipment in Dundee. During this period (i.e. 1965 to the present) whole jute mills and weaving plants were sold second-hand (in Africa, Bangladesh and Thailand) and much of the equipment was relatively new.

Earlier we pointed out the different conditions which could lead to an increase in the supply of old machinery. These conditions are of importance to buyers, because the quality of the machinery that becomes available depends partly on the reasons why the original owners want to get rid of it. This is borne out to some extent by the study we made in Dundee. For example, before the Second World War, machinery supply was mainly determined by deterioration of equipment (which was subsequently replaced by machinery of more or less the same kind). Much of this equipment was very old indeed and of dubious quality. At that time the second-hand merchants were obliged to do a considerable amount of reconditioning, which has been much less common since the war as relatively newer equipment has been sold off because of innovation and replacement, and declining markets. [1] It is certainly true that much of the equipment sold off in the late 1960s as a result of declining markets was almost new, and was technically up-to-date.

However, the study also showed that the relationships between the underlying reasons for increases in the supply of machinery and its quality and potential usefulness in the developing countries are not always the same. For example, rove spinning frames that were replaced in 1948 by the innovation of slip-draft spinning were often very old, and many of them were in poor condition. So innovation does not necessarily make relatively new machines available; what counts is the rate of innovation and how long it is sustained. Another problem is that even when there is a large amount of excess capacity, machine owners quite often retain old machinery for long periods. There may be a number of reasons for this. For example, the owners may hope for an eventual recovery of demand, or they may hope that it will be possible to use old machinery in new ways—for example by using jute looms to weave synthetic substitute materials. Or they may keep the equipment in hopes of a particularly profitable second-hand sale, perhaps by selling off the jute mill as a whole instead of in bits and pieces. At all events, this form of "stocking" is quite common in Dundee; it is helped along by the low opportunity cost of factory space in a depressed area. Consequently, some of the machinery that has been sold off as a result of declining markets is older than might have been expected. The second-hand circular looms which we studied in Kenya had been kept idle in Dundee for about four years.

Organisation of the market

The second important result of the study in Dundee has to do with the organisation of the market for second-hand equipment. It relates directly to assumptions about the factors that determine the prices of second-hand machines, which underlie a good deal of the theoretical argument discussed in the first part of this paper.

To start with, we found that decisions to discard old machinery are hardly ever affected by prospects of a second-hand sale. Nearly all replacement decisions in Dundee are made independently of the market for second-hand

---

[1] Another reason for this may be that rising labour costs in the United Kingdom and increasing skills in the developing countries have meant that the comparative advantage in reconditioning has passed to the developing countries.

machines. The main reason for this is that the market is very thin. Buyers only appear intermittently. They enter the market not in response to information about the availability of old machinery in the United Kingdom but because the firm in the developing country has decided on an investment programme. The communication of information on used machinery for sale is imperfect; moreover, sellers are usually in the market for only a limited time before they scrap their machinery. Whether or not a transaction takes place depends, therefore, on there being a coincidence in timing between the decision of buyers to hunt out second-hand machines and replacement decisions in Dundee. When these coincidences occur, they usually involve only one buyer and one seller.

Some of these effects have been modified from time to time by the stock-piling of unused equipment after a fall in demand, and also by dealers who sometimes supply a number of potential buyers with information about the availability of equipment. Generally speaking, however, the market remains highly imperfect. There is simply no assured equilibrium price for used machines which machine owners can take into account when they make replacement decisions.

A more important consequence of this market organisation is that nearly all transactions involve bargaining between the maximum price that the purchaser is prepared to pay and the minimum that the seller can accept (which may be near to scrap value). The buyer is normally at a disadvantage in the bargaining process. First, it is often difficult for him to know what is the maximum price he should pay. Apart from problems of predicting future innovations (which buyers often know less about than sellers), there is the difficulty of guessing how transport and re-assembly may affect the performance of the machinery. Secondly, the seller normally has more information and knowledge about the machine than the buyer; he is better informed about its past history and its technical limitations. In these circumstances, prices may be driven up well above the minimum acceptable price to the seller; they may end up quite close to the buyer's maximum price. Indeed, as we shall see in the factory study, the buyer can make serious mistakes about what his maximum price should be, and the price at which the bargain is struck can be well above the ceiling which a more realistic buyer would have set.

Role of dealers

A third subject of inquiry in the Dundee part of the study was the role of second-hand dealers. The *raison d'être* for dealers in this industry is that sellers and buyers know very little about one another. It would not be possible for the Dundee firms to meet the overhead costs of keeping themselves informed about potential buyers, since each firm comes on to the market only very occasionally. Similarly, their customers would have to meet the costs of keeping themselves informed about the availability of machinery in Dundee, and that would be uneconomic for them. The dealer concentrates transactions and is able to meet these costs, although the market is not large enough to support more than a very few dealers anyway.

The dealer has other sources of advantage. The most important is that he is in a much stronger position than his customers to assess the technical quality of machinery in the Dundee factories. This is partly because he has access to better technical advice than his customers, and anyway has much more experience than they of making this kind of judgement. But probably more important, the dealer is likely to know much more than his customers about the way in which particular pieces of machinery have been treated during their working life in Dundee.

Finally, the dealers carry out some particular functions for their customers. They provide the labour to lift the old machines, pack them and transport them to the shipment point; they do the legal and commercial paper work of the transaction; they provide some assurance of spare parts, and sometimes they are able to find engineers and technicians to help with re-assembly and installation.

When it comes to the actual transaction, the knowledge the dealer has about the quality of the machinery and the practices of the firm which is making the sale is obviously useful in bargaining: it gives the dealer an advantage over his customers. But he has other advantages also, particularly because he can conduct the bargaining process in a way which his customers in developing countries simply cannot copy.

For example, many of the bargaining situations involve playing for time. The dealer keeps his offer down and waits, hoping that the owner will be placed in a position of accepting his price or else selling less profitably for scrap. This procedure is likely to work best in cases where the company wishes to free factory space for new equipment, when there are costs involved in waiting for the dealer to increase his offer.

In this type of market it is difficult for a buyer to know whether it is better to work through a dealer or to go it alone. The customer has to meet some stringent conditions before he can be sure that he will be able to do better by making a direct purchase: he must have a way of informing himself about availability of old machinery (perhaps over quite a long period); he must be sure that his bargaining capacity is at least as good as the dealer's; he must have detailed knowledge of the particular machines he is buying; and he must be able to obtain spare parts. Usually, the customer would have to incur costs to meet these conditions, so that he has to be sure that these costs plus the actual selling price are lower than what he would have had to pay the dealer. The dealer is essentially a second-best solution, but since the market is imperfect it may be quite difficult to find a better solution unless the government in the advanced country is willing to step in to improve the flow of information and knowledge in the market.

COMPARATIVE PERFORMANCE IN KENYA OF JUTE PROCESSING EQUIPMENT BOUGHT NEW AND SECOND-HAND

The real test of the economic value of second-hand machines in developing countries is whether or not they are privately and socially profitable in operation: their advantages and disadvantages do not appear until they are used to produce output.

This part of the chapter relates to a Kenyan factory, and involves a comparison of the performance of second-hand machines with that of machines which were purchased new by the factory owners. [1]

The comparison covers two types of jute machinery: first, circular looms, which are used to make tubular cloth for grain bags, and secondly, 80-spindle slip-draft spinning frames used to spin jute yarn. The factory uses both new and second-hand machines in each category. The comparison is based on output and costing data which we collected from factory records and from extensive talks with the factory management over a period of two to three months. The factory managers were extremely co-operative and helpful and gave us unusually free access to their records and knowledge.

The first part of the comparison is about the technical performance of the machines: the second deals with economic performance.

Technical performance

Circular looms produce tubular cloth which is subsequently cut into bag lengths and sewn. In Kenya the looms are used to make what are called "combination grain bags"—"combination" because the cloth is made from a combination of fibres: the weft is sisal and the warp is jute. The Kenyan bags are made from very heavy yarns: the weft is 1,100 tex and the warp 420 tex. Circular looms are rather complex by comparison with most other types of jute machinery.

---

[1] In the interest of brevity, the machines that had been purchased new are sometimes referred to in this section as "new machines", even though they were no longer new at the time when the data were collected.

There are four kinds of circular looms in the sample, which consists of: 8 machines (1949), designated J2 Mk I 49; 12 machines (1954), designated J3 Mk I 54; 4 machines (1963), designated J3 Mk II 63; 5 machines (1966), designated J3 Mk I 66. The 1949 machines are the ones that were bought second-hand.

There are some technical differences between these types of machinery, but as it turns out, these differences do not affect our main conclusions.

We obtained output data on a shift-by-shift basis for each loom over a period of 20 days. However, because we were suspicious of some of the readings for the first 10 days, we mainly used the data for the second half of the period. The average output per shift from the looms for this latter part of the period was as follows:

Type of loom	Average output per shift (metres)
J2 Mk I 49 (second-hand)	68.1
J3 Mk I 54	316.5
J3 Mk II 63	335.3
J3 Mk I 66	275.4

All these average outputs differ significantly from one another (at the 1 per cent level), except those for 1963 and 1954. The results show that the second-hand machines were much less productive than the machines which had been purchased new. Both in our discussions with engineers in Dundee (some of whom had been responsible for these very machines during their working life in Dundee) and with factory management, we examined a number of possible explanations for this extremely poor performance. Using Kenyan yarn the machines were operating at a factory-rated efficiency of 10-15 per cent; with yarn of comparable weight in Dundee they had operated at about 50 per cent efficiency.

Some of the factors that contribute to this poor performance are not specific to second-hand machines. However, the most strongly favoured explanations, and the ones which are held to account for most of the loss of output, are specifically related to the fact that the machines were second-hand. The Dundee engineers (with whom we had extensive discussions) mentioned the two main factors. First the J2 looms had been standing idle in Dundee for a long time—probably about four years. They had certainly been looked after during this period, but nevertheless they must have suffered considerably. Secondly, because these were old looms that had stood in the same Dundee factory for many years, they had "settled" in their original factory space. Lifting, transport to Kenya and re-assembly must have caused many problems of alignment. The performance of the machines strongly suggests that these problems have not been solved. The machines were in fact installed by factory personnel in Kenya. They were purchased more or less directly by a senior manager from the Kenyan factory, with minimum support from the second-hand dealer. There was no technical assistance with installation.

At this stage, we took a look at comparable output data for the spinning frames.

The spinning frames in the Kenyan factory were all 80-spindle slip-draft frames weaving jute yarns. We have already discussed how spinning is done. An important point about it is that the machinery is simple and robust; it is very different from circular looms from this point of view. There are 16 spinning frames in the sample. Six of them are Fairburn-Lawson frames bought new in 1954 (referred to below as FL 54s). Six are Mackie frames bought new in 1968. Four are Mackie frames made in 1963 and bought second-hand in 1971. The output from these frames is indicated below.

Type of frame	Output per shift (kg)
FL 54	222.1
M 63 (second-hand)	300.0
M 68	335.2

The first point is the poor performance of the 1954 machines. They actually produce less per shift than the second-hand 1963 models. The main reason for this is one of design: the machines are not necessarily being used inefficiently —they are simply technically inefficient machines. Secondly, although the average output from the second-hand machines is lower than from the technically comparable 1968 machines, it is not much lower; the difference is only about 10 per cent. In the case of spinning frames the second-hand machines compare much more favourably with machines bought new than in the case of circular looms. Part of the reason for this may be that the second-hand spinning frames are not as old as the second-hand looms, nor did they stand idle in Dundee for as long as the looms. But the main reason is almost certainly that the spinning frames are much simpler machines than the looms. They are built on a massive metal framework, which reduces re-alignment problems after lifting and transport, and the moving parts—per spindle at least—are simpler and more robust than in the case of the circular looms.

The lower average output on second-hand frames compared to that on frames bought new is probably due to mechanical wear on rollers and other moving parts, which can lead to more frequent yarn breakages. If the spinner is reasonably skilled a yarn breakage on the spinning frames can be corrected quite rapidly without stopping the frame. However, if yarn breakages are frequent, there may be delays in rectifying them, since the worker may have some difficulty in keeping up, and may fail to notice all the breakages immediately.

A second stage in the study of technical performance was the analysis of sources of variation in the output of machines. We considered two hypotheses. First, it seemed likely that there would be greater differences between the performance of individual machines of a given model as the model became older. If this was so, it meant that purchasers of second-hand machines had to solve the problem of choosing the better machines from a batch which might be very variable in quality. So we wished to know whether there were greater differences between the machines in the second-hand group than between the machines that had been bought new. Secondly, we thought that the output of particular machines might vary more from day to day and from shift to shift as the machines got older. Loom output could vary because the efficiency of the weavers (and other workers who supported them) varied. Our argument was that older looms were more sensitive to the efficiency of the worker than new looms. Newer looms were stopped for more or less routine reasons like shuttle changes or minor yarn breakages, which could be fixed up fairly easily by semi-skilled workers. Semi-skilled workers probably took about the same time as skilled ones to get the loom into operation again. Consequently, while there might be some variation in the output from the loom it would not be affected very much by the fact that some workers were less skilled than others.

Older looms, however, particularly second-hand ones, were likely to suffer from mechanical problems and possibly more severe yarn breakages (because of alignment problems). In other words the ailments of older machines were different from those of newer ones. It might need skill to cure these ailments quickly. The production time that was lost on older machines might depend on whether the man who tried to solve the problem was skilled or not, to a greater extent than with newer machines. Since the workers were not uniformly skilled, this should show up in a greater variation in the output of the older looms.

The data on circular looms supported the first of these arguments. There are no significant differences among machines of the newer 1966 model, but there are among those of other models. Also the relative variation in output increased with the age of the machines. The coefficient of variation was 0.015 for 1966 machines and 0.984 for the second-hand machines.

On the other hand the spinning frame data did not support the first hypothesis. There were highly significant differences among machines of both the 1968 and the second-hand 1963 model, but the coefficients of variation among machines were more or less the same for both models.

Once again we were led to the conclusion that the spinning frames had not deteriorated with age to the same extent as the circular looms, nor did they suffer as much from the ailments typical of the second-hand machines, mainly because the spinning frames were robust machines and relatively uncomplicated.

The second hypothesis suggests that the older second-hand machines may be more sensitive to lack of operator skills than the new ones. The second-hand circular looms show significant variation between shifts whereas the newer models do not. Also coefficients of variation are greater for the second-hand looms. In the case of the spinning frames, we cannot distinguish between the new and the second-hand models in this way. The hypothesis is confirmed for the circular looms but not for the spinning frames.

These comparisons of technical performance can be related to the key assumptions in the argument about second-hand machines discussed in the first section. In the case of the circular looms, it is clear that there were important miscalculations about the attainable rate of output—miscalculations which are specifically the result of the second-hand origin of the machines. For these machines, the uncertainties about output performance were seriously under-estimated. The spinning frames, on the other hand, do not give rise to this kind of problem. The relatively large differences among the second-hand circular looms, and their relatively greater sensitivity to worker skills, also point towards an important kind of uncertainty. Once again, there is much less of a problem with second-hand spinning frames.

Taken together, the results demonstrate the difficulties of generalising about second-hand machinery. However, they also suggest that it is possible to distinguish between kinds of machinery that are more susceptible to the problems usually raised about second-hand equipment, and kinds that are less so.

Economic performance

To compare the economics of machines bought new or second-hand respectively, we have to examine the rationality of the investment decision which resulted in the firm's buying second-hand in the first place. This involves a comparison of the *ex ante* total unit costs of production on each type of machine. To do this the following kinds of information are needed:

(a) a list of the various types of machine (including the second-hand ones) that were available to the firm at the time it made the decision to buy second-hand;

(b) the rates of output from each type of machine;

(c) the price of each type of machine [1] (at the time of the investment decision), anticipated economic life and the rate of interest; and

---

[1] To be precise, we need to know the total costs of installing each type of machine, of which the price of the machine itself is only a part. We describe total investment costs in more detail later on.

Table 21.   Circular looms: investment costs per loom
(EA shillings)

Type of loom	Total investment costs	Of which: purchase price of loom
J2 Mk I 49 (second-hand)	41 414	5 849
J3 Mk II 63	121 324	65 248
J3 Mk I 66	125 459	71 528

(d) variable costs of production and working capital required on each type of machine (we will specify these more completely later on).

A number of assumptions were made in order to set up these data for the circular looms and the spinning frames. [1] First we assumed that the firm could choose between the second-hand machines it actually installed and new machines of the type it had already installed under earlier investment programmes. This is admittedly a bit arbitrary since it leaves out some other possibilities that might have been open to the firm, but it is justified on grounds of data availability. Secondly, we made estimates of what machine prices had been when the firm had installed the second-hand ones (i.e. in 1970). These estimates were based as far as possible on quoted market prices at the time. Thirdly, we made estimates of working capital costs and of various smaller components of variable costs. Fourthly, we calculated annual capital costs on the basis of the conventional annuity formula, using the estimates of machine life made by factory personnel.

The results of the cost comparison are summarised below. We look first of all at total investment costs of the circular looms and the spinning frames, and then at the total unit costs of production.

Table 21 summarises some components of total investment costs for the circular looms (at 1972 prices). It is a picture of some of the investment alternatives which were available when the firm bought second-hand J2 looms in 1971. The table shows how deceptive the low f.o.b. price of second-hand machines can be as a guide to investment costs because of all the other costs that are involved, and that remain the same whether a machine is bought new or second-hand. The total investment costs for a J3 Mk I loom are about 1.7 times greater than the f.o.b. cost of the loom itself. Total investment costs for the second-hand looms were 7 times the f.o.b. cost of the loom alone. Consequently, although the second-hand looms were bought at only 8 per cent

---

[1] These assumptions are discussed in detail in C. Cooper and R. Kaplinsky, op. cit.

Table 22. Spinning frames: investment costs per frame
(EA shillings)

Type of frame	Total investment costs	Of which: purchase price of frame
FL 54	194 861	155 290
M 63 (second-hand)	73 745	29 706
M 68	196 179	150 156

of the price of a new J3 Mk I loom, their total investment costs were 33 per cent of the investment costs of a J3 Mk I.

Most of the damage is done by the high costs of installation for circular looms. The machines are mounted on a specially constructed wooden platform about 10 feet high. Costs of installation were a good deal more than twice the cost of the second-hand looms, though only about one-fifth of the costs of new looms.

In general it is clear from this data that a good deal of the price advantage of second-hand machinery can be whittled away by these other elements in investment costs. If machinery is expensive to transport or to install or if a lot of working capital is needed to operate it, the argument for buying second-hand is considerably weakened.[1]

Table 22 gives comparable data for spinning frames.

Once again, the price advantage of second-hand machines is cut down by other fixed elements of investment cost, but the effect is much less marked than with the looms. If the figures for the Mackie 1968 machines bought new are compared with those for the Mackie 1963 machines bought second-hand, it will be seen that total investment costs of frames bought new are about 1.3 times the machines price (as against 1.7 times for looms). Total investment costs for second-hand frames are 2.5 times the machine price (as against 7 times for looms). The purchase price of second-hand frames is 20 per cent of the price of a new frame (for looms the proportion is 8 per cent) and their total investment costs are 37 per cent of total investment costs for new machines. Obviously two factors are at work. One is that to start with the price of second-hand frames is a quite high proportion of the price of new ones (higher than in the case of the looms). The other, more important factor is simply that transport and installation costs are not as big in relation to

---

[1] This is certainly true from the point of view of the private businessman. From a social point of view, the foreign exchange cost of the investment is an important consideration, though this is not as yet a major concern in Kenya. If the local currency is seriously overvalued, the premium on foreign exchange will to some extent offset high local costs of installation and working capital, in favour of the second-hand machines.

Table 23. Cost of weaving 1 metre of tubular cloth on circular looms
(EA shillings)

Type of loom	Machine costs	Working capital costs	Labour costs	Power costs	Indirect variable costs	Total cost
J2 Mk I 49 (s/h)	.071	.014	.167	.007	.003	.265
J3 Mk II 63	.031	.007	.054	.007	.003	.101
J3 Mk I 66	.044	.009	.058	.007	.003	.120

machine costs as they are for circular looms, so that advantages of low machine price for the second-hand spinning frames are not whittled down to quite the same extent as are the advantages of low-priced circular looms.

The data on looms, and to a somewhat lesser extent on spinning frames, suggest that assumptions about transport and installation costs are very important to the argument about second-hand machinery. Essentially, if there are important elements of investment costs which are the same for both new and second-hand machines, this will tend to reduce the maximum price which buyers in developing countries can afford to pay for second-hand machinery and so reduce the range of purchase prices in which second-hand machinery is useful to them.

Table 23 summarises the total unit costs of production in these various types of looms. These data allow us to draw conclusions about the decision to install second-hand looms.

Total unit costs on the second-hand 1949 machines are much greater than on the new machines. The correct decision would have been to install J3 Mk II machines (or possibly some more advanced loom which we have not taken into account). The rate of output from the second-hand looms would have to be more than 2.5 times what was actually achieved to make them an optimal choice.

It is important to remember that there is a good deal of hindsight embodied in the data. We know what rate of output was actually achieved on the second-hand looms; we also know that it is very low mainly because of the problems of lifting, transporting and re-installing the looms. It is precisely this information which was not available to the firm when it made the decision to buy second-hand. The decision was almost certainly based on a higher anticipated rate of output than was achieved in practice. The factory-rated efficiency of the second-hand machines is exceedingly low—about 12 per cent or so. It would not have been absurd to expect the machines to operate at about 35 per cent efficiency. Since at this rate of efficiency the second-hand looms would certainly have been the correct choice, the main lesson to be drawn is that there are very considerable risks in buying complex equipment second-hand.

Table 24. Costs of spinning 1 kg of yarn
(EA shillings)

Type of frame	Machine costs	Working capital costs	Labour costs	Power costs	Indirect variable costs	Total cost
FL 54	.095	.006	.111	.029	.026	.267
M 63 (s/h)	.027	.006	.063	.032	.027	.153
M 68	.062	.006	.067	.029	.027	.190

There is a second lesson implicit in the data, which is that the case for buying second-hand looms depends much more on the anticipated rate of output from the looms than on the price that is paid for them. In fact, while machine price accounted for about 15-20 per cent of the total unit costs of machines bought new, it accounts for only 4 per cent of the unit costs of second-hand machines. Even if it had been possible to beat down the price of the machines by half, the effect on total unit costs would have been extremely small: they would have been lower by about .005 of an East African shilling. A small increase in average output—say 10 per cent—would reduce unit costs five times more than cutting the machine price by 50 per cent. So the machine purchaser would have done well to put most of his efforts into finding out about the mechanical performance of the looms, and the likely effects of moving them, rather than bothering too much about the price he had to pay for them. This is a good general rule about second-hand machinery for:

(a) labour-intensive processes with which investment costs as a whole do not contribute much to total unit costs; and

(b) both capital-intensive and labour-intensive processes in which case transport and installation account for a major part of total investment costs.

Table 23 also leads to the interesting conclusion that because of the very low rate of output obtained with the second-hand looms, they should probably be scrapped and replaced with new J3 Mk II machines or with more modern looms altogether, at least from the point of view of private profitability. This conclusion follows from the fact that unit variable costs on the second-hand looms are considerably higher than unit total costs on the J3 Mk II looms.

Table 24 gives the figures on unit costs of production on spinning frames. The table shows clearly that it was a good decision to buy the second-hand spinning frames. They have considerably lower total unit costs than equivalent new machines (represented by the Mackie 1968 model). So in the case of spinning frames the second-hand machines are the most profitable investment. We can go further: the total unit costs of production on the second-hand machines are in fact lower than unit variable costs on the Fairburn-Lawson machines, and it would have been profitable to replace the Fairburn-Lawson machines with second-hand 1963 ones.

We also examined the way total unit costs of production varied among machines of the same type. Our expectation was that they would vary more among older second-hand machines than among relatively new ones. This was borne out in the case of the circular looms, in respect of which both the absolute variation in unit costs among machines and coefficients of variation were substantially larger for the second-hand machines than for the new. However, there is a further aspect. We found earlier that the variation in rates of output were also bigger on the second-hand machines than on the new. However, the results show that the differences between new models and the second-hand ones were much greater in the case of variations in unit costs than in the case of rates of output. This is essentially a reflection of the shape of the cost curves: the second-hand machines are operating in a range of outputs where the unit cost curve slopes sharply downwards (because of fixed costs); the new machines, on the other hand, are operating at outputs where the unit cost curve is much flatter (i.e. at the optimum point of production). Consequently, a given absolute variation in output among second-hand machines is reflected in a much greater variation in total unit costs of production, whereas on new machines a given variation in output is reflected in a smaller variation in unit costs. We think this is an important result because it underlines the difficulties that can arise in predicting the performance of individual second-hand machines of a hiven model and year.

Similar results were obtained for spinning frames. Even though we found little difference between new and second-hand frames from the point of view of variations in rates of output, there were important differences between them from the point of view of variations in unit costs. This again is because the second-hand frames—even though comparatively much closer to new machines in their output performance than the second-hand looms—are nevertheless operating in a range where the unit cost curve slopes steeply downwards. So even in the case of these comparatively high-quality second-hand machines, there are much greater absolute and relative variations in unit costs among individual machines than in the case of new ones.

Once more these results can be related to the assumptions underlying the arguments about the usefulness of second-hand machinery. One assumption which is frequently made is that the rates of output obtained from machines in the advanced countries are a good guide to the rates that will be attainable in developing countries. The case study on output from circular looms has already shown how misleading this can be. In particular in the case of circular looms there are factors which are specific to machines bought second-hand and which resulted in a very serious shortfall in output performance in Kenya compared with that in Dundee. The performance of spinning frames was less affected by the fact that they were second-hand, largely because they are more robust machines than circular looms.

The analysis of economic performance shows just how important the rate of output obtained from second-hand machinery can be. Unit costs of production are very sensitive indeed to rates of output, much more so than to machine prices themselves, although a good deal of the discussion about the second-hand machines centres on the price problem. This sensitivity to the rate of output is an important point of uncertainty about second-hand machinery, particularly in cases like that of circular looms, in which there are reasons to expect that the attainable outputs from second-hand machines in the customer firm may be more unpredictable than output from new machines. This economic analysis underlies the problem of risk and uncertainty.

The problem of uncertainty is the greater because differences among second-hand machines in regard to output performance are sometimes greater than differences among new machines. This was the case with circular looms though not with spinning frames. More important, however, is the fact that these differences among machines in terms of technical performance tend to appear as even greater differences in economic performance. Even second-hand spinning frames show more variation in total unit costs of production from one machine to another than new ones, although the output variations were much the same as for new machines. This is a further reason for suspecting that there are greater uncertainties in investment in second-hand machinery than in new.

It is difficult to generalise from the results of a single case study: there are no grounds for asserting that the performance of circular looms in this case is more characteristic of what happens when firms buy second-hand machinery than the performance of the spinning frames; there is not even a basis for guesswork about the question. However, it can be said that the case study demonstrates the dangers of having a definite opinion either way about the usefulness of second-hand machines in general. This conclusion, of course, is not particularly comforting for either side in the debate.

The theoretical argument that lies behind the idea that developing countries should use second-hand machines is compelling if one leaves aside such factors as transport costs and differences in factor productivity between advanced and developing countries. It is not difficult to show that the maximum price which a firm in a developing country can afford to pay for a used machine is higher than the minimum price at which a machine owner in an advanced country will be prepared to part with it (leaving out the problem of scrap value for the moment). Another way of putting this is: the value in use of second-hand equipment is higher in developing countries than in advanced ones, because of factor prices differences. As a result, there is a private and social incentive to use second-hand machines in developing countries, so that it looks as if such machines constitute an important source of efficient and relatively labour-intensive technology.

Of course, we have shown that it is quite inadequate to assume away transport costs, installation cost and the like. It is also inappropriate to leave out differences in factor productivities. For one thing, there are important differences in labour skills. For another, machine efficiency may deteriorate in the process of moving the equipment from its original factory to a new one. These points alone, among a large number of others that have been made in the course of this analysis, are quite sufficient to justify a certain reticence about second-hand machines. When practical technological problems are taken into account, the dogmatic advocacy of the use of second-hand machines in developing countries begins to appear a little absurd. However, the opposing view is absurd too, i.e. the idea that second-hand machines are always inefficient, always have high maintenance and spare-part costs and a low rate of output, and are always on the point of economic obsolescence. The second-hand spinning frames in the case study are efficient and socially and privately optimal; they use one-third of the investment per workplace needed for new machines. They were obviously a very good investment. Generally speaking, proponents and opponents of the use of second-hand machines both miss the point. The arguments about the merits and demerits of second-hand machines are carried out in terms that obscure the main problem, which can be put simply: any investment involves risks and uncertainties; the problem about investment in second-hand machines is that the risks and uncertainties are nearly always much greater than in the case of new machines. It is very easy to make mistakes; the story of the second-hand circular looms shows just how easy.

The fact that it is easy to make mistakes is important for two reasons. First of all, it leads to a wasteful use of investible resources, because mistakes are usually associated with low output-investment ratios. Also, a mistaken investment in second-hand machines might nevertheless have the advantage of providing a relatively large amount of employment per unit of investment in the short run, but this is achieved at the cost of reducing the re-investible surplus and the potential growth in employment and output. [1] These are, of course, the conventional arguments against the use of inefficient technologies. Secondly, there is a point that is left out of the conventional argument. If the performance of the second-hand machines is poor enough, it can happen that unit variable costs on them will be actually higher than unit total costs on new machines. In this case, profit-maximising firms will be strongly disposed to scrap the

---

[1] If the second-hand machinery is inefficient—like the circular looms—there is, of course, a sacrifice of output per unit of investment in the short run too.

machines and substitute new ones (or possibly much better second-hand ones). The trouble is that replacement means the expenditure of more investible resources without generating more employment. If replacement investment is taken into account, a mistaken investment in second-hand machinery can lead to a situation where investment per workplace is ultimately higher than it would have been if the investment had been made in new machines in the first place. So the high risk of making mistakes about second-hand equipment is a very serious problem not only because of the usual disadvantages that follow from using inefficient technology but more particularly because it can lead, paradoxically, to very high investment costs per workplace.

CONCLUSIONS

It is not easy to devise policies to deal with the problems of risks and uncertainty, or at least with the particular kinds of risk and uncertainty that characterise second-hand machinery. However, some fundamental rules about buying second-hand are suggested by the analysis, and could provide a starting-point for the elaboration of a general policy on the matter.

The first rule is that the risks of buying second-hand are much greater with some types of machine than others. A simple, robust machine like a spinning frame is a better proposition than a complex, sensitive machine. To repeat a point made earlier, second-hand machinery is not just old machinery, it is old machinery which has been lifted, transported and re-installed. Machine makers do not generally design machinery with this in mind. Some types of machine stand up to the experience better than others, and this should be an important consideration in deciding whether to use new or second-hand machinery.

A second rule is to find out why the machinery is for sale. Machinery may be sold off because its variable costs have risen with age and are too high to justify its use in production. It is risky to buy such equipment. Although it may seem that the machinery still has a value in operation in developing countries (because of lower wage rates), the risks of running out of spare parts, or of incurring heavy maintenance costs, or of deterioration during re-installation, are generally high.

Machinery that is sold because innovation has made it economically obsolete is often still in good working order. The problem in this case is that although the machines may be reliable and economically efficient at developing country factor prices, their economic life may be rather short: there may be further innovation which will make the machinery obsolete even at low wage rates. But there is another, potentially more difficult problem about this type of machinery. After an innovation, it often happens that machinery manufacturers stop making the older type of machine. If the innovation results in widespread replacement of the older types of machines in the advanced countries, the manufacturers may stop making spare parts for the older machines. Thus there can be considerable difficulties in obtaining spare parts.

Some machinery is sold because producers in the advanced country face a declining demand for the product. Generally speaking, machinery that becomes available because of changes in product markets in the advanced countries is probably the kind to buy. In such a situation the machines sold off are quite often (though not always) comparatively new.

In addition, it is important to check up on the production history and maintenance records of used machinery. In particular, if the machines have been idle for any length of time, it is essential to find out what was done about keeping them in working order.

A third principle is to give consideration to likely future changes in technology and in factor prices. In industries in which there is a high rate of innovation, the economic life of second-hand machinery may be very short. It may still be sensible to use it (indeed one consequence of a high rate of innovation is that the second-hand machinery may be comparatively new); but because of the short economic life of the equipment, the capital-saving effect of buying second-hand may be a good deal less in the long run than it seems to be in the short.

Fourthly, it is important not to be misled by very low machine prices into excessive optimism about savings in investible resources. The case study shows how important other elements in investment cost can be in comparison to the cost of the machine. Transport and installation costs need to be considered carefully.

Fifthly, it is very often more important to worry about how well the machines will perform when they are re-installed than about small reductions in their second-hand price. This is particularly true in labour-intensive processes, and where machine price is a small proportion of total investment costs. This is a difficult problem. It is inherently rather difficult to know what effect transport and re-installation will have on machine performance, although the effect will be more marked with complex and sensitive equipment; as we have seen, it can result in very large reductions in output.

The sixth rule is to bear in mind that second-hand machines differ from one another more than new machines in terms of output performance. This results in large relative variations in the unit costs of output between machines.

Finally, there are two rather obvious problems, which can, however, be overlooked: there is the problem of the availability of spare parts and the problem of technical assistance with installation.

Clearly, it would also be nice to have a lot more information about second-hand machinery and its problems. For example, it would be useful to collate technological and economic information about how different types of machinery deteriorate with age and are likely to be affected by lifting and re-installation. It might be possible for technologists to produce some sort of generalisation on the question of which kinds of machines are likely to suffer most through ageing and removal particularly in terms of output performance.

It would also be valuable to know much more about how the markets for different types of second-hand machinery actually operate. Is the thin market peculiar to the case of jute machinery because of particular historical circumstances, or are there similar markets? Are the markets more competitive in cases in which there is a substantial demand for second-hand machinery in the advanced countries themselves? How does the existence of an internal market like this affect prices and so the usefulness of used machines? For example, dealers might have much more room to manoeuvre in fixing prices in thin markets than in more smoothly functioning ones. How efficient is the flow of information in the market? These are only a few of the questions that call for an answer.

It is of course very doubtful whether even a massive programme of empirical research will produce clear categorisations of "good" and "bad" types of second-hand machinery. However, some level of generalisation about empirical realities, higher than what has already been achieved, is almost certainly possible and is probably worth striving for.

Even in the absence of further information, there are a few points to be made about policies on the use of second-hand machinery. A first point is to recognise what is probably an important general limitation. Widespread use of second-hand machinery may reduce investment costs per worker in the short run, but in the long run it will not solve the employment and income distribution problems that are caused by technological changes biased towards increased capital intensity; so the use of second-hand machinery is not an alternative to the search for more "appropriate" technologies in the laboratories and workshops of developing countries. It is of course possible that a policy of using old machinery in developing countries might actually contribute towards their technological capacity to create new "appropriate" methods of production. For example, it might create greater demands on local machine shops and start a learning process there. Usually, however, these external economies will be achieved only if the productive sectors are specifically organised to capture them, which often calls for government intervention. [1]

---

[1] An interesting example in the case of Kenya is the possibility of using the large East African Railway Workshops, as well as the small machine shops, as a technical support service for industries in which second-hand machinery is widely used. If this were properly planned, it might be a way of generating new types of technical knowledge in the railway workshops, and, of course, it might contribute to making second-hand equipment more viable in Kenya.

A second point is that existing policies in developing countries in general on the use of second-hand equipment are rather undiscriminating. We have not been able to survey them in any detail, but our impression is that most developing countries seem to fall into one of two extremes: they either ban second-hand machinery altogether, or they allow it to be imported without paying any attention to the particular problems and risks that are associated with it. Furthermore, no explicit distinctions are drawn between the different types of institutional setting in which transactions in second-hand machinery take place. A transaction between a dealer in a reasonably competitive second-hand market and an independent firm in the developing country is one thing; an intra-company transaction in which a foreign enterprise pays for equity in a subsidiary company by supplying old equipment from headquarters is quite another. The first case is the one we have studied, and raises the problems we have discussed. The second raises many more problems, particularly because it may be very difficult indeed to assess the value which the foreign enterprise places on the machinery; prices are rather arbitrary in an intra-company transaction.

Neither an open door policy nor a policy of total prohibition is likely to be optimal for business deals between independent enterprises: the open door policy ignores the possibility that imperfect markets, imperfect communications between buyers and sellers and technical weaknesses on the part of buyers could lead to quite serious mistakes, wastage and misallocation of investible resources and foreign exchange; whereas the policy of total prohibition ignores the fact that there are economic and social gains to be derived from the use of second-hand machinery, if only one knows enough about it to reduce risks.

One can imagine circumstances in which there would be a case for total prohibition. For example, where there is a serious lack of technical knowledge, so that one cannot compensate easily for the lack of technical skills in the enterprises by creating advisory services, or one can do so only at very high social cost, total prohibition of second-hand machinery may be the only way out. [1] However, it is rather doubtful whether existing policies of total prohibition are really based on such a rational evaluation.

In most cases an optimal policy would probably be a policy of discriminating import of used machinery, based on a knowledge of how various second-hand markets are organised and on the accumulation of knowledge about success and failure in the use of various types of old machines. The rules we listed above provide some indication of the factors which it is neces-

---

[1] Even here, however, it may be that the net benefits of an open door policy, after allowing for the misallocations it may produce, would be greater than those of a total prohibition. Obviously, this is an extremely difficult evaluation to make.

sary to be discriminating about. Discrimination in this sense involves a good deal more than an act of will. It is probably only possible to implement this kind of policy if the skills required in making an investment decision about second-hand machinery are properly organised. This aspect will be touched on in a moment. It is worth noticing, however, that neither of the extreme policies discussed makes it possible to build up the skills and knowledge that are needed in order to implement a policy of discriminating use of old machinery.

It is rather hard to generalise about the skills that are needed to purchase old equipment successfully. At the most general level, it is probably true that any improvement in the quality of management decision making is likely to contribute something to the capacity for making successful use of old machines. More specifically, the technical skills available in the productive enterprises are plainly important too. Still more specifically, government intervention to improve information and knowledge about the advantages and dangers of second-hand equipment would certainly be needed to support any discriminating policy.

In some cases it might be possible to accumulate expertise about old equipment through existing research and technical organisation related to the industries concerned. Better organisation of the machine shop sector, and perhaps the encouragement of specialised units dealing only with old machines, could also provide a basis for building up the knowledge required.

Finally, it seems that the flow of market information between buyers and sellers is highly imperfect in a number of sectors. On the face of it, this is a problem which could be tackled by co-operation between governments, possibly through an appropriate part of the United Nations. There is also the possibility of setting up international classifications of machine quality, which might reduce risks to the buyer. This would be hard to implement, but might be considered as part of aid programmes. Generally speaking, the organisation of trade in second-hand machinery has not been given much attention.

THE CHOICE OF TECHNIQUE AND EMPLOYMENT IN THE TEXTILE INDUSTRY

6

by H. Pack [1]

The implications of substituting labour for capital in manufacturing processes has been a major topic in work on economic development published over the past 20 years, and growing urban unemployment in less developed countries has recently led to a resurgence of interest in the question. Until recently there were few empirical studies of substitution possibilities, discussion being mainly concerned with the relative desirability of labour-intensive and capital-intensive production methods, on the assumption that there was in fact a choice. For a variety of reasons existing estimates of statistical production functions, though suggestive, are not well suited for resolving the issue. [2] In particular, even relatively disaggregated studies, for example at the three-digit level, may embody differences in product mix over time or across regions that mask the relevant production relations. Moreover, the competitive assumptions underlying the estimation of production functions are rarely justified in less developed countries. [3] Since most policy discussions focus on individual projects within a development plan, such relatively aggregative studies provide decision makers with insufficiently refined analyses. Thus it is necessary to undertake analyses with a different orientation, akin to that of process analysis, if some of the outstanding issues are to be resolved. In the past few years a few inquiries of this type have been conducted using data obtained primarily in less

---

[1] Associate Professor of Economics, Swarthmore College, Pennsylvania.

[2] See the review of empirical research on capital-labour substitution possibilities in Chapter 2 of this volume.

[3] A brief survey of the problems encountered in obtaining useful information from existing studies is given in H. Pack: "The employment-output trade-off in LDCs: A micro-economic approach", op. cit. A recent article that attempts to surmount some of the difficulties of analysing the textile industry runs into other obstacles; see Peter Isard: "Employment impacts of textile imports and investment: A vintage-capital model", in *American Economic Review*, Vol. LXIII, No. 3, June 1973.

developed countries from firms currently in production. [1] The case considered
in this chapter differs in so far as it is based on data from the United Kingdom,
which nevertheless are quite relevant for less developed countries.

EVIDENCE OF SUBSTITUTION POSSIBILITIES
IN OLDER TECHNIQUES

In practice a major difficulty in calculating substitution possibilities in a
particular case is the scarcity of useful engineering comparisons of alternative
techniques ("technique" here denoting a production method with a specified
capital-labour ratio: a set of techniques generates an isoquant). There are, of
course, many potential sources of information on technical characteristics
of production, including textbooks, trade publications and the material
published by equipment producers. However, such publications are not very
useful for the problem at hand; they are of course not intended for the construction of isoquants, and thus omit essential data. For example, in engineering
texts a technique is discussed and then presented as optimal and its costs
including labour requirements are specified, but no comparison is given with
other techniques which exist (and perhaps were described as optimal in the last
edition). Moreover, there is very often no discussion of labour, all the author's
attention being devoted to the technical processes. Trade journals and the
descriptive material put out by equipment producers consider one or two
machines in detail (including comparisons with earlier models) but provide
little information on any required changes in other stages of the production
process: for example, a discussion of such properties as the speed of a machine
and its output per hour may be extensive, but there will be no consideration of
the implications of increased output per hour for the next step in the production
(will more or faster machines be required at that stage, or are other alternatives
possible?). While the data needed can be pieced together from the information
given, there are substantial possibilities of error, particularly in consistency.
Fortunately, in textiles—a field of considerable importance because of the major
role assigned to it (perhaps incorrectly) in many development plans—a recent
study by the Textile Council of the United Kingdom presents process-wide
comparisons of alternative production methods in cotton textiles. [2] Since the
study was carried out to discover the ability of the United Kingdom textile

---

[1] For a survey of much of this material see David Morawetz: "Employment implications
of industrialisation in developing countries: A survey", in *Economic Journal*, Vol. 84, No. 335,
Sep. 1974.

[2] Textile Council: *Cotton and allied textiles*, 2 vols. (Manchester, 1969). Two earlier
studies resulting from the interest of Indian planners in textiles are those of A. S. Bhalla:
"Investment allocation and technological choice: A case study of cotton spinning techniques",

industry to meet import competition in the future, particularly from low-wage countries, much of the study focuses on the differential labour input of the available techniques.

Although the techniques considered are primarily ones that are actually in use in the United Kingdom, or known but not used there, they are of considerable interest for an analysis of the choices facing less developed countries because in such countries most of the techniques considered by a company or a planning agency for starting production or replacing equipment, while capable of adaptation, would belong to the spectrum of developed country techniques. [1] A fundamental question, then, is whether older techniques available in developed countries allow efficient substitution, by comparison with new methods, of labour for capital and, if this is so, whether such substitution is economically efficient for the less developed countries. The older methods would, in some instances, depend on the purchase of used equipment, although in some weaving activities machines are still being produced that utilise older production methods.

The analysis of the Textile Council covers the three major divisions of the industry, namely spinning, weaving and finishing. Most of their analysis focuses on the first two sub-sectors, and we shall confine our analysis to these two activities. For each of these activities, a number of production methods were considered. In spinning, the methods available correspond to different ages of machines which differ primarily in such attributes as speed: the processes available consist of production lines with all-1950 equipment, all-1960 equipment or all-1968 equipment, whereas in weaving there are Elitex (Airjet), Sulzer, Lancashire, Battery, and Battery looms combined with a Unifil machine, the major difference among these combinations being the method by which the weft is supplied within the loom. For each of the processes in spinning and weaving, a complete list of the kind of machines (and the required numbers of each), power requirements, material wastage, space use and labour requirements is provided. [2] Thus it is possible to trace out the

---

op. cit., and Amartya Sen: *Choice of techniques*, op. cit., Appendix A. In both the focus is on the efficiency of handicraft as compared with industrial methods as well as the implication of each technique for generating savings.

[1] Advertisements of equipment producers in Indian textile trade publications emphasise "Western" features of their equipment, including the labour-saving devices. For examples see the *Indian Textile Journal*.

[2] For example, with regard to spinning, the study gives the number and type of machines in the blowroom, cardroom and ringroom that would be required to produce a given output per week using a 1950, 1960 and 1968 technology. The number and type of workers consonant with each combination are given. In addition, such information is also provided for each vintage for several different qualities of yarn. Spinning includes the preparing of fibres as well as spinning proper. Though it is possible to disaggregate the data to allow for differences in the development of the preparing and spinning activities, we think the added detail would obscure some of the important, more aggregated relationships.

substitution possibilities among the inputs as well as to examine the economic efficiency of such substitution at a variety of relative factor prices.

In effect, we shall be constructing isoquants with three, four or five techniques, a typical linear programming isoquant. One cannot rule out the possibility that, even within the range of capital-labour ratios considered, other technically efficient alternatives exist, such as combinations of machines of various vintages within the spinning process. These combinations would, of course, correspond to additional techniques that would have to be analysed along with the original ones. Interestingly, the Council considered some mixed techniques in cotton spinning such as 1950 ring frames combined with 1968 cards. Though we cannot be sure that all relevant combinations of this type were considered, the presentation and careful analysis of a number of these mixed techniques increases our confidence that the processes presented provide a reasonable description of the choices actually available.

Spinning

In spinning, two major technical trends have developed; newer machines use much less labour per unit of output and the output per spindle is higher: "replacement of 1950 by 1968 conventional machinery leads to a 50 per cent improvement in output per spindle hour and a 100 per cent improvement in labour productivity". [1] This performance does not necessarily imply the economic superiority of new machinery unless the increase in the price of machinery required to realise such gains is such as not to offset its higher productivity. Unfortunately, it is exceedingly difficult to obtain estimates of how much it would cost to produce 1950-vintage equipment at the current time; even in less developed countries with internal capital goods production capacity, production of this type is not undertaken. Thus, in this section we consider the profitability of various techniques on the assumption that the only source of older-style equipment is used equipment. For many purposes this is a perfectly proper point of departure since the quantity of such equipment in the developed countries is quite large and since, at the probable rates of discarding, such equipment should be more than sufficient to meet the needs of the less developed countries. [2] Our analysis will also provide estimate of the prices at which

---

[1] Textile Council: *Cotton and allied textiles,* op. cit., Vol. I, p. 61.

[2] On the rationality of using second-hand equipment, see A. K. Sen: "On the usefulness of used machines", op. cit. The most important factor in practice is the percentage of the new price for which used equipment will sell. Although the production of new but old-style equipment in a local capital goods industry may be warranted for other reasons, existing data suggest that the supply of used equipment will be considerable in relation to the needs of the less developed countries. Thus it is estimated in a recent UNIDO projection—UNIDO: *The textile industry* (New York, 1971)—that a maximum of 28 million spindles will be needed by the less developed countries over the 1967-80 period. If one adds half this amount for

new models of older design would have to be available to constitute an economically rational choice.

The basic data for the 1950-68 comparison are shown in table 25. [1] These show the costs incurred in producing 20 S cotton in dollars per round. Although there are some differences in costs for cotton yarn of other degrees of fineness, most of the analysis below holds, with some relatively small changes in magnitude, for these others as well. The capital charges for 1968 reflect assumed borrowing and lending rates of either 9 or 19.5 per cent per annum and an expected service life of 10 years. The former rate is that used by the Council, whereas for most of the less developed countries a considerably higher rate would seem appropriate, and the rate used here increases annual capital charges by 50 per cent. [2] The additional capital charges could also be viewed as a correction for the failure of most less developed countries to impose a tariff on investment goods; almost all other products benefit from tariff protection, while imports of equipment are made excessively cheap by overvalued exchange rates. The capital charge for 1950 assumes that equipment of this

---

replacement, there will be a total requirement of 42 million spindles. In 1963 there were over 61 million spindles in the United States, Canada, Japan and the developed countries of Western Europe (OECD: *Modern cotton industry* (Paris, 1965), p. 91). Many of them will probably come on to the world market as more labour-saving equipment becomes available. Presumably the same will be true of equipment for other parts of the spinning operation, such as opening, scutching and carding machines. The same sources indicate that loom demand by the less developed countries might be about 652,000 over the same period, whereas there were about 1,226,000 looms in place in 1963 in the developed countries.

It is obvious that many of the spindles and looms in place in 1963 would be among the more modern types discussed in this chapter. Thus the less developed countries, if they rely on equipment discarded in the developed countries, will gradually be modernising themselves, though continuing to use more labour-intensive equipment than the developed countries, which will presumably continue to discover still more labour-saving equipment. This may force the less developed countries to modernise more quickly than they should, and if that were to occur there would be a strong argument either for the developed countries to produce their older equipment or for the less developed countries to undertake such production themselves. On the other hand, sufficiently rapid progress on the part of the developed countries might require a wage that would be lower than the alternative cost of labour in less developed countries (see, for example, table 30, line 3). If this were the case, it would be desirable to follow the trail blazed by the developed countries but to modernise slowly. The data presented later suggest that, in spinning, remaining at any one stage of technological advance may lead to values of w^* (break-even wage, in dollars, at which the 1950 and 1968 processes yield the same average cost of production) below subsistence, whereas in weaving it will be possible to retain the Lancashire loom for a considerable time.

[1] These differ in several ways from the calculation of the Textile Council: their method of calculation uses actual labour costs rather than the underlying hour requirements; differential space costs resulting solely from assumed differences in capacity utilisation are here ignored; and our capital data assume similar rates of capacity utilisation, usually 168 hours per week, for all techniques.

[2] The present discounted value of an annuity of one dollar for 10 years at an interest rate of 0.09 per cent is 6.4176, whereas at an interest rate of 0.195 it is 4.2657. As the capital cost of a machine is divided by these numbers to obtain the equivalent yearly cost of using a machine, the assumption of the latter interest rate yields an increase of approximately 50 per cent in the annual cost.

Table 25.   Costs per pound of 20 S carded cotton
(1968 dollars)

Costs	Date of equipment	
	1950	1968
Power	.0119	.0145
Waste	.0343	.0219
Capital, if—		
$r = .09$	.0023	.0229
$r = .195$	.0035	.0344
Average total, if—		
$r = .09$	.0485	.0593
$r = .195$	.0497	.0708
Operative hours per pound	*.0781*	*.0355*

Note: r = interest rate per annum.
Source: Textile Council: *Cotton and allied textiles*, op. cit., Vol. I, p. 64.

vintage is available at 10 per cent of the price of new, 1968-type equipment. The sensitivity of our results to this assumption will be examined, but the assumption does not appear to be unreasonable. The technical reports underlying the Council's calculations indicate little resale value within the United Kingdom. This is to be expected, since at current wage rates the average variable costs of the 1950 plant exceed the average total costs of the 1968 plant. On the other hand, it is not clear why the scrap value should have been quite so low in recent years, since countries with lower wages could have used such equipment quite profitably. [1]

It may appear questionable whether used equipment should be given the same life expectancy and the same maintenance costs as new equipment, an assumption of our calculations. It is often argued that used equipment breaks down more often and is increasingly difficult to repair because of a growing absence of spare parts as the original production date recedes further into the past. This view overlooks the oft-cited ingenuity, obvious to any visitor, shown by repairmen in less developed countries in keeping older machines and vehicles in use for extraordinary lengths of time; similarly it disregards the ability of small machine shops (or the repair shops of industrial

---

[1] The Cooper and Kaplinsky study in this volume (Chapter 5) provides detailed estimates of the actual cost of installing used equipment in Kenya, including both transport and installation costs. These costs do not vary with the type of machine, and will thus cause a greater proportionate increase in the cost of a used machine than in that of a new one. On the other hand, given the relation between variable and total costs, 0.10 overstates the acquisition cost *per se* and may be viewed as including these other costs. The calculations below also provide estimates of the maximum amount that could be paid for used equipment at a variety of wage rates, and these calculated values should be viewed as including shipping and installation costs.

companies) to duplicate required parts. Moreover, the absence of spare parts seems to be a spurious issue: there is a large, flourishing international trade in such parts, and three or four well financed specialists in spare parts within a country such as Kenya have been able to make the absence of parts largely a myth. Repair may also be considerably less expensive in less developed countries because the wages of skilled repairmen are only a fraction of those of their counterparts in developed countries. Finally there is some evidence, however subjective in origin, that very modern machines (e.g. numerically controlled machine tools) are more difficult to service than older ones because the basic scientific understanding of repairmen would have to be increased. Thus in Kenya a firm utilising both automatic and manually controlled injection moulding equipment has to send its repairman to the United Kingdom for further education as well as specific training to service automatic equipment, whereas the manually controlled equipment was repaired by these relatively uneducated men after a few months of training on the job. Though none of the above evidence can prove that assumptions of similar life (at least 10 years, the relevant planning horizon) and maintenance costs are correct, it would seem that they are not unreasonable.

Consider a company facing the choice of an optimum production process, choosing between 1950 and 1968 equipment. The first question considered is at what hourly wage, w^*, the two processes will yield equal average total costs. At any wage less than w^* the 1950 process will be less expensive, since it is more labour-intensive. Total non-labour costs per pound using the 1968 plant (assuming $r = .09$) are $0.0593, while they are $0.0485 with the 1950 plant. Operative hours per pound are shown at the bottom of table 25. w^* is found by solving the following expression:

$$\$0.0485 + 0.0781 \, w^* = \$0.0593 + 0.0355 \, w^*$$

and $w^* = \$0.255$ per hour in 1968 or about $530 per annum, if 2,080 hours of work per annum are assumed to be average, including paid vacation. By comparison, average hourly wages in Indian and Pakistani spinning companies were about $0.175 per hour or $364 per annum in this period, and even these rates are likely to exceed the shadow price of labour. Thus, if the costs of acquiring 1950 equipment is about one-tenth of the price of new equipment and if the interest rate is 9 per cent, the use of older equipment will certainly be justified for at least some less developed countries. This result is shown in the first column of table 26. In a similar manner the break-even wage can be calculated when capital costs are 50 per cent higher ($r = 0.195$) and amount to $978 per annum. The maximum amount (as a percentage of the cost of new, 1968 equipment) that may be paid for a used machine is shown in the third and fourth columns, a wage of $300 being assumed.

Table 26. Alternative factor prices required for a 1950 spinning plant to be optimal (assumed and *calculated* values)

Values assumed						
α	.100	.100			.500	.500
r	.090	.195	.090	.195	.090	.195
w			300	300		
Values calculated (given assumed values above)						
α			.300	.550		
w^*	530	978			87	366

Key: α = the percentage of the 1968 new plant price which is or can be paid for a 1950 plant.
    r = interest rate per annum.
    w = wage per annum in dollars.
    w^* = break-even wage, in dollars, at which the 1950 and 1968 processes yield the same average cost of production.

α is 0.30 and 0.55, the latter being more useful given the more realistic shadow rate of interest. Thus, if reliance upon newly produced equipment of old design were to become widespread, its availability at prices considerably below that for new equipment of current design would be of critical importance.

While the first two columns of table 26 assume the price of a 1950 plant to be 10 per cent of that of a 1968 plant and the third and fourth columns indicate the maximum price that could be paid with a wage of $300 per annum, it is of interest to consider the question of the relative economic efficiency of equipment of older design if it could be acquired only at higher prices than have been indicated so far. If it could be produced for half the price of new 1968 equipment [1], the new value of w^* (fifth and sixth columns), calculated as before, yields a break-even wage of $0.04 per hour, or about $87 per annum, when an interest rate of 9 per cent is used. The role of shadow prices becomes evident when a shadow rate (19.5 per cent) is used, w^* now being $366 per annum, a market price of labour in many less developed countries and surely above the shadow price, whereas $87 may well fall below the shadow price for most such countries.

More generally, table 26 shows the assumed values of two of three variables (price of used equipment as a percentage of new, the interest rate, wage) and the calculated maximum value of the third that will result in the choice of 1950 plant rather than 1968 one. Each line of the table thus provides a different solution to the equation

$$rk_{50} + wl_{50} = rk_{68} + wl_{68}$$

where k is the capital-output ratio (cents per pound of cotton) and l the labour-output ratio (hours per pound of cotton); k_{50} will be some fraction, α, of k_{68}.

---

[1] This would reflect the savings on the various devices installed on new-design equipment that reduces its operating costs, particularly labour-saving ones. Such a price would be more likely if less developed countries, with their lower skilled labour cost, produced such equipment themselves.

Thus, for any column, an increase in r or a decrease in w^* or α will augment the profitability of 1950 equipment in relation to that of 1968.[1] It should be noticed that the former yields more than twice as much employment per unit of output as the latter, but we leave further discussion of this aspect of the matter to a later stage.[2]

One other possibility relevant to our own interest was calculated by the Textile Council, namely modernisation of both 1950 and 1960 plants including "new opening and cleaning machines, modernised cards, new draw frames, new speed frames and modernised ring frames".[3] Less extensive modernisation of a 1960 plant was also examined. Interestingly, the latter course proved most economic, given United Kingdom labour costs: though a partly modernised 1960 plant uses more labour, it saves enough capital by comparison with the other two choices to make it the optimal choice if modernisation is undertaken. When factor prices prevailing in less developed countries are used in the modernisation calculation, as long as the wage is less than \$375 the basic unmodernised 1950 plant remains the most profitable one.

We have seen that the 1950 plant employs more than twice as much labour as a 1968 plant. We have argued elsewhere that the adoption of labour-intensive production methods is likely to consist primarily of replacing auto-mated material movement within a plant by the use of simpler, labour-intensive movement methods: for example, barrels of paint may be moved by men instead of through an expensive piping system.[4] Textiles, however, appeared to be one industry in which material movement of the type described offered only minor opportunities for additional employment, so that if additional employment was to be generated in this industry it would have to be through the use of more labour-intensive primary processing equipment. The very nature of the spinning process, which takes raw cotton and makes it longer while strengthening and straightening it, suggests that the process will be much more efficient if material is fed automatically from stage to stage. The figures in table 27 reveal the expected pattern. Equipment of 1950 provides more jobs, mainly for direct production workers rather than for ancillary personnel: while 1950 equipment uses about 125 per cent more labour than the 1968 equipment (53 additional jobs for an output of 96,000 pounds of spun cotton per week), only four of the

---

[1] More strictly, a term representing depreciation should be included; i.e., instead of r, $(r + d)k$ should represent capital costs. To keep the notation simple, r alone is used in the rest of this chapter.

[2] See pp. 164-165 below.

[3] A potential purchaser of textile equipment has the option of buying some combination of new and used equipment, a possibility that provides a technique between the 1950 and 1968 ones.

[4] For a number of such examples, see H. Pack: "Employment and productivity in Kenyan manufacturing", in *Eastern Africa Economic Review*, Vol. 4, No. 2, Dec. 1972.

Table 27. Number of workers per shift required to produce 48 tons per week of cotton yarn

Year of plant	(1)	(2)	(3)	(4)
	Direct production workers	Ancillary workers	Total	(2)/(3) (Percentages)
1950	78	17	95	17.9
1960	46	15	61	24.6
1968	29	13	42	31.0

Source: Textile Council: *Cotton and allied textiles*, op. cit., Vol. II, pp. 64-66.

additional jobs are in ancillary activities. But the major implication is the sub-stitutability of labour for capital even where most workers are involved in the primary producing activity.

Weaving

Whereas, in spinning, differences in equipment are best denoted by date of manufacture, in weaving the major differences among equipment types depend on the type of loom. Five types of looms will be compared—modernised Lancashire, conventional modern automatic (Battery), modern automatic with Unifil, Elitex (Airjet) and Sulzer. Most of these types have been available for quite a time, although improvements on the basic procedures have of course been made. Recent developments—

have been concerned with the method of weft supply. There have been progressive improvements in this field. At one extreme is the Lancashire loom where a good proportion of the operative's time is taken up with replacing the small weft pirns in the shuttle. At the other extreme are looms of the Sulzer and Elitex type where the weft supply is taken direct from large packages. In addition, the newer looms generally operate at higher speeds. A further factor working in favour of the Sulzer looms, and possibly also the Elitex, is that yarn breakage rates tend to be lower, since the yarn is subjected to less harsh treatment during the weaving process. The main effect of these changes is to reduce the labour content per yard of cloth, since operatives are able to look after greater numbers of the newer types of loom than of the old.[1]

These changes in labour productivity have been achieved by increasingly sophisticated machinery that costs more, and the desirability of utilising these looms thus very much depends on relative factor prices. There are substantial differences in the cost of weaving different materials. We first present a detailed calculation for shirting, then consider the costs of producing other types of cloth. Table 28 shows the basic data on costs per 100 yards (in dollars) for producing 40-inch cotton-polyester shirting.

---

[1] Textile Council: *Cotton and allied textiles*, op. cit., Vol. I, p. 68.

Table 28. Costs of weaving shirting material with looms of different types
(dollars per 100 yards)

Cost	Lancashire	Battery	Battery-Unifil combination	Elitex (Airjet)	Sulzer
Space	.198	.152	.152	.093	.130
Power	.385	.397	.397	.385	.250
Weft waste	.035	.035	.035	.408	.290
Pirning	.525	.350			
Capital					
$r = .09$	.110	1.097	1.435	1.225	1.980
$r = .195$	.165	1.645	2.153	1.838	2.970
Average total					
$r = .09$	1.253	2.031	2.019	2.111	2.650
$r = .195$	1.308	2.579	2.737	2.724	3.640
Hours per 100 sq. yds.	*.0436*	*.0222*	*.0229*	*.0163*	*.0101*

Source: Textile Council: *Cotton and allied textiles*, op. cit., Vol. II, pp. 83, 86-90.

Table 29 is similar in construction to table 26. In the first set of calculations, those of w^*, for the values assumed of two factor prices in the first two columns, the maximum wage that would allow the Lancashire process to break even is shown in the third column. Any value less than the assumed or calculated one would strengthen the profitability of the Lancashire loom in relation to that of the other methods. What is conspicuous about the first set of calculations, those of w^*, is the quite high wage rates that would still allow the Lancashire loom to be as desirable as the others; with $r = 0.09$ the lowest value of w^* being $653 (for the Elitex) and the highest being $1,141 (Sulzer). With $r = 0.195$ the range changes to between $1,084 and $1,826. This suggests that even at market prices for labour, most less developed countries would be well advised to choose the Lancashire loom, if the quantities of used ones available on the world market are sufficient so that their price is not more than 10 per cent of the price of new Battery looms. Of course, to the extent that market interest rates are below 9 per cent in a particular country the wage would have to be lower to induce a company to adopt this loom, but even at interest rates of 6 per cent the value of w^* would not lie below market wage rates in many of the less developed countries. On the other hand if companies had to pay interest rates that were close to 19.5 per cent, the wage at which it would still pay to adopt the Lancashire loom, assuming it was available at 10 per cent of the Battery loom price, would be almost at the wage level of developed countries.

It will be noticed in table 28 that the labour intensity per yard is more than four times greater with the Lancashire loom than with the Sulzer loom, which is the most labour-saving of those considered. Moreover, the additional labour

Table 29. Alternative combinations of factor prices required for Lancashire looms to be optimal in production of shirting material (assumed and *calculated* values)

Type of loom compared with Lancashire	First set [1]			Second set [1]		
	α	r	w*	α	r	w
Battery	.10	.090	*755*	*.48*	.090	300
	.10	.195	*1 233*	*.65*	.195	300
Battery-Unifil	.10	.090	*770*	*.36*	.090	300
	.10	.195	*1 441*	*.58*	.195	300
Elitex (Airjet)	.10	.090	*653*	*.42*	.090	300
	.10	.195	*1 084*	*.62*	.195	300
Sulzer	.10	.090	*1 141*	*.61*	.090	300
	.10	.195	*1 826*	*.74*	.195	300

[1] For explanation see text.

Key: α = the percentage of the 1968 new plant price which is or can be paid for a 1950 plant.
  r = interest rate per annum.
  w = wage per annum in dollars.
  w* = break-even wage, in dollars, at which the 1950 and 1968 processes yield the same average cost of production.

required on the former is not more highly skilled, contrary to the argument that more modern equipment is a substitute for skilled labour and is therefore suitable for less developed countries. On weighting the hours of each type of operative by the one-shift wage paid to operatives on the Lancashire loom [1], the average hourly wage on the Lancashire loom process was found to be $1.05 per hour and that on the Sulzer $1.06 per hour—hardly a differential to suggest much skill replacement.

The second set of calculations in table 29 is that of the maximum amount that could be paid for the Lancashire loom as a percentage of the value of each other type of loom, if the Lancashire loom is to be competitive with it, assuming a wage of $300 per annum. The calculated values of α are generally quite high; when $r = .09$ the adoption of a simple new Lancashire loom would be warranted if it could be obtained at about 36 to 61 per cent of the price of the more advanced types of loom, this range changing to between 58 and 74 per cent with an interest rate of 19.5 per cent.

It is worth noting that even with the new looms currently available there is no sense for a less developed country in adopting something as modern and labour-saving as the Sulzer. For example, compare the Battery and Sulzer looms, the former using more than twice as much labour per unit of output. At $r = 0.09$ the value of w* at which the Battery would be equally profitable is $2,086 per year, and at $r=0.195$ this value would be $3,132—in other words,

---

[1] The higher wage levels reported by the Textile Council for workers on other looms reflected shift differentials.

Table 30. Wage required for Lancashire loom to be as inexpensive as other looms in production of sheeting material

Type of loom	Values		
	Assumed		Calculated
	α	r	w^*
Lancashire compared to			
Battery	.10	.090	189
	.10	.195	395
Battery-Unifil	.10	.090	46
	.10	.195	283
Sulzer	.10	.090	418
	.10	.195	792

Key: α = the percentage of the 1968 new plant price which is or can be paid for a 1950 plant.
r = interest rate per annum.
w^* = break-even wage, in dollars, at which the 1950 and 1968 processes yield the same average cost of production.

far above the wages paid in less developed countries. The argument often put forward that the newest equipment is dominant in the sense of using less labour and capital per unit of output is not supported by these calculations.

One problem easily overlooked in abstract discussion of choice of technique is the possibility of producing a range of products with any given set of machines. Unless product specifications are very similar, there is no *a priori* reason to assume that a given loom will exhibit the same relative efficiency in all types of weaving operations. A loom that is efficient in producing one fabric may not be sufficiently flexible to allow others to be made as efficiently. While it is widely acknowledged that some labour skills or machines are specific to an industry, the possibility that such specificity is of importance at a process level is very rarely mentioned. To illustrate these considerations, we analyse the production of cotton sheeting.

Table 30 shows that when the Lancashire loom is compared with the Battery and Battery-Unifil, the break-even wage becomes $189 and $46 respectively when using a 9 per cent interest rate and $395 and $283 when using a 19.5 per cent interest rate. Of these, $46 is almost certainly less than the shadow rate of wages in probably all countries, though even this may not be relevant in so far as w^* becomes $283, instead of $46, when a more realistic interest rate is used. Thus despite the economic efficiency of the Lancashire loom in producing shirting material, it might not be economical to use it to produce sheeting.

Given the differences in relative efficiencies in producing different fabrics, it is interesting to see the values of w^* for four major fabrics. These are set out in table 31 in which part A shows the values of w^* at two interest rates and part B ranks the values of w^* by fabric. For all materials except sheeting, the Elitex (Airjet) turns out to be the loom most competitive with the Lancashire,

Table 31. Break-even wage for four types of woven material (dollars)

	Lancashire compared with:			
	Battery	Battery-Unifil	Elitex (Airjet)	Sulzer
A.				
$\alpha = .10$ $r = .090$				
Shirting	755	770	653	1 141
Sheeting	189	46	Not feasible	418
Zephyr	1 090	1 258	439	972
Printing	1 073	1 012	347	913
$\alpha = .10$ $r = .195$				
Shirting	1 233	1 441	1 084	1 826
Sheeting	395	283	Not feasible	792
Zephyr	1 716	2 169	863	1 641
Printing	1 654	1 789	715	1 539
B.				
	Ranking for each material			
$\alpha = .10$ $r = .090$				
Shirting	2	3	1	4
Sheeting	2	1	–	3
Zephyr	3	4	1	2
Printing	4	3	1	2
$\alpha = .10$ $r = .195$				
Shirting	2	3	1	4
Sheeting	2	1	–	3
Zephyr	3	4	1	2
Printing	3	4	1	2

though even here the values of w^* are such as to suggest that its adoption would be unwarranted if used Lancashires are available. Only in the production of sheeting does the Lancashire ever require a wage likely to be below the shadow price of labour in most of the less developed countries.

Some of the problems of making an optimal choice of technique are nicely illustrated in table 31. Suppose that a semi-developed country with an opportunity cost of labour of $1,000 and using an r of 0.195 is trying to choose among different types of equipment. The Airjet loom would not be economic if considerable amounts of shirting were to be produced, since the break-even wage relative to the Lancashire is $1,084, and the use of the Lancashire would be optimal. The Airjet would be the correct choice for zephyr and printing materials; on the other hand it cannot produce sheeting at all. Thus two fabrics are produced most economically with the Airjet, one cannot be produced with it at all and the fourth can be most cheaply produced with a Lancashire loom. Though the example, particularly the wage alternative, is arbitrary, it forcefully illustrates our earlier point, often lost sight of in the published material on choice of techniques, that most plants (and machines) produce a considerable range of products, in each of which there may be a different relative efficiency. It is perfectly easy to minimise cost when the relative desirability of various products is specified, by using linear programming; but it is not possible to make choices solely on the grounds of factor proportions and factor prices without having a specification of product choice. In the case considered here the optimal technique cannot be ascertained without knowledge of the product mix to be produced.

There are a number of issues that we have so far intentionally ignored but which are of some importance to the results given above. The value of w^* is determined not only by differences in relative factor proportions but also by differences in intermediate material costs. Formally, this can be seen in the method of calculating w^*. Denoting the labour-intensive technique by the subscript 1 and the capital-intensive technique by 2, w^* is derived by equating the costs of production of the two techniques:

$$rk_1 + p_m m_1 + w^* l_1 = rk_2 + p_m m_2 + w^* l_2$$

where m is the input of materials per unit of output, and p_m the cost per unit of these materials. w^* is thus equal to

$$\frac{p_m (m_1 - m_2)}{l_2 - l_1} + \frac{r (k_1 - k_2)}{l_2 + l_1}$$

or the marginal rate of substitution of intermediate materials for labour plus the marginal rate of substitution of capital for labour. If intermediate input requirements are the same for different types of machine ($m_1 = m_2$), then w^* will depend solely on the marginal rate of factor substitution. On the other hand, differential intermediate requirements will change the value of w^*.

It is often claimed that new machines are more efficient in their utilisation of material and that this constitutes one of their attractions, at least partly explaining their adoption in some less developed countries despite their higher capital costs and lower labour utilisation. In textiles, the evidence on this question is mixed: while newer spinning processes do reduce waste considerably, the more recent loom designs result in higher waste levels than does the Lancashire [1]; in both activities the difference in waste levels is an important determinant of the level of w^*, this being particularly important in spinning, where the absolute differences in waste costs are quite large.

In power utilisation there is no consistent trend: newer weaving machines use less power than the Lancashire, while 1968 spinning mills use considerably more power than the 1950 one. The power costs used in our calculations were those of the United Kingdom, and to the extent that power costs in the less

---

[1] Qualifications are necessary in so far as different levels of waste occur in weaving different types of material.

developed countries differ, the calculated value of w^* will have been raised in spinning and lowered in weaving if the United Kingdom price is lower than the correct (shadow) price in less developed countries, and vice versa if the United Kingdom price is higher. Given the high values of w^* in weaving, the general nature of our results will not be affected in this activity, though in spinning the effect might be more significant. Without a detailed analysis of power costs in the less developed countries it is difficult to evaluate the significance of any such changes in prices.

AGGREGATE EMPLOYMENT EFFECTS

The evidence reproduced above demonstrates the economic sense of not using the most modern production processes technically applicable to cotton textile manufacturing. It was seen that until high wage levels were reached, the most labour-intensive equipment in use in the United Kingdom was optimal in the sense of allowing the lowest average cost of production. We have yet to consider the aggregate implications of the differences in capital-labour ratios associated with the various techniques. Table 32 shows these ratios measured as initial equipment cost per man-year of labour engaged in direct production. Although equipment is measured as a stock, conversion into a flow would lead to proportional changes for all types of equipment if equal lives and identical interest rates are assumed. It should be remembered that the capital-labour ratios for the 1950 spinning plant and for the Lancashire looms reflect assumptions about the cost of used equipment by comparison with new in so far as the older-style equipment is no longer being produced.[1]

A 1968 spinning plant has about 21 (4.4) times the capital-labour ratio of the 1950 spinning plant on the assumption that equipment of the latter costs 10 (50) per cent of new, 1968-type equipment. The differences in factor intensity in weaving show an even greater range, the extreme being that a Sulzer loom requires about 90 (18) times as much capital per worker as the Lancashire loom when the latter's price is 10 (50) per cent of that of a new Battery loom.[2] Such magnitudes imply very large employment differentials. The employment effects of choosing between the techniques are of two kinds—

(a) the direct effect of producing a given output level with one technique rather than another; this effect is measured as $(l_1 - l_2) Q^*$ where 1 denotes the labour-intensive technique and 2 the capital-intensive one; and

---

[1] An intensive canvas of a number of engineering consultants and equipment producers to obtain estimates of the current cost of producing older-style equipment was fruitless. As mentioned earlier, the production of such equipment would presumably cost much less than that of current models, given the simpler designs.

[2] The cost of the Lancashire loom is assumed, as in the preceding section, to be directly related to that of the next most labour-intensive loom, the Battery loom.

Table 32.   Capital-labour ratio in spinning and weaving

Type of equipment	(1) Equipment (1968 $)	(2) Man-years of labour	(3) Capital/labour ratio (col.(1) ÷ col.(2))
A. Spinning inputs per ton of cotton			
1950 plant [1]	44	.0781	566
1960 plant [2]	220	.0781	2 830
1968 plant	442	.0355	12 462
B. Weaving inputs per million square yards of shirting			
Lancashire [3]	7 164	21.8	329
Lancashire [4]	35 820	21.8	1 645
Battery	71 635	11.1	6 454
Battery-Unifil	92 803	11.5	8 106
Airjet (Elitex)	78 777	8.2	9 665
Sulzer	150 063	5.1	29 715

Note: Some multiples are not precise because of rounding of underlying calculations.
[1] Assumed to cost 10 per cent of 1968 plant.   [2] Assumed to cost 50 per cent of 1968 plant.   [3] Each Lancashire loom assumed to cost 10 per cent of Battery loom—more Lancashire looms are needed for a given output.   [4] Each Lancashire loom assumed to cost 50 per cent of Battery loom.

(b) an indirect effect resulting from the saving in initial capital expenditure due to the lower capital-output ratio of the less capital-intensive technique. This latter effect can be expressed as $\dfrac{(k_2 - k_1) Q^*}{(K/L)_1}$ where k_i is the capital-output ratio for technique i. While direct employment gains are usually taken into consideration in the published work on the subject, indirect gains have been largely ignored. This neglect presumably stems from the assumption that the lower price of used equipment accurately reflects a shorter remaining physical life, and that capital costs on a yearly amortised basis will thus be the same for both new and used equipment. However, if the lower price represents economic obsolescence within the developed countries whereas expected physical lives are about the same, the lower initial price of used equipment represents a net saving in capital that is equivalent to an increase in the real level of national saving. The expression for the indirect increase in employment assumes that any capital saved can be used to generate further employment that requires no higher a capital-labour ratio than the labour-intensive textile techniques. Moreover, it assumes that the resultant output growth equal to $(k_2/k_1 - 1) Q^*$ can be sold without lowering profitability, a not unreasonable assumption for many small, less developed countries.

To obtain an indication of the number of jobs involved in a medium-size, partly industrialised country, a calculation using these formulae was made using the approximate output levels of the Philippines for 1970, about 42,000 tons of spun cotton and 227 million square yards of woven cloth. These levels are considerably lower than those of the more developed countries of Latin America, but greater than the magnitudes of almost all African countries. The calculations indicate the benefits in terms of employment (and by implication in terms of output) to be derived from pursuing an economically rational choice of technique. Needless to say, such gains should not necessarily be assumed to have been made (or to be capable of achievement) in the Philippines, which already has in place equipment which may differ from that used as the norm in the calculations. Table 33 presents the results of the calculations. In both spinning and weaving alternative assumptions about the cost of used equipment are made: (1) for spinning, a 1950 plant is assumed to cost either 10 or 50 per cent of a 1968 one; (2) for weaving, a Lancashire loom is assumed to cost either 10 or 50 per cent of a Battery loom. The differences in assumptions have two effects, namely they change both the saving in capital costs resulting from the choice of used equipment, and the capital-labour ratio by which this saving is translated into additional employment. The figures in the first column indicate that the potential gains in direct employment are quite large. Thus, selection of a 1950 spinning plant and Lancashire looms instead of a 1968 plant and Battery-Unifil looms would increase employment from 4,086 (1,487 + 2,599) to 8,224 (3,275 + 4,949), or by 100 per cent. If weaving was done by the most modern technique, the Sulzer, the former figure would be 2,633. The addition of roughly 4,000 or 5,700 jobs (101 and 212 per cent) in one sector of manufacturing can hardly be dismissed as being of little consequence. Although the basis for the calculation is Philippine output levels, proportional results would occur with other levels. Still more impressive, however, are the indirect gains. Again using the 1950 or 1968 spinning comparison and the Lancashire or Battery-Unifil comparison, the combined direct and indirect employment difference is 91,988 when the price of old equipment is put at 10 per cent of the new price and 15,197 when it is put at 50 per cent. These are, of course, huge increases, and stem largely from the indirect effects. The latter would be decreased if all the less developed countries followed a used equipment policy, since they would bid up the price of the equipment.

The estimates in table 33 exclude other investment costs, such as in buildings and inventories, that would be needed as complements to machines and labour: allowance for the need to finance these would perhaps halve the indirect employment gains to be attributed to a policy of utilising used equipment. Nevertheless, the added employment is so large that it is perhaps not excessive to assert that if similar results were to hold in a number of other industries that

Table 33.   Increase in employment from use of labour-intensive techniques

Type of equipment	Employment		
	Direct	Additional indirect	Total
Spinning			
1968	1 487	0	1 487
1950 [1]	3 275	28 761	32 036
1950 [2]	3 275	3 195	6 470
Weaving			
a. Battery	2 520	0	2 520
Lancashire [3]	4 949	44 483	49 432
Lancashire [4]	4 949	4 943	9 892
b. Battery-Unifil	2 599	0	2 599
Lancashire [3]	4 949	59 089	64 038
Lancashire [4]	4 949	7 864	12 813
c. Airjet (Elitex)	1 850	0	1 850
Lancashire [3]	4 949	49 411	54 360
Lancashire [4]	4 949	5 928	10 877
d. Sulzer	1 146	0	1 146
Lancashire [3]	4 949	98 596	103 545
Lancashire [4]	4 949	15 765	20 714

[1] Assumed to cost 10 per cent of 1968 plant.   [2] Assumed to cost 50 per cent of 1968 plant.   [3] Each Lancashire loom assumed to cost 10 per cent of Battery loom.   [4] Each Lancashire loom assumed to cost 50 per cent of Battery loom—more Lancashire looms are needed for a given output.

are likely to permit efficient factor substitution, much of the urban unemployment problem would be relatively easy to solve, although the adoption of appropriate policies might encounter strong opposition for a variety of reasons.

LIMITED USEFULNESS OF INTERMEDIATE TECHNOLOGY

In recent years there has arisen, in response to the unemployment problem, an important stream of thought that suggests the use of funds to finance an intensive effort to generate new or "intermediate" techniques that will have lower capital-labour ratios than the most modern ones, yet be efficient in the neo-classical economic sense. Presumably this suggestion has its origin in the argument that there is as yet no alternative to existing capital-intensive techniques, or in the difficulty of obtaining spare parts for the equipment used with known older techniques. Yet the evidence in this chapter indicates that existing older-style equipment does offer efficient labour-intensive possibilities (even though little research on this type of equipment has been carried out in recent years) and is available in large quantities in used form. Although it is not possible *a priori* to rule out the potential generation of still more efficient labour-intensive techniques if sufficient research funds were available, it is far from clear that the effort is warranted by a cost-benefit calculus. Resources devoted to developing new techniques would, if successful, yield a prototype machine only after a number of years; the actual beginning of substantial production of this new textile equipment would take still longer, and a period of adjustment would be necessary in actual textile production; thus potential employment gains, if any, would accrue only a number of years later. Alternatively, the same resources could be devoted to disseminating information on existing techniques, to help firms to make efficient use of more labour with older-style equipment and to facilitate the storage and distribution of spare parts. While it is not certain that the action just described is inherently superior to the intermediate technology approach, the potential immediate employment gains from utilising existing labour-intensive techniques, as compared with the delay and uncertainties attendant upon the generation of entirely new techniques, suggest that in this industry the latter is not likely to be optimal.

CONCLUSIONS

The theme of this chapter is that it would be sensible to use older, used equipment when undertaking investment in textile industries in the less developed countries. The analysis presented suggests that, at factor prices relevant for many poorer countries, the choice of used equipment would be optimal. Nevertheless, before recommending such a strategy without reservation, it would be necessary to obtain other types of information. It was assumed throughout the paper that the productivity of the various types of equipment would be the same in a less developed country as it is in the United Kingdom.[1] While this is plausible *a priori*, some economists would claim that a major advantage of newer equipment is that it can be a substitute for scarce skilled labour. The absence of such labour could lead to poor performance with older equipment, and thus the gap between what can be achieved in the United Kingdom and in a less developed country would grow with the age of equipment: the gap between potential and actual output would be greater with the Lancashire than with the Sulzer, and hence there is an upward bias in our calculations of the break-even wage and of the proportion of the price for new plant that can economically be paid for older plant. On the other hand, even if there were such differentials in actual productivity (and this is far from certain), the maintenance difficulties due to the greater technical

---

[1] See the careful study of these and related questions by Cooper and Kaplinsky in Chapter 5.

sophistication of the Sulzer loom might result in more down-time when repairs become necessary. The net effects on machine productivity could go either way, and an intensive analysis of the ability of less developed countries to absorb different techniques is an important area of further research.

Similarly, detailed analyses of the effect of transport costs, licensing fees and other costs, as well as of potential innovations that may be introduced once used equipment is in place [1], should be brought into the calculations. Even if such additional evidence is marshalled and supports the desirability of using older equipment, it will still be necessary to adopt policies that encourage such decisions. Although changes in relative factor prices are an important element of an over-all policy in this matter, they are unlikely to be sufficient without better information and support. Unfortunately, few governments have yet shown an interest in altering the economic environment to provide better signals, and still fewer have supported the types of institutions that could perform essential functions connected with the identification and correct use of used equipment. Yet on the evidence put forward in this chapter the benefits of such an effort could be considerable, in terms both of employment and of output.

---

[1] See Pack, op. cit., for examples from Kenyan industry, and Gustav Ranis: "Industrial sector labor absorption", in *Economic Development and Cultural Change*, Vol. 21, No. 3, Apr. 1973, for examples from Japanese experience.

SUGAR PROCESSING TECHNIQUES IN INDIA

7

by C. G. Baron [1]

The encouragement of labour-intensive technologies and industries in the manufacturing sector has frequently been recommended in recent years as an important element in employment-oriented development strategy. In particular each of the ILO's comprehensive employment strategy missions [2] under the World Employment Programme has to a greater or lesser extent proposed measures, of various kinds, to influence factor prices and to promote research and development so as to encourage labour intensity and thus a faster rate of growth of employment in manufacturing. Nevertheless, because this is largely virgin territory for the economist, the existence of technically efficient labour-intensive alternatives to modern, large-scale production methods has been a matter of some controversy and even scepticism.

However, an increasing number of analytical studies have in recent years tended to justify the place of the choice of techniques in employment-oriented development policy. Some of these studies are contained in this volume. The present chapter extends our knowledge of technological choice as it exists in practice. It is the result of field work in India [3] on methods of sugar production,

[1] International Labour Office. Thanks are due, for many helpful discussions in Delhi, to officials of the Ministry of Food and Agriculture, and to the National Federation of Co-operative Sugar Factories. An earlier draft has benefited considerably from the comments of the steering group of the ILO's technology and employment project, namely Gustav Ranis, Amartya Sen, Gerard Boon, Charles Cooper, Bruce Johnston, Howard Pack and Paul Strassmann. The author is particularly indebted to Mr. M. K. Garg of the Planning Research and Action Institute, Lucknow, for data on the small-scale sugar plants described here, in the development of which he has played a pioneering role. Responsibility for the comparison nevertheless rests entirely with the author.

[2] That is, the ILO employment missions to Colombia (1970), Sri Lanka (1971), Iran and Kenya (1972), and the Philippines (1973). The last-named developed policy recommendations in these directions in particular detail.

[3] The author visited India in 1972 and collected much of the data presented here in Delhi, in the course of conversations with officials connected with the sugar industry. He also made several visits to the sugar producing areas in Uttar Pradesh, and plants typifying the alternative technologies were seen, as described in Appendix II. A good deal of information

(footnote concluded overleaf)

and its main theme is the comparison between a small-scale intermediate technology for making white sugar, and a larger and better established capital-intensive production process. The over-all conclusion tends to bear out the desirability of intermediate technology. However, the study also illustrates the difficulties that arise, first in establishing to what extent a choice of technology can be said to exist, and secondly in analysing the implications of such a choice in a manner which takes into account all the issues involved. Product choice is found to be highly relevant.

The arrangement of this chapter is therefore as follows. The two technologies are first briefly described, as well as their operating characteristics and the political and economic setting in which the cane sugar agro-industry operates. Information is then supplied on the factor and material inputs required under each technology, and the choice between the two technologies is then analysed with reference both to private profitability and to the social costs of production, drawing upon the considerations on shadow pricing in the surplus labour economy suggested in several works, most notably in the UNIDO *Guidelines for project evaluation.* [1] After a summary of the findings, the choice of technologies and of products is discussed within the framework of decision making and planning in India.

THE TWO TECHNIQUES USED IN WHITE SUGAR PRODUCTION

Several different forms of sugar are produced in the world. The processes for the production of brown and raw sugar differ slightly from that for the production of white. The present discussion relates only to white crystal sugar, which is the variety produced by both the large and the small [2] plants under review.

The production of white sugar from sugar cane involves a sequence of the following processes: crushing of the cane to extract a juice, settling and filtration of this juice to separate out non-sugar components, boiling of the filtered juice to concentrate for crystallisation, and centrifuging to separate sugar crystals from molasses. Small quantities of chemicals such as sulfur are added to aid the process. Bagasse, the cane fibre, is the by-product of crushing

---

about the large-scale sugar industry is also contained in the reports of various government bodies, notably reports of the Sugar Enquiry Commission (1965), the Committee on the Rehabilitation and Modernisation of Sugar Factories in India (1965), the Tariff Commission (1969), and the Second Central Wage Board for the Sugar Industry (1970).

[1] UNIDO: *Guidelines for project evaluation,* op. cit.

[2] Some summary technical and engineering data on the small-scale plants is given in Appendix III below.

and is used as a fuel for the later stages of production. Other by-products, especially molasses which can be used in distilleries, emerge at the last stage.

White crystal sugar was first manufactured in sugar mills in India in the 1920s. The passage of the Sugar Industry (Protection) Act in 1932 enabled the industry to expand rapidly. There are now about 220 mills having an average crushing capacity of 1,350 tons of sugar cane per day. Altogether these mills have in the last three years produced an annual average of 3.8 million tons of sugar. Two or three new mills open each year, the rate of establishment being determined by the Government's licensing policy. The machinery installed has since the late 1950s been manufactured in India itself. For the most part there is not a wide technological choice, since the chemical processes referred to above are standard. Certain operations, such as cane unloading, can be mechanised, but this has been avoided in Indian sugar mills because of trade union and social pressure to maintain the level of employment. The choice of machinery, and the ways in which it is added to and improved from time to time, is determined mainly by two considerations, namely the need to reduce costs by making more sugar out of the same quantity of sugar cane (i.e. raising the recovery rate), and the need to meet quality standards set by the Government. Partly because wages constitute only a fraction of total costs and also because of trade union pressure against any reduction in employment, there is little incentive to substitute machines for men in the operations that can be mechanised.

However, in recent years a choice of techniques has emerged in the following way. After trials at experimental plants [1] over several years sugar units began to be established from 1960 onwards in Uttar Pradesh, the state that produces the largest quantity of sugar cane in India. In 1965 there were reported to be 114 such units [2], and in 1972 they were said to number 550 in Uttar Pradesh alone, without counting the units in other states. The small units appear to be commercially successful and their contribution to employment in Uttar Pradesh and to the total output of sugar is substantial [3], so that an economic comparison of the small plants with the mills seems worth while.

---

[1] These pilot plants were supervised by the Planning Research and Action Institute in Lucknow, aided by the National Sugar Institute in Kanpur. The Khadi and Village Industries Commission, located in Bombay, has also actively encouraged the establishment of small-scale sugar plants.

[2] Government of India, Ministry of Food and Agriculture: *Report of the Sugar Enquiry Commission, October 1965* (Delhi, 1965).

[3] No single government office is responsible for compiling statistics about the small-scale units. The *Report of the Sugar Enquiry Commission ... 1965*, op. cit., mentions (p. 50) 114 units "using the sulfitation process of clarification", mainly in Uttar Pradesh. It is reported by the Planning, Research and Action Institute that about 550 units were operating in Uttar Pradesh in the 1971-72 season. The output and employment of these units can be estimated only very roughly, by making an assumption about the average crushing capacity and associated employment creation. A conservative estimate would be that the annual output is of the order of 200,000 tons and that total employment in these units is about

(footnote concluded overleaf)

The small units are technically known as open pan sulfitation (OPS) units, to distinguish them from the mills, which use the vacuum pan sulfitation process. The technology employed is a simple version of the chemical processes in a sugar mill, but with some characteristics of the traditional *khandsari* [1] sugar cottage industry. In favourable circumstances the sugar produced in the small units is of the same technical standard (in terms of colour, size and dryness of the crystals) as average mill sugar, although it is more generally not quite up to that standard. [2] There are, however, no difficulties in selling it, usually at a price 5 or 10 per cent below the free market price of mill sugar. Moreover, as we shall see, the OPS units are labour-intensive and capital-saving in comparison with the mills; these are considerable advantages in a developing country such as India with a shortage of capital and a serious rural employment problem. On the other hand, their recovery rate of sugar from sugar cane is lower than that in the sugar mills, which constitutes a serious disadvantage. It is interesting to speculate about how much more employment would have been generated had the OPS technology been developed at a much earlier stage. Apparently this possibility was considered before the introduction of the Sugar Industry (Protection) Act in 1932, but a decision was then taken in favour of the modern machinery.

THE POLITICAL AND ECONOMIC SETTING

It is necessary to outline the broader context within which the mills and OPS units are to be compared. This is complicated by the fact that only 30 per cent of the annual output of sugar cane is used for making sugar by these two types of plant. The remainder is used for making *gur* [3] and *khandsari*, the traditional sweetening agents, apart from the 10-15 per cent of the cane used for replanting

---

75,000 persons, permanent and seasonal. These magnitudes are respectively 5 and 40 per cent of the totals for all sugar mills in India, the second figure, for employment, being particularly remarkable. Another 200 units are said to be operating in other states in India, most notably Andhra Pradesh with 120. A field survey is really required to ascertain the precise facts about these units and to estimate the associated output and employment more precisely.

[1] The sugar produced by the *khandsari* process is yellow to whitish in colour and opaque and dull in texture. It is either powdery or in small crystals, and is valued for its digestibility and nutritional qualities.

[2] A sugar merchant in Kanpur said that the distinction between mill sugar and OPS sugar was one to be drawn only by a specialist. A technologist at the National Federation of Co-operative Sugar Factories agreed that some of the OPS sugar was "very good". On the other hand, the extent to which OPS sugar deviates from the best standards of white, hard, dry crystals is certainly dependent upon the diligence and skill of the supervisors in the small-scale units, which vary. In the mills, on the other hand, the chemical processes—boiling times and temperatures, etc.—are well known and documented.

[3] Obtained by boiling and evaporation of the clarified juice of sugar cane. Like *khandsari*, it is claimed to have a greater nutritional value than white sugar.

and for animal feed. Of the two traditional products, *gur* is by far the more important, accounting for about 50 per cent of cane consumption in an average year. The output of traditional *khandsari* is now at a very low level.

The sugar cane cycle is responsible for many of the current difficulties of the sugar industry. The cycle derives from the fact that in India, unlike other sugar producing countries, which have large plantations, sugar cane is grown by individual cultivators who make their own choice of the crops to be sown each year: a cultivator switches from sugar cane to food grains or a vegetable crop according to his estimate of profitability. The time lag between planting and harvesting is between 12 and 18 months, and it is this time lag that generates the cycle. In a year of deficiency in sugar cane production, prices rise as *gur* producers and the sugar mills compete for cane supplies. Cultivators respond by planting more cane, so that two or three years later cane is plentiful. Prices fall despite the fact that the mills must pay a minimum price for cane—now about 8 rupees per quintal [1]—because the *gur* producers are not so restricted and account for most of the demand. Prices having thus fallen, production decreases in the next two years, so completing the four-year cycle. The sugar cane cycle causes recurrent deficiencies and surpluses in the supply of sugar, as well as being responsible for under-utilised capacity in the sugar mills in most years, and higher costs of production. Several measures to damp down the cycle have been proposed: for example, the Sugar Enquiry Commission (1965) suggested a large buffer stock for sugar. [2] To this the Tariff Commission Report (1969) added the idea of a government support price for *gur*, which would naturally support the cane price. However, no such policy has been implemented.

Economic analysis is complicated first by the existence of the cycle and secondly by the various control measures employed by the Government to stabilise the supply of sugar to consumers. Under the system of partial decontrol [3] the Government buys 60 per cent of the sugar produced by the mills at a fixed price, reselling it to consumers after levying an excise duty. This sugar supply is termed "levy sugar". The remaining 40 per cent of sugar produced, "free" sugar, sells on the open market at a much higher price. This system leads to many abuses. Nor is it the only restriction on the mills: the minimum price of sugar cane is fixed, as mentioned above; wages are fixed by a wage board, and labour forces in the mills are held constant by trade union pressure to maintain the level of employment; finally, the growing sector of the sugar industry, the co-operative sector, is indirectly controlled in the public interest, as a consequence of its being financed by the public authorities and by semi-public

---

[1] A quintal is one-tenth of a ton.

[2] *Report of the Sugar Enquiry Commission ... 1965*, op. cit., p. 163.

[3] Prevailing in July 1972. The form of control over supplies exercised by the Government varies from time to time according to the condition of the market.

bodies. The OPS units in Uttar Pradesh, on the other hand, are not subject to controls: their sugar may be sold freely on the open market, and labour is not unionised; there is no restriction on the price they pay for their cane; a low rate of excise duty is levied, but on their capacity rather than on output, since with so many units operating only a capacity tax is administratively feasible.

The combined development of the modern technology of the sugar industry in India is ensured by technologists connected with the National Sugar Institute at Kanpur and the two factory associations. Most of the machinery for new factories and for the modernisation of old ones is made within India by about a dozen firms. This specialist industry still has some links with firms in developed countries, but was set up because of the foreign exchange crisis of 1957 to avoid the necessity of importing machinery as the sugar industry developed. Since then, however, it has been making equipment for other industries also. The machinery for the OPS small-scale units is made by smaller firms in provincial towns, and the expansion of the OPS units in Uttar Pradesh is said to have stimulated local machine-building capability. Although the technology of the OPS units has been locally developed, it is possible that white sugar may at one time have been made with similarly simple equipment in other countries now more economically advanced.

As regards the demand for sugar, consumption is low in India by comparison with other countries; in 1969 it was 5.8 kg per head, as against 24.9 kg in Sri Lanka, 17.8 in the Philippines, and over 50 kg in advanced countries such as the United Kingdom, the United States and Canada. There would therefore appear to be plenty of room for a growth in demand. Most of the underlying demand for sweetening agents is met by the traditional product, *gur*, which is often considered to be more nutritious than sugar, but the urban population in India tends to prefer sugar, especially with tea, the consumption of which is increasing also. *Gur* consumption still predominates in rural areas and among the poorer people in the towns. A significant lacuna in the information required to plan the sugar industry's future development is the lack of evidence concerning the relationship between the demand for sugar and the price levels of *gur* and sugar respectively. [1] Even if the necessary statistics existed, however, these would not represent true demand and true market prices because of the measures of control of sugar supply adopted by the Government. It can be argued that this intervention in the market for sweet-

---

[1] An attempt to explain sugar consumption econometrically was attempted in S. C. Gupta, S. N. Sinha and M. Prasad: "Relation between consumption and price of sugar to deal with the problem of surplus production", in *Indian Sugar* (New Delhi), Vol. XXI, No. 2, May 1971. The price effect was not found to be significant in this work. Micro-studies of price and income elasticities of demand would be useful in planning the future growth of sugar production.

ening agents implies a choice of products by the Government, a choice which may be just as relevant to the over-all employment level as the choice of techniques. Since each increment in sugar output in sugar mills or in small-scale units displaces *gur* output, employment is to some extent diminished since *gur* is much more labour-intensive. [1] The question of whether or not the Government is intervening too strongly to protect the sugar industry and encourage its future growth is, however, very difficult to answer, not only because of the great short-term fluctuation in supply but because the income elasticity of sugar and *gur* demand is unknown. The determination of the long-term equilibrium demand would, moreover, depend upon assumptions about the urban-rural income distribution.

COST OF LABOUR AND OF CAPITAL

We start our comparison of the two techniques for the production of white sugar by examining the cost of labour and of capital. Table 34 gives various basic statistics in respect of representative units of the two technologies. The first series relates to a modern mill of the type now being built, with a crushing capacity of 1,250 tons per day, a fixed capital investment (in land, buildings and machinery) of 28 million rupees, and a labour force, permanent and seasonal, of 900 persons. The second series relates to a standard OPS plant with a crushing capacity of 80 tons per day, a fixed capital investment of 540,000 rupees and a labour force of 171 persons. The fixed capital per workplace amounts to 31,100 rupees in a modern mill and 3,160 rupees in an OPS plant.

The figures given in table 34 for employment include all employees, skilled and unskilled, permanent and seasonal. The measure of employment versus output might therefore be considered rather a rough one. However, it does in fact reflect the labour-output ratio even when seasonality is taken into account. In a typical OPS plant the permanent/seasonal breakdown is as shown below:

Permanent staff, for 12 months (250 working days)	9
Seasonal staff, for 5 months (150 days)	19
,, ,, ,, 4 months (120 days)	3
,, ,, ,, 100 days	140
	171

---

[1] At its very simplest, *gur* production requires the employment of two bullocks and four persons to process about 15 tons of cane in a 100-day season. The labour input is about 50 man-days per ton of *gur*, as opposed to about 10 man-days per ton of sugar in the sugar mills.

Table 34. Capital and labour requirements per unit of output

Item	Modern mill				OPS plant
	9.5 per cent recovery [1]		11 per cent recovery [2]		8 per cent recovery [3]
	135-day crushing season [4]	200-day crushing season [5]	135-day crushing season [4]	200-day crushing season [5]	
Output per season (tons of sugar)	12 825	19 000	14 850	22 000	640
Fixed capital per ton of sugar output (rupees)	2 183	1 474	1 885	1 273	843.7
Output (tons of sugar) per employee	14.25	21.11	16.50	24.44	3.74
Employees per 100 tons output	7.02	4.74	6.06	4.08	26.7

[1] This has been the all-India average for sugar mills in recent years. [2] This higher average corresponds to the level reached in Maharashtra State. [3] In the OPS plants, 8 per cent is a reasonable average; 8.5 per cent may also be technically feasible. [4] Average length of the season in Uttar Pradesh in recent years. [5] Full capacity utilisation.

By adding together the number of man-days worked in each category of staff (as shown above) an employment-output ratio of 31 man-days per ton of sugar can be derived. The corresponding measures for the mills were as follows:

	Man-days per ton of sugar [1]	
	Highest	Lowest
Uttar Pradesh:		
West	16.5	5.1
East	20.7	9.3
Maharashtra	10.3	2.6
All-India average	10.42	

The all-India average figure given here is about one-third of the figure of 31 man-days per ton calculated above for the OPS units. Even with the seasonal nature of employment taken into account, therefore, the OPS units generate about three times more employment per unit of output than the sugar mills.

There are several reasons for this difference in labour intensity. First, in the small units more watchers are needed for each chemical process, in place of the scientific measuring gauges used in the mills. Secondly, in the mills the juice obtained from the cane in crushing is moved between processes by power pumps; in the OPS units pumps are used, but manual power also plays its part, particularly in certain operations during which the juice has to be stirred.

---

[1] Figures from the report of the Committee on the Rehabilitation and Modernisation of Sugar Factories in India (1965). They are for the season 1962-63, when production was rather below average, during a trough in the sugar cycle. These figures are therefore probably rather higher than the average over all years. This does not detract from the argument in this section, but rather reinforces it.

For the most part, the unskilled labour force employed in the mills in the crushing season is used on cleaning operations; in the OPS units unskilled labour has rather wider functions.

The figures regarding the fixed capital requirement in table 34 show that for any level of capacity utilisation the investment cost per unit of output—or per workplace for that matter—is higher for a modern mill than for an OPS unit. This is true even when output is high because of a high recovery rate of sugar from sugar cane. In conjunction with the findings on the labour input requirements, this shows that the two technologies are in a relationship of efficiency with respect to each other: at no feasible level of capacity utilisation do the capital-output and labour-output ratios for the modern mill both fall below the corresponding ratios for an OPS unit.

Working capital, representing stocks of raw materials and finished product, is likely to bear a higher ratio to fixed capital in a labour-intensive plant than in a capital-intensive plant processing the same material. This consideration may significantly affect the comparison of rates of surplus, as has been pointed out and illustrated in one case by Sen. [1] The extent of this effect depends upon the time lag between the purchase of raw materials and the sale of the final product; it will thus be especially significant in an industry in which most of the cost is for the raw material. Nevertheless, in the Indian sugar industry, working capital is probably higher for the mills than for the small-scale OPS units, because the Government carefully regulates the sale of sugar by the mills, to ensure that supply is reasonably evenly distributed over the year. The OPS units are not subject to such regulation, and may therefore sell their sugar in the course of the crushing season. If this were not so, and the mills and OPS units were competing on an equal basis, with the same time lag between production and sales, then the ratio of total fixed and working capital to output would be higher for the OPS units, relative to the mills, than is suggested by the ratios for investment per unit of output in table 34.

The full implications of the difference in the factor inputs required under each technology are brought out by comparing the output and employment arising from the same initial investment, 28 million rupees, in one modern mill and in the corresponding number, 52, of OPS units. The output figures for the mills are calculated on various possible assumptions about the length of season (100 to 200 days) and recovery rates (9.5 to 11 per cent). On that basis it will be found that the modern mill employs 900 persons on a permanent and seasonal basis, and the corresponding OPS units 8,892 while the total output ranges from 9,500 to 22,000 tons for the modern mill and amounts to 33,280

---

[1] See "Technique for the cotton-weaving industry in India", in Amartya Sen: *Choice of techniques*, op. cit., Appendix C.

tons, assuming an 8 per cent recovery rate, for the OPS units. The OPS units together therefore provide about ten times the employment per unit of output generated by a modern mill (as long as we do not discriminate between permanent and seasonal staff); and for the same investment they generate between one-and-a-half and three times the output of one mill, the exact ratio depending on the recovery rate and length of season in that mill. If therefore it is desired to maximise either output per unit of investment or employment, the OPS units appear to be the best choice, even if we compare a very efficient modern mill in Maharashtra, with an 11 per cent recovery rate and working a 200-day season, with the equivalent number of OPS units. This discussion has, however, concerned only the capital and labour inputs to production, regardless of the real opportunity costs of these inputs, and the other elements in costs. To these we now turn.

THE OVER-ALL COST STRUCTURE

The detailed data on costs in a standard OPS unit with a crushing capacity of 80 tons per day are given in Appendix I. The main items in recurrent costs are (1) sugar cane, (2) conversion costs, including minor material inputs, fuel and power, (3) wages and salaries, and (4) depreciation, interest charges on working capital, overheads, etc. In the OPS units, which work a standard 100-day season, the calculation of costs is easier than for the mills because capacity utilisation does not vary greatly. The sum of the last three items above is 286,000 rupees per season, or 45 rupees per quintal of sugar produced. To this must be added the cost of cane, the price of which varies considerably. For the purpose of this study, we assume a price fluctuation of between 8 and 12 rupees per quintal, which for an OPS recovery rate of 8 per cent implies a cane cost of between 100 and 150 rupees per quintal of sugar.

The structure of costs of production in the sugar mills is more complex. Every mill is different, so much so that one suggestion current in sugar circles in India is that for levy sugar the fixed price to be paid by the Government should be estimated separately for each mill. A report of the Tariff Commission for India in 1969 [1] gave data on average costs throughout the country for the season 1966-67. The actual average conversion cost, equivalent in composition to the figure of 45 rupees per quintal quoted above for the OPS units, was 28 rupees per quintal. This average is therefore rather less than the estimate for an OPS plant for the season 1971-72; on the other hand conversion costs in the mills have increased considerably since the 1966-67 season.

---

[1] Government of India, Tariff Commission: *Report on cost structure of the sugar industry and the fair price for sugar, Bombay, 1969* (Delhi, 1970).

Table 35.   S mill: conversion costs per quintal of sugar
(rupees)

Costs	1965-66	1966-67	1967-68	1968-69	1969-70	1971-72
Salaries and wages	5.02	7.86	14.20	11.26	9.01	17.30
Fuel and power	1.84	1.60	2.05	2.09	1.14	6.08
Repairs and maintenance	1.26	3.04	4.04	2.80	1.84	5.13
Depreciation	14.70	13.48	16.65	10.53	7.66	10.09
Overheads	0.60	2.47	3.90	4.26	2.91	4.13
Total	23.42	28.45	40.84	30.94	22.56	42.73
Capacity utilisation (%) [1]	*52*	*45*	*34*	*59*	*98*	*54*

[1] Capacity is defined for the purpose of this table as output at maximum daily crushing capacity in a 200-day season with a 10 per cent recovery rate.

The accounts of one sugar mill visited, the S mill as we shall term it, serve to illustrate this point, and also several others. The S mill is quite a new co-operative sugar factory in Uttar Pradesh, having a crushing capacity of 1,270 tons of cane per day. It is described in more detail in Appendix II. Conversion costs in this mill were as shown in table 35.

From this table it is apparent that conversion costs are generally rather higher than the 28 rupees per quintal mentioned in the Tariff Commission report. An upward trend can be detected in several items despite the low costs of the 1969-70 season, so that an estimate of costs for a 120-150 day season might be 40 rupees per quintal of sugar. This approaches the OPS plant's conversion cost of 45 rupees at 1971 prices, if the recovery rate is 8 per cent.

The conversion and capital costs together therefore account for only a fraction of the total cost of sugar production. The cost of cane constitutes by far the greater expense—between 60 and 70 per cent of all costs—and here the mills have an important advantage over the OPS units. Their higher recovery rate means that they need less cane per unit of sugar produced: whereas the sugar mills need between 9.1 and 10.5 tons of cane to produce 1 ton of sugar [1] (these figures corresponding to recovery rates of 11 per cent and

[1] The figure of 1.05 tons of cane per quintal of sugar is based on an average of 9.5 per cent recovery. This is subject to variation according to mill efficiency and the quality of cane, as the table below indicates.

State	Average recovery rates per season			Number of factories
	1967-68	1968-69	1969-70	
Maharashtra	11.27	11.28	10.72	40
East Uttar Pradesh	9.62	9.42	9.24 ⎫	
West Uttar Pradesh	9.44	8.91	8.93 ⎭	71
All India	9.87	9.50	9.33	215

9.5 per cent respectively) the standard OPS unit requires between 11.8 and 13.3 tons of cane (for recovery rates of 7.5 to 8.5 per cent). Although the difference depends upon which specific sugar mill and which OPS unit are being compared it is significant enough in any case for the small-scale units to be accused by the mill associations of being "cane wasters". This turns out to be an argument of limited validity, because the extra cane requirement of the OPS units is offset by the additional capital requirement of the sugar mills; but it may nevertheless be an argument of wide application, since processing agro-industries typically have high raw material costs. Thus in developing countries where wage rates are low, the choice of technology may be more dependent upon the price of the principal input in relation to that of capital than upon the relative factor/price ratio between capital and labour.

At a naive level of choice criteria the optimal sugar production method is not clear. On the one hand the engineer would prefer the more capital-intensive technology because of the higher extraction and quality, but if employment maximisation is the objective the OPS units are preferred, as they are also if output is to be maximised with investment resources as the constraint; and if agricultural production is thought to be constrained, then the mills are preferred because they require less cane per unit of output. There is therefore a conflict of arguments which will be considered together in the analysis following.

SOCIAL COSTS

In this section we shall look more closely at the social costs and benefits of the two technologies in order to determine in what circumstances the expansion of large-scale sugar production is justifiable. Two difficulties stand in the way of such an analysis. The first is that the real benefit of sugar production is not easily measurable. Because of the sugar cane cycle the price of free sugar fluctuates despite the regulation of most of the supply by the Government; thus what is actually paid for the average ton of sugar is not known, and a price elasticity cannot be estimated. Moreover, as we have noted above, there is a quality difference between OPS and mill sugar, although not apparently a very significant one to consumers: the OPS sugar tends to sell at a price 5 to 10 per cent below the free market price of mill sugar. To avoid the first difficulty in a cost-benefit analysis we shall simplify matters by assuming an identical product and so an identical benefit from each method of production. The benefits thus being regarded as the same, the analysis can concentrate on the social costs of production alone.

The second difficulty has been referred to before. It is that each sugar mill operates with a different schedule of costs, and with varying technical efficiency. The data used here are at best reasonable averages.

The real opportunity costs of production by either technology have a present value

$$PV(C) = \sum_{o}^{n} \frac{C_t}{(1+i)^t} + I.P_i$$

where C_t is the total of running costs each year (weighted as necessary by shadow prices), I is the initial capital investment requirement, P_i is the shadow price of investment funds, and i is the social rate of discount. This expression is simplified by assuming [1] that a one-time investment is followed by a constant stream of costs each year, and also by assuming a long life for machinery (as is reasonable in this case). Then

$$PV(C) = \frac{C}{i} + I.P_i$$

This will be converted to an annual equivalent cost per unit of output to allow a comparison between the two technologies. The annual equivalent cost will in fact be our criterion of choice: the technology for which it is less is that which economises most in the use of scarce resources per unit of output.

The shadow price of investment, P_i, is estimated by following the method of Sen and Datta-Chaudhri in a previous evaluation of an industrial project in India according to the formula

$$P_i = r(1-a)/(i-ar)$$

where

a = rate of re-investment of profits
r = rate of return on investment in the economy
i = social rate of discount.

Following these authors we shall assume that $a = r = 0.20$, i.e. that the rate of return on investment is 20 per cent, as is the proportion of profits re-invested.

The social rate of discount is both a parameter and an unknown in the calculation. [2] However, in an economy such as India's in which consumption per head is increasing only very slowly the social rate of discount will be on the low side. In the first calculation presented below, i has been taken to be 10 per cent; in further calculations its value will be varied, with significant effect, as we shall see. With i at 10 per cent, and a, r taking the values suggested, P_i equals 2.67, this figure indicating the present value of the aggregate consump-

---

[1] A. K. Sen and M. Datta-Chaudhri: "Durgapur Fertilizer Project: An economic evaluation", in *Indian Economic Review* (Delhi), Vol. V, No. 1, Apr. 1970.

[2] This is the approach adopted in UNIDO: *Guidelines for project evaluation*, op. cit., p. 167. The importance of the value assigned to the social rate of discount can be determined by sensitivity analysis.

tion forgone elsewhere in the economy owing to the investment of 1 rupee in sugar production now. P_i increases as the social rate of discount decreases, reaching the value 16 when i equals 5 per cent.

Turning now to the details of the computations, in the first term of $PV(C)$ above, C/i, the costs included are *(a)* sugar cane, *(b)* minor material inputs, chemicals, fuel and power, *(c)* repairs and maintenance, *(d)* overheads and contingencies, and *(e)* salaries and wages. These are easy to estimate for the OPS units, in which the length of season is fairly fixed. For a modern mill it is necessary to estimate the costs, some fixed (i.e. *c*, *d*, *e*), and others variable (i. e. *a*, *b*) from the time series for the S mill (which is in fact as typical a modern mill as one could find). This has been done for three levels of capacity utilisation, corresponding to seasons of 200, 150 and 100 days. By assumption, the price of cane in these seasons is 8, 10 and 12 rupees per quintal.

Depreciation has been excluded from recurrent costs in the modern mill and the OPS unit because it represents only a nominal transfer in favour of the eventual replacement of machinery and equipment which is likely to be long-lasting, and for which repair and renewal costs are allowed. [1]

Each item of costs can be valued in social terms. For example, the real social cost of a marginal unit of electricity may be higher than the tariff value, in so far as increments in electricity supply involve the utilisation of scarce foreign exchange and investment resources. Shadow pricing considerations of this kind have not been taken into account in the present value calculations in respect of the minor inputs. However, the two major items in recurrent costs, wages and salaries and sugar cane, deserve special attention.

A typical modern mill pays each employee four times more per annum (2,450 rupees) than does the average OPS unit (684 rupees), so that despite the greater labour intensity of the OPS units their labour costs per unit of output are similar to those in a sugar mill. To be precise, the labour cost in the S mill equals that in an OPS unit at a level of capacity utilisation in the mill corresponding to a season of 135 days [2], which is equal, by coincidence, to the average length of the season in Uttar Pradesh in recent years.

---

[1] See UNIDO: *Guidelines for project evaluation*, op. cit., p. 299.

[2] Details of comparison of labour costs:

	S mill			OPS
Total salary and wage costs (millions of rupees)	2.2			0.117
Number of workers	900			171
Salary and wage costs per worker (rupees)	2 450			648
Length of season (days)	100	150	200	100
Salary and wage costs per ton of sugar (rupees)	232	154	116	183

The wage and salary bill for the modern mill given here is estimated from the S mill accounts for 1969-72, and is consistent with other data on sugar mill costs. The OPS salary and wage bill figure is that also shown in Appendix I.

In economic theory the debate on choice of technique has usually been pursued in the context of an assumption that the wage rate for labour is the same for either technique. We see here that this is not the case. A sugar mill is in fact an enclave of the manufacturing sector, although physically located in an agricultural area. Thus the wage rate paid, and associated conditions, are related to those in other manufacturing industries. The net payment to labour in the sugar mills is determined from time to time by reference to the level of consumer prices, the concept of a "minimum need-based wage rate" for industrial workers, and the capacity of the employers to pay. [1] Allowance is not made for the fact that if a sugar mill did not exist its workers would be earning very much less than they are in agricultural employment, if indeed they were employed at all. A further subsidy to sugar mill workers is the provision of housing at a low rent in the vicinity of the mill. This provision represents the major element in the social cost of transferring labour from the agricultural sector to the manufacturing sector.

The practical effect on the calculation of the present value of social costs is that since wage costs per unit of output are more or less equal under the two technologies, the attachment of a shadow price factor to these costs will make no difference to the choice between the technologies. No adjustment of this kind has therefore been made for the purpose of the present comparison, although such an adjustment would be advisable if we were comparing the social costs of producing sugar with those in another industrial project.

There is one final but possibly quite crucial point to be made in relation to wages and employment. As we have seen, in each method of production, wage costs per unit of output are roughly equal. However, in a mill a relatively high wage is being paid to a few, in comparison with the OPS units in which a much lower wage is paid to more workers, but for only a part of the year. In social welfare terms is it better to withdraw a few persons from the agricultural sector and pay them a complete wage, or subsidise the agricultural incomes of a greater number in the OPS units but only seasonally? Upon this essentially value judgement the optimal choice of technology may depend. In view of the incidence of rural poverty in India it is arguable that the average social utility of the OPS wage payments is higher than that of the mill wages.

Equally we might consider attaching a shadow price to sugar cane, to take into account the labour element in its cost. Such a procedure would favour the OPS units, since they require more cane per sugar output. Since the purchase price of sugar cane varies from year to year as growers attempt to maximise

---

[1] See Government of India, Ministry of Labour, Employment and Rehabilitation, Department of Labour and Employment: *Report of the Second Central Wage Board for the Sugar Industry, 1970* (Faridabad, Government of India Press, 1970).

Table 36.   Present value of the costs of sugar mill production
(rupees)

1. *Length of season (days)*	*200*	*150*	*100*
2. *Tons of cane crushed*	*200 000*	*150 000*	*100 000*
3. *Tons of sugar output*	*19 000*	*14 250*	*9 500*
4. Price of cane per ton	80	100	120
5. Total cost of cane	16 000 000	15 000 000	12 000 000
6. Total wages and salaries	2 203 000	2 203 000	2 203 000
7. Total cost of material inputs other than cane	850 000	637 500	425 000
8. Repairs and maintenance	489 000	489 000	489 000
9. Total running costs (addition of lines 5-8)	19 542 000	18 329 500	15 117 000
10. Present value of total running costs (discount rate 10 per cent)	195 420 000	183 295 000	151 170 000
11. Investment cost	28 000 000	28 000 000	28 000 000
12. Investment cost weighted by the shadow price of investment [1]	74 670 000	74 670 000	74 670 000
13. Present value of all costs [2] (addition of lines 10 and 12)	270 090 000	257 965 000	225 840 000
14. Present value of all costs per ton of output (line 13 ÷ line 3)	14 215	18 103	23 773
15. Annual equivalent of line 14, expressed per quintal of output	142	181	238

[1] $P_i = 2.67$.   [2] $PV(C)$.

Note: To obtain the annual equivalent per quintal the previous line is twice divided by 10: *(a)* to convert from per ton to per quintal; *(b)* to translate from present value to annual equivalent by multiplying by the rate of interest ($i=10$ per cent) (see text).

their income over a range of crops the opportunity cost of sugar cane is difficult to gauge, although it may indeed be rather less than the minimum price payable by the sugar mills. However, this possibility is insufficient justification for a shadow pricing procedure, the only concrete reason for which would be a clear-cut value judgement in favour of rural incomes at the expense of aggregate consumption. [1]

The present value calculations at a social rate of discount of 10 per cent are set out in detail in tables 36 and 37. For the purpose of comparability, each present value has been reduced to an annual equivalent figure per unit of output, i.e. $PV(C)$ has in each case been multiplied by i/Q.

At the assumed 10 per cent social rate of discount, the mills appear to have a clear advantage only when seasons last for an average of 200 days; 150 days

---

[1] Such a value judgement assumes that the social utility of an increment in a (low) rural income is greater than the social utility of the same increment in a (higher) urban income. The implications of this value judgement are discussed in depth in a recent study—Deepak Lal: *Men or machines: A Philippine case study of labour-capital substitution in road construction* (Geneva, ILO, 1974; mimeographed).

Table 37.   Present value of the costs of sugar production in an OPS plant [1]
(rupees)

1. *Length of season (days)*	*200*	*150*	*100*
2. Price of cane per ton	80	100	120
3. Total cost of cane (8,000 tons) [2]	640 000	800 000	960 000
4. Fixed costs [3]	220 080	220 080	220 080
5. Present value of running costs (lines 3 + 4), discounting at 10 per cent	8 600 000	10 200 000	11 800 000
6. Investment cost	540 000	540 000	540 000
7. Investment cost weighted by the shadow price of investment	1 440 000	1 440 000	1 440 000
8. Present value of all costs (lines 5 + 7)	10 040 000	11 640 000	13 240 000
9. Present value of all costs per ton of output (line 8 ÷ 640 tons)	15 687	18 187	20 687
10. Annual equivalent of line 9 expressed per quintal of output	157	182	207

[1] For fuller details see Appendix I.   [2] Crushing capacity 80 tons per day.   [3] Repairs and renewals 17,020 rupees; fuel, power and chemicals 65,460 rupees; overheads 20,600 rupees; contingencies 10,000 rupees; wages and salaries 117,000 rupees.

is closer to the actual average in all India, and for that average length of season the present value of social costs is practically equal for the two techniques. For a short 100-day season the OPS units produce sugar more cheaply. Thus, these figures confirm what is observed in practice, that the OPS units are growing faster numerically in northern states, and not in Maharashtra, where the seasons are longer and recovery rates significantly higher.

The influence of this last factor is more fully brought out in figure 3, which is the result of the same present value calculations carried out for a range of social rates of discount and for two levels of recovery rate of sugar from cane, 9.5 per cent (the national average) and 11 per cent (the Maharashtra average). Since the social rate of discount is normally considered to be low in the surplus labour economy, its significant range in these diagrams is 0-10 per cent, and the following remarks concern this range. Diagram I indicates that if seasons are only 100 days in length the OPS units clearly have lower social costs of production. With a season of intermediate length, Diagram II indicates two switching values of the social rate of discount. If the rate is considered to be below 9.6 per cent the small-scale units have lower social costs per unit of output than mills with a recovery rate of 9.5 per cent: if the social rate of discount is very low—less than roughly 6 per cent—the OPS units are preferable in social cost terms to even the most efficient mills. Diagram III confirms that if the seasons last 200 days the social costs of sugar mill production are lower for either recovery rate. Apart, therefore, from the difference in recovery rates among regions, the degree of capacity utilisation is vital. Sugar mills are not justifiable if they crush cane for less than 150 days per season: above 150 days

Figure 3. Annual equivalent of present value of cost of sugar production per ton

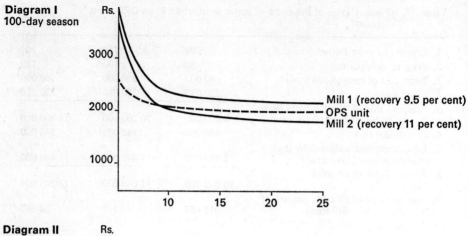

Diagram I
100-day season

- Mill 1 (recovery 9.5 per cent)
- OPS unit
- Mill 2 (recovery 11 per cent)

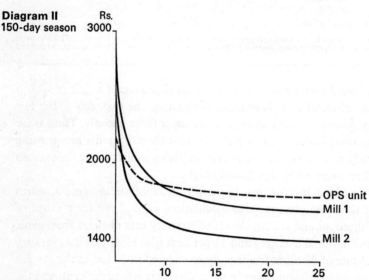

Diagram II
150-day season

- OPS unit
- Mill 1
- Mill 2

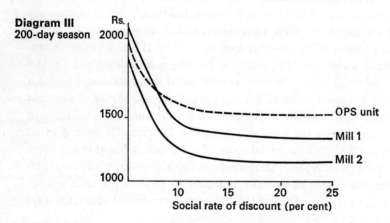

Diagram III
200-day season

- OPS unit
- Mill 1
- Mill 2

Social rate of discount (per cent)

they become justfiable as long as weight is not attached to the value of a wider income (and employment) distribution. (However, the fact that the small-scale OPS units are intended only to work a 100-day season each year renders investment in them risk-free in a way that investment in the large sugar mills is not since the latter are more dependent for their profitability on cane supplies lasting beyond this 100-day period.)

PRIVATE PROFITS AND RE-INVESTMENT

The comparison of private profitability between the two techniques, large-scale and small-scale, is confused by market conditions and the variability in supply from year to year. OPS sugar tends to sell at a price 5 to 10 per cent below the "free" price of mill sugar, but the mills are forced to sell most of their output at a lower controlled price. The latter is said by the mill owners to have been fixed at an unprofitable level, at least for mills in the north. Certainly in years of short supply the OPS units enjoy the private advantage of being able to sell their sugar at a higher price than they might otherwise be able to secure if there were not a system of supply control.

The small-scale units also profit when sugar cane is in surplus, for then, although the selling price of sugar may be low, the price of cane is also low; moreover, the OPS need not pay the minimum price fixed for cane by the Government, and may thus enjoy a purchasing advantage over the mills. Their cane purchasing position is also made easier because, individually, they require less cane and their demand can be met by the small growing area around them. On the other hand a large-scale mill requires good communications to assure capacity utilisation, because sugar cane must be crushed within 24 hours of cutting; even with such good communications, additional transport costs are incurred that may add as much as 1 rupee to the cost per quintal of cane. The choice of techniques may then be effected by location, and it is for this reason that the small-scale units have received the support of a leading sugar technologist in India. [1]

The seasonal variation in output, and the consequent variation in costs and prices, render the calculation of rates of return for each technique almost meaningless. As to the investible surplus, in the case of the modern co-operative sugar mills it will be returned to the financial sources from which the initial

---

[1] See the speech of Mr. Gundu Rao at the opening of the Fourth Technical Seminar on the Open Pan Sulphitation Process, Lucknow, 1966. In the course of that speech, Mr. Rao noted that the small-scale unit "situated very close to the cane areas has the advantage of fresh cane. There are vacuum pan factories which are losing 0.5 to 1 per cent in recovery on account of staleness."

investment is drawn, these sources including central and state governments and the Industrial Finance Corporation. OPS profits may be consumed or invested by their owners in rural industry and agriculture. Unfortunately, no figures are available to sustain fully the thesis that rural areas benefit from such re-investment, although this seems probable. A further long-term social and economic advantage offered by the small-scale units is that they allow the mobilisation and development of mechanical and entrepreneurial skills in rural areas.

SUMMARY OF FINDINGS

Setting aside for the moment the slight difference in quality involved, the conclusions that we can draw from the preceding sections about the choice of technology in sugar production are as follows. It was established that neither technology minimises costs over the entire range of costs of labour and capital. Then a form of social cost-benefit analysis was applied so as to indicate the parameters which determine the optimal choice between the technologies, i.e. that choice which minimises real social costs, given the objective of maximising aggregate consumption discounted over time. In fact this discount rate emerged as a key parameter, on which depends the social value attached to a unit of investment. Other major parameters of crucial importance in the comparison between large-scale and small-scale production are the recovery rate of sugar from cane and the average length of the crushing season. Interestingly enough, the general level of wage rates makes little difference as long as the wage rate in the mills does not increase in relation to that paid by the small-scale OPS units.

These findings are confirmed by the fact that the sugar mills in Maharastra have the advantage over OPS units because of their higher recovery rates and longer crushing seasons. On the other hand, small-scale production has developed rapidly in the northern states of India because the mills obtain lower recovery rates and are handicapped by the fact that the crushing seasons are of uncertain and often short duration.

IMPLICATIONS FOR TECHNOLOGICAL CHOICE

It would be facile to conclude from the findings of this study that sugar mills should be built in Maharastra and OPS units in the northern states. What the analysis really demonstrates is the need for careful examination of proposals for new mills, wherever these are to be located, in the light of the existence of the small-scale technology. Two refinements in methodology are however required. The first of these is that a value judgement should be made on whether or not a premium is to be attached to rural income and consumption at the expense of aggregate consumption. If this were done the situation would be more favourable to the OPS units, especially in so far as the surplus they earn is re-invested locally. Only if such a premium is applied will the widespread concern in India about rural poverty be reflected in the project evaluation. The second refinement concerns the data: there is a clear need to verify the information so far available on small-scale production costs, and to determine the cost of overheads (extension services, research and development, etc.) entailed in their expansion. [1]

Two arguments in defence of the mills have already been touched upon, but are so important as to merit a concluding comment. The first is the assertion that the OPS units are "cane wasters" by comparison with the mills, because of their lower recovery rate. This is a valid argument but clearly a very limited one, for the mills may equally well be dubbed "capital wasters", especially in seasons of low cane production.

This leads us to the second line of defence of the mills, which is that their unit costs would be reduced if only the sugar cane cycle could be stabilised. Investment in new sugar mills would then be rendered no more risky than investment in OPS units. However, it is not at all clear how stabilisation in sugar cane production is to be achieved, given the nature of the market, except by raising the minimum price payable for cane to an otherwise unacceptable level.

Beyond the choice of technique in sugar production lies the more difficult choice of products. In sugar production in India there are alternative technologies resulting in similar, though not identical, products. The existence of another, very labour-intensive sweetening agent, *gur*, also derived from sugar cane, further complicates the question of the choice of products. The present case study cannot therefore achieve the same clarity of conclusion as Timmer's study [2] of rice milling, in which five principal methods of production of varying capital intensity resulted in the same end-product. Yet it can be said that if there is a case in a country like India, with a very low income per head, for a more equitable distribution of consumption goods, then rather than producing a high-quality product such as sugar for consumption by the few, it may be preferable to produce twice the quantity of *gur*, which is a cheaper but reputedly more nutritive food. A less extreme argument would be that the labour-intensive production of low-grade sugar is preferable, on similar grounds, to

---

[1] A field survey of the OPS units for this purpose has already been suggested (see footnote 3 to page 177 above).

[2] C. P. Timmer: "Employment aspects of investment in rice marketing in Indonesia", in *Food Research Institute Studies in Agricultural Economics, Trade, and Development* (Stanford University), Vol. XI, No. 1, 1972.

the capital-intensive production of a lesser quantity of high-grade sugar (given the restraint on investment). Why then has government policy been so preoccupied with a product the quality of which is almost unduly high in view of the poverty-weighted income distribution in India?

The answer to this question may in part lie in history. The sugar industry was established behind a tariff wall during the colonial era. Possibly the consumption needs of either the metropolitan country or the expatriate community in India at that time played some part in the decision. Had there then been research into small-scale methods of production the industry might have developed differently.

Beyond that point in history there are other reasons for the continuing expansion of capital-intensive and large-scale production. There is first a strong and vocal vested interest in such expansion. This consists of professionals and businessmen connected with the two employers' associations, as well as chemists, engineers, accountants and managers in mills throughout the country. These people are proud of the modern technology of the industry, the more so because all new machinery is now made in India itself, and have little interest in seeing small-scale production increase at the expense of mill expansion. Altogether, this may be considered as another example of the institutional urban bias in rural development planning in India. [1]

Capital intensity is also favoured, as we have seen, by the application of what has been termed the "engineering criterion" of choice of technique. [2] (This phrase is due in the first place to Timmer [3], and the idea was also developed by Wells [4] in a case study comparison of economic and engineering rationalisations of technological choice in Indonesia.) In the past the engineering criterion has been defended in the sugar industry on the grounds that high quality is required in production for the purpose of export. However, this argument is of doubtful weight because so small a proportion of production—at most 10 per cent in recent years—is actually exported.

A final reason why the possibility of expanding small-scale production has not been more carefully considered in India is that there is no single body responsible for evaluating investment possibilities in the two technologies

---

[1] The phrase is due to Lipton: "Strategy for agriculture: Urban bias and rural planning in India", in P. Streeten and M. Lipton (eds.): *The crisis of Indian planning* (London, Oxford University Press, 1968).

[2] The preference for the engineering criterion is revealed by the contents of the sugar industry journals, in which articles on engineering and chemical problems and the agronomics of sugar cane cultivation have pride of place. What economic articles do occasionally appear are at a simple, descriptive level.

[3] C. P. Timmer, op. cit.

[4] Louis T. Wells: *Economic man and engineering man: choice of technology in a low-wage country,* paper presented at the Ford Foundation Seminar on Technology and Employment, New Delhi, 21-24 March 1973.

together. New sugar mills licensed by the Government obtain loans at a rate of interest of 7 per cent (other elements of the investment cost usually being interest-free) from the Industrial Finance Corporation. The latter appears to interpret its function in a fairly narrow sense: it verifies the financial viability of the proposals made to it for new sugar mills without taking into account the existence of the alternative technology and without analysing the related conse-quences for income distribution and employment. However, because of the diminishing number of regions producing enough cane to justify new large sugar mills, a policy of dual development, small-scale and large-scale, may become increasingly appropriate in the future.

THE ECONOMICS OF SMALL-SCALE SUGAR PRODUCTION

The figures given here summarise the detailed data supplied by Mr. M. K. Garg of the Planning Research and Action Institute in Lucknow. They refer to a small OPS plant with a crushing capacity of 80 tons per day in two-shift working.

	Thousands of rupees
Land	15
Buildings and construction	215
Machinery and fittings	310
Working capital	60
Total capital investment	600
Chemicals and other minor inputs	24.26
Fuel and power	41.20
Permanent staff	61.00
Unskilled labour (at 4 rupees per man per day)	56.00
Contingencies	10.00
Depreciation	
— on buildings, etc., at 5 per cent per annum	11.45
— on machinery at 10 per cent per annum	31.10
Repairs and renewals	17.02
Overhead charges (covering office expenses, entertainment, etc.)	10.60
Interest on seasonal working capital at 12 per cent	24.00
Total of working expenses, excluding the cost of cane	286.63
Cost of sugar cane (80,000 quintals in a season) at 5.25 rupees per quintal	420.00
Excise duty and income tax	136.53
Working expenses (as detailed above)	286.63
Total annual running costs	843.16

This total cost must be deducted from income to give the net profit. The income from the sale of sugar and minor by-products, assuming a selling price for the sugar of 160 rupees per quintal, will be 908,800 rupees. The net profit will amount to 65,640 rupees (908,800 — 843,160). Thus the rate of return on a capital investment of 600,000 rupees is 10.9 per cent. This figure is very sensitive to year-to-year variation, however, because it depends so greatly on the prices of sugar output and the sugar cane input. (The estimates of these prices used above are lower than those used in the text of the chapter.)

DESCRIPTION OF A TYPICAL MODERN MILL
AND A TYPICAL OPS UNIT

The S mill is a modern co-operative factory established in 1966 in West Uttar Pradesh. The main building is over 200 metres in length and three or four floors in height. There is a complexity of machinery. Around the factory itself is a small housing estate. Most of the workers, seasonal or permanent, live in these quarters at a fairly nominal rent. It will not be long before all the factory personnel are housed in this way. The community is self-contained, the nearest town being 15 miles away, so that most shopping is done on the factory estate itself. Some of the workers are said to be saving from their wages to buy houses elsewhere.

The book value of all the fixed capital stock, mainly purchased before 1965, is 18.9 million rupees. (This compares with the cost of a mill being built in 1973 of 28 million rupees.) Of this sum about half is a loan from the Government of India, the Government of Uttar Pradesh and a co-operative society of growers. The growers receive a perpetual return to their investment in the form of rather higher prices for cane in the area around the mill than prevailed before its existence.

The senior staff have an air of professionalism. They are qualified engineers, chemists and accountants. The quality of sugar has been high since the factory opened. On the other hand, the mill has had considerable difficulty in obtaining sufficient cane, so that this expensive and modern plant is poorly utilised.

The labour force at the S mill numbers 285 permanent staff and 630 seasonal labourers who are paid a retainer allowance when not working. Because of trade union pressure no consideration is given to the possibility of reducing costs by diminishing the labour force. The mill has none the less decided to install some expensive cane unloading machinery. This machinery will be more efficient than the labour now employed on this task, which will not however be declared redundant.

A typical open pan sulfitation unit in the same region was visited. This unit has a daily crushing capacity of 60 tons of cane. It has been financed and is closely managed by a large landowner who is to lose land in the coming land reform in India and has therefore decided to experiment in increasing the family income by means of a small-scale rice mill and an OPS unit. The landowner is well satisfied with the results of these experiments. The OPS unit has made a profit in every year in the last four, and not only in years of sugar scarcity.

One difficulty is the *karigars*. These are the semi-skilled workers who are responsible for the principal chemical processes in the manufacture of the sugar. The *karigars* may achieve the right mix, timing and temperatures for a good-quality sugar one day, but there is no guarantee that they will be able to repeat the performance on the next; for this reason the quality of sugar is variable. Nevertheless, a sample of the sugar from this OPS unit was very good, although a little moist by comparison with mill sugar.

The OPS unit provides employment for about 150 persons during the crushing season from December to March. These men live locally, mainly in a nearby village, and have to leave in April to harvest other crops. For this reason the unit does not generally work for more than about 100 days.

OPEN PAN SULFITATION TECHNOLOGY AND EQUIPMENT [1]

Summary description

The process is a simplified form of the single-sulfitation process employed in vacuum pan factories.

The cold raw juice from the crusher is received alternately in two tanks in the mill house. It is pumped to a sulfitation tank where milk of lime of 15° Baume is added to the juice to bring the pH to between 8.2 and 10.4. The sulfur is burnt in a special furnace with the help of a compressor and the SO_2 gas thus produced is made to bubble through the juice in the sulfitation tank. The bubbling is continued till the pH of the juice is between 6.8 and 7. The juice is then run into the sulfitation bel where it is boiled. After sulfitation the juice comes out of the tank at a pH of 4.5 to 5.0. It flows by gravity to the sulfured juice heating bel, which consists of a gutter pan and a circular flat-bottomed pan at a lower level. The juice flows from a tank into the gutter, where it is more or less neutralised with milk of lime and is heated simultaneously. It is then run into the round pan where final neutralisation is carried out and the juice is brought to the boil. In order to avoid charring, a small quantity of juice, just sufficient to cover the bottom of the gutter pan, is always left behind. After the juice has been boiled in the round pan it is pumped to the settling tanks. The clear juice from these tanks flows by gravity into the gutter pans of the standard bel while the muddy juice is sent to bag filters. The filtrate from these, together with the washing, is also sent to the gutter pans of the bels. After the other pans of the bels are filled, the juice is concentrated to *raab* in the usual manner.

The hot concentrated product from the bel is transferred directly without any aeration to the crystallisers fitted with stirrers. Crystallisation in motion takes place, and after the *raab* has cooled for two days it is ready for centrifuging. This gives first sugar and the first heavy and light molasses. These molasses are re-boiled and put into crystallisers similar to those for the first *raab* but smaller in capacity. The second *raab* is also ready for centrifuging on the third day. The light and heavy molasses obtained from the second *raab* are again boiled, yielding a third *raab*. This is aerated, cooled and allowed to crystallise at rest in masonry tanks for four to five weeks. On centrifuging, this gives third sugar and a final molasses. The sugar obtained is dried in the sun or in the dryer in the usual way.

---

[1] Condensed from Planning Research and Action Institute, Lucknow: *Open pan sulfitation process of khandsari sugar manufacture* (1965). The Institute, which is located at Kalakankar House in Lucknow, can supply further information to interested readers.

Machinery

The machinery requirements suggested in the context of India may be modified elsewhere. The basic components in an OPS unit crushing 60 tons of cane per day are—

1 crushing unit (6-roller hydraulic crusher fitted with cane carrier and cutter);
2 pumps (open impeller type fitted with motor—2-in. delivery);
1 compressor (20 cu. ft. air displacement capacity per minute, fitted with motor and air receiver);
4 centrifugals ($1\frac{1}{2}$ ft. × 1 ft., fitted with motor);
1 dryer (complete);
1 weighbridge (3 ton capacity);
1 platform balance (to weigh up to 2 quintals); and
7 electric motors, including starters, switches, etc.—
    1 60 h.p. motor (for 6 roller-crushers),
    1 10 h.p. motor (for cane carrier and cutter),
    1 7.5 h.p. motor (for cane carrier and cutter), and
    4 5 h.p. motors (for crystallisers, etc.).

The simpler equipment listed below may be manufactured in local workshops with fairly simple equipment.

2 raw juice tanks — 8 ft. × 2 ft. × $1\frac{3}{4}$ ft. — made up of $\frac{1}{8}$ in. steel sheets;
2 sulfitation tanks — $2\frac{1}{4}$ ft. diameter × 7 ft. deep, fitted with lime-adding tanks;
1 sulfitation bel consisting of the following:
    2 long pans — 8 ft. × 3 ft. × $1\frac{1}{2}$ ft.,
    1 round pan — top diameter $6\frac{1}{2}$ ft., bottom diameter 5 ft., height $1\frac{1}{2}$ ft.,
    1 round pan — top diameter $5\frac{1}{2}$ ft., bottom diameter 5 ft., height $1\frac{1}{2}$ ft.,
    1 round pan — top diameter $4\frac{1}{2}$ ft., bottom diameter 4 ft., height $1\frac{1}{2}$ ft.;
1 sulfur furnace (casted);
1 scrubber — 10 in. diameter, $2\frac{3}{4}$ ft. high;
12 settling tanks — 3 ft. × 2 ft. × 4 ft.;
12 bag filters — 4 ft. × $1\frac{3}{4}$ ft. × 4 ft.;
2 filter presses — $1\frac{1}{2}$ ft. × $1\frac{1}{2}$ ft. — 30 plates;
4 standard bels consisting of the following:
    2 rectifying pans — 8 ft. × 2 ft. × 1 ft.,
    1 round pan — 6 ft. diameter, $7\frac{1}{2}$ in. deep, of $\frac{3}{8}$ in. thickness,
    1 round pan — 5 ft. diameter, $7\frac{1}{2}$ in. deep, of $\frac{3}{8}$ in. thickness,
    1 *parchha*, casted — $3\frac{1}{2}$ ft. diameter, $7\frac{1}{2}$ in. deep, and
    1 fourth pan — $4\frac{1}{2}$ ft. diameter, $7\frac{1}{2}$ in. deep, casted;
4 standard molasses bels consisting of the following:
    1 round pan $3\frac{1}{2}$ ft. diameter, $7\frac{1}{2}$ in. deep, casted,
    1 round pan 3 ft. diameter, $6\frac{1}{2}$ in. deep, casted, and
    1 round pan $2\frac{1}{2}$ ft. diameter, $5\frac{1}{2}$ in. deep, casted;
3 recuperators fitted with either wet-bagasse or horseshoe furnaces; and
24 crystallisers:
    18 for first *raab* — 5 ft. × $3\frac{1}{2}$ ft. × $3\frac{1}{4}$ ft., and
    6 for second *raab* — 4 ft. × 3 ft. × $3\frac{1}{4}$ ft.

In addition various pipe and electrical fittings are required.

MANUFACTURE OF CEMENT BLOCKS IN KENYA [1]

8

by F. Stewart [2]

Cement blocks are probably the most important building material in urban Kenya. They are made by mixing cement, sand and small stones together and forming the mixture into blocks of varying size. The blocks are then used as bricks in the construction of buildings. Block manufacture was selected as an example to illuminate the question of choice of techniques for two reasons—first because it was believed that there was in use in Kenya a range of techniques ranging from hand-operated machines which could with justice qualify as "intermediate technology" to fairly sophisticated mechanised machines, and secondly because it was believed that the product resulting from these different methods was homogeneous and hence that the obvious problems emanating from non-homogeneity would be avoided. The second reason was a bad one in two respects: first, and less important, the product turned out not to be completely homogeneous; secondly, a comparison of techniques that produce homogeneous products limits the investigation and makes the results most misleading. [3] In this case attention is focused on the making of blocks to the exclusion of other substitute building materials such as mud bricks, coral, stones or mud. Consideration of some of these methods would extend the choice of techniques, in the sense of technically efficient, labour-intensive methods of production which meet the country's requirements, far more than would an examination of cement block manufacture alone.

Until recently, detailed empirical studies have been very rare. [4] The present

---

[1] This research was partly conducted with finance from the Ministry of Overseas Development of the United Kingdom (now the Overseas Development Administration). I am grateful for their permission to publish this study.

[2] Senior Research Officer, Institute of Commonwealth Studies, University of Oxford.

[3] This point is discussed theoretically in F. Stewart: "Choice of technique in developing countries", op. cit.

[4] See G. Jenkins: *An annotated bibliography of empirical studies on the choice of techniques* (Oxford, Institute of Commonwealth Studies, 1974) for references to other studies.

study is designed to throw some light on the question of whether an efficient range of techniques exists or whether the later more capital-intensive techniques invariably dominate all other techniques by virtue of using less of all inputs per unit of output. For this purpose the information needed was the quantity of each input required to produce a given amount of output, not the money costs, i.e. an engineering production function instead of a cost curve. Actual costs of inputs become relevant only when an entrepreneur decides which technique to use, or a planning office which technique would be optimal. However, it is sometimes impossible conceptually, as with capital, and sometimes impossible through lack of information, to disentangle money costs and the quantity of input used. Other questions illuminated by the study are the relationship between skill requirements and the capital intensity of techniques; technique and product quality; and the process approach to choice of techniques.

THE FIELD SURVEY

In 1969 23 different organisations were interviewed, each of which was operating one or more block-making machines. These organisations were not intended to be a representative sample of block-makers in Kenya: the aim was to find a number of examples of each of the different ways of manufacturing blocks rather than to give any accurate representation of the industry as a whole. Of the 23 organisations, 19 were located in or very close to Nairobi; the remaining four came from Thika, Machakos, Nyeri and Nanyuki. The 19 in Nairobi included all the main block suppliers, but the rest of the country was extremely sketchily represented. The organisations to be interviewed were selected in a somewhat ad hoc way. The main types of organisation using block-making machines—quarry operators, builders, local government authorities, and some building material firms—were known. [1] The survey included all the quarry operators who produced blocks in and around Nairobi. Every building firm listed in the yellow pages of the Nairobi telephone directory was contacted and if it owned a block-making machine the firm was requested, and in most cases granted, a personal interview. The other organisations interviewed in Nairobi were contacted through the advice of some of the other firms questioned. The organisations outside Nairobi were selected in a

---

[1] For this and other information, including a comprehensive list of quarry operators, I am extremely grateful to E. J. Wells who, in part in co-operation with E. Rado, was conducting a major study of building materials, with particular reference to quarries, in Kenya, at the Institute for Development Studies, from 1966 to 1969. See E. J. Wells and E. Rado: "The building industry in Kenya", in G. Smith (ed.): *Industry and labour in East Africa* (Nairobi, East African Publishing House, forthcoming); and E. J. Wells: *The production and marketing of ballast in Kenya* (Nairobi, Institute of Development Studies, 1970; mimeographed).

much more arbitrary way. I was informed by others working in connected fields that various firms operated block-making machines [1] outside Nairobi, and I contacted them accordingly.

The 23 covered a number of different types of organisation: 10 operated quarries and themselves provided an outlet for some of the products of the quarries while benefiting from proximity to the most bulky raw material (ballast or small stones) for block making. Eight were building firms. Whereas the quarry operators produced blocks for sale, all the builders produced exclusively for their own building operations and not for sale. Most of the builders relied on commercially sold blocks for a good deal of their operations—primarily those in Nairobi—and used their own block-makers only for out-of-town jobs. Thus their requirements, in terms of scale of output and mobility of the machine, differed substantially from those of the main commercial producers; this fact had important implications for the choice of techniques. Three of the firms that supplied building materials but did not operate quarries also made blocks for sale. One of these firms was a major cement producer, cement providing an obvious link with block making. Another produced building materials such as tiles and had stopped production of cement blocks, perhaps temporarily, because it was less profitable than the production of other items. Representatives of two councils, the Nairobi City Council and the Machakos County Council, were interviewed. Like the builders, they produced for their own use exclusively. In Nairobi all the blocks for a major housing scheme were being produced, and the Machakos block-makers provided virtually all the blocks used by Machakos County Council. Seventeen of the organisations were owned and run by Asians: the remainder, with the obvious exception of the two councils, were European. All the quarries were Asian. The interviews were carried out in the midst of the implementation of the Kenyanisation policy and the Asian exodus. This had some relevance to the questions under survey since it appeared that in some cases short-lived assets were being chosen in preference to long mainly because of the peculiar uncertainties at that time.

The 23 organisations between them owned and operated 40 block-making machines. These machines could be classified into three broad categories: (a) hand-operated machines; (b) vibrating stationary machines powered by electric motors or diesel engines; and (c) electrically powered mobile laying machines. Within each of these categories (which will be described in more detail later) there were significant distinctions. The hand-operated machines were basically of similar design, though they differed in age, in cost when bought, and in manufacturer. Within the second category there were

---

[1] In particular by E. J. Wells and by John Anderson, also then on the staff of the Institute for Development Studies in Nairobi.

major differences between imported machines and locally manufactured machines in terms of price, durability and repair and maintenance requirements. This category has therefore been split into two—locally manufactured and imported machines. Within the latter sub-category there were substantial differences between the machines according to the scale of production for which they were designed. Where scale is relevant the sub-category has been split accordingly. The housing scheme in Nairobi used a locally produced machine which had been especially adapted to the needs of the scheme according to the design and under the supervision of the resident engineer. That machine, therefore, is in a category of its own. Each of the machines in the third category (the laying machines) differed in age and scale of production. Thus generalisation about the group as a whole is not very meaningful. In the rest of this chapter the techniques have been categorised as follows:

Category	Technique
1	Hand-operated
2	Locally produced stationary vibrating machine, of which—
2a	normal machine
2b	special adaptation
3	Imported stationary vibrating machine, of which—
3a	small
3b	large
4	Laying vibrating machine, of which—
4a	small
4b	large

Table 38 indicates that quarries are less well represented among organisations using hand-operated machines (category 1, accounting for 30.8 per cent) than among organisations using all types of machines (52.5 per cent); they are particularly well represented in categories 2 and 3, the stationary vibrating machines. The table also suggests that firms in Nairobi generally use proportionately fewer hand-operated machines than firms outside Nairobi; 54 per cent of hand-operated machines were in Nairobi, compared with 85 per cent of the electrically or diesel powered machines. If we include the one Mombasa firm interviewed with the Nairobi firms as "urban" firms, this conclusion is reinforced, with over 93 per cent of the powered machines being urban as compared with 54 per cent of hand-operated machines. Table 39 gives a breakdown by type of organisation separately for Nairobi and outside Nairobi.

It will be seen that no quarriers used hand-operated machines in Nairobi, whereas each of the quarriers outside Nairobi did. In Nairobi building firms were the only organisations to use hand machines. All the builders operating in

Table 38.   Users, location and origin of the machines covered [1]

Item	Categories					
	1	2a	2b	3[a]	4	All
Type of organisation:						
Quarrier	4 (2)	8 (5)	— —	7 (6)	2 (1)	21 (14)
Builder	7 (6)	3 (3)	— —	2 (2)	— —	12 (11)
Council	2 (1)	— —	1 (1)	— —	— —	3 (2)
Other	— —	— —	— —	1 (1)	3 (2)	4 (3)
Location:						
Nairobi	7 (6)	10 (7)	1 (1)	9 (8)	3 (2)	30 (24)
Other	6 (3)	1 (1)	— —	1 (1)	2 (1)	10 (6)
Origin of machine:						
Imported	12 (8)	— —	— —	10 (9)	5 (3)	27 (20)
Local	1 (1)	11 (8)	1 (1)	— —	— —	13 (10)
Sample total	*13 (9)*	*11 (8)*	*1 (1)*	*10 (9)*	*5 (3)*	*40 (30)*

[1] Some of the organisations operated more than one machine. The figures in parentheses show the number of organisations. Since some organisations were responsible for more than one type of machine, the total of the figures in parentheses is greater than the total number of organisations.   [a] Category 3 includes both large and small imported vibrating machines. All the large machines (of which there were five) were located in Nairobi; quarries operated four of them, a builder the fifth.

Table 39.   Location of the techniques sampled and kind of organisations using them

Location and kind of organisation	Machine categories			
	1	2 and 3	4	All
In Nairobi:				
Quarriers	0	14	2	16
Builders	7	5	—	12
Other	0	1	1	2
All	7	20	3	30
Outside Nairobi:				
Quarriers	4	1	—	5
Other [1]	2	1	2	5
All	6	2	2	10

[1] No builders were interviewed outside Nairobi.

Nairobi used hand-operated machines only for small or out-of-town jobs. As we shall see, this major difference between urban and rural choice of technique, which is paralleled elsewhere [1], is to be attributed to questions of scale, product requirements and relative factor costs.

[1] See my inquiry into maize grinding in Kenya, in a paper entitled "Employment and the choice of technique: Two case studies in Kenya", in Dharam Ghai and Martin Godfrey (eds.): *Essays on employment in Kenya* (Nairobi, East African Literature Bureau, forthcoming).

Although altogether the organisations contacted were directly responsible for 40 machines, data has not been included for all 40 in what follows. In some cases firms owned identical machines and gave identical answers to all questions for two or more machines. Such answers have been treated as applying to a single case. In many interviews incomplete answers were obtained—e.g. no estimate of repair costs, or of mixture used. In general, incomplete answers have been included except where virtually no information, beyond a statement of the existence of the machine was provided. The detailed data analysed in this paper therefore cover 33 cases, broken down as follows:

Category	Number of examples
1	10
2a	8
2b	1
3a	4
3b	5
4	5

BLOCK-MAKING MACHINES

Block making is a process of mixing materials—cement, sand and ballast—and then forming the mixture into blocks. The block-making machine, which is the subject of this investigation, is the machine which converts the mixture into blocks. Hand-operated machines consist basically of a box the size of the block; the mixture is put into this block and is compressed by the pressure of the lid (sometimes lowered several times), which normally has some system of springs so as to increase the pressure applied. In some examples of hand-operated machines the mixture was hit with sticks to compress it before the lid was put on. After the mixture has been compressed the block is removed and left to dry for two weeks or so, after which it is ready for use. Blocks may be of various sizes, and either hollow or solid. Generally, the hand-operated machines were less versatile than the mechanical ones, though some of them could produce blocks of more than one size, either hollow or solid. Hollow blocks have two advantages over solid ones: they use less materials and are lighter to carry. On the other hand they are also inclined to be weaker, are more difficult to make and collapse more often.

With the stationary vibrating machine the mixture is put into moulds (which can be changed to form solid or hollow blocks of different sizes); the machine then vibrates vigorously for about 30 seconds, after which the blocks are removed and left to dry. The smallest vibrating machines take one block of $9 \times 9 \times 18$ in. at a time, or two of $6 \times 9 \times 18$ in. The rate of output thus varies substantially according to the size of block. Generally, the hand-operated machines can take only one block at a time. Hollow blocks take longer than solid ones because of the higher rate of breakage and the greater care required. [1]

The locally produced stationary vibrating machines were all designed to produce either one 9×9 in. block or two 6×9 in. blocks at a time. Within the imported category some machines also had this output, while others produced at roughly twice the rate with two 9×9 in. or four 6×9 in. blocks. Where this difference affects the results, category 3 has been split into 3a and 3b (small and large imported stationary vibrating machines respectively). Laying machines also produce blocks on the vibrating principle, but they are mobile and lay rows of blocks on the ground as they move. The number they lay in each row depends on the size of the machine, which varies substantially, and on the size of the blocks. Daily output also depends on the speed of operations or the number of rows laid per day. Since they lay the blocks directly on the ground it is not necessary to transport the blocks from the machine to a drying area as with the stationary machines. However, the mobility also creates problems: the mixture has to be carried to the moving machine, and the ground on which the blocks are laid has to be very smooth if a smooth block is to be achieved. On the other hand, pallets, on which the blocks from stationary machines are formed and carried away, are not needed. Pallets cost between 2 and 4.50 East African shillings each, and between 2,000 and 4,000 are needed.

The vibrating process not only saves the use of labour to exert the necessary pressure but also produces a stronger block of more uniform quality. It was claimed that the quality of blocks produced by hand-operated machines was more variable: some easily met the minimum requirements of 400 lb/sq. in. crushing strength, whereas others did not. The uniformity and strength of the blocks are also affected by the method of mixing and of transferring the mixture from mixer to machines. In principle any method of mixing and transfer is compatible with any type of block-maker; in practice, however, this is not so, primarily because of the scale factor. The tendency for a weaker block can be compensated for by increasing the proportion of cement used in the block. However, cement is by far the most expensive material: cement in Nairobi was said to be around 245 shillings a ton, compared with 15 or 20 shillings a ton for sand and a similar price for ballast; the figures given for Nyeri, about 80 miles

---

[1] One firm (using a stationary vibrating machine) suggested that faulty blocks occurred at the rate of 6 out of 250 for solid blocks and 20 out of 250 for hollow.

Table 40. Ratio of cement to other materials in cement blocks in individual cases, by category of block-making machine

Category of machines				
1	2a	2b	3	4
1:7 [1]	1:6 [2] 1:10 [4]	1:11	1:5 [3]	1:8
1:9	1:7	—	1:7	1:8
1:10 [5]	1:8	—	1:10	1:8
1:12	1:8	—	1:11	1:12
1:12 [6]	1:12	—	1:12	1:12

[1] This was a double-storey building and included stones. For a single-storey building the firm in question used a ratio of one part of cement to six parts of dust plus sand, which it obtained free of charge. Two hand machines were used in this case. [2] No ballast. [3] This very strong mixture was justified explicitly. It was claimed that it produced a very strong block, and that breakages were only 2 per cent, compared with an estimated 15 per cent for a lighter mix. Since the firm concerned was a builder producing blocks for own use, it may have been more concerned about avoiding damaged blocks than other firms. [4] With ballast. [5] Apart from cement, the mixture is described as consisting entirely of sand or whatever local materials are available. [6] A local electric vibrating machine had also been used, with a mix weakened to 1:15.

from Nairobi, were 400 shillings a ton for cement, 25 shillings a ton for sand and 10 shillings for stones (estimates of material costs varied according to the location and nature of the producers, who tended to price their own products at cost price). As a result of these prices, an increase in the proportion of cement used substantially increases the cost of the blocks. An official in the materials department of the Ministry of Public Works suggested that the correct proportions for cement blocks were one part cement to nine of other ingredients (three of sand and six of stones), but ratios of up to 1:12 (which produce a weaker block) are quite common.

Table 40 shows the ratio of the mixture adopted by firms which supplied information on this subject. There is no obvious tendency for a stronger mixture to be adopted when hand machines are used, although one firm which had operated both types of machines used 1:12 for hand machines and 1:15 for vibrating machines. The required strength of a block depends on how much weight it is going to have to support: for one-storey buildings the required strength is less than for multi-storey buildings. All the hand-operated machines were producing for one-storey buildings only: most of the firms argued that hand-operated machines were not suitable for multi-storey buildings, although one operator claimed that they would be suitable if the mixture were strengthened. The mixture used by the modified local machine on the Nairobi housing scheme was particularly interesting in this connection. One of the modifications the engineer had introduced had been double vibrating, on the grounds that a weaker mixture could then be used to secure a block of the same strength. He had carried out a number of experiments showing how the uniformity and predictability of block strength increased with the increasing rate of vibration.

He also claimed that the modifications secured a more uniform and smoother block which reduced building labour requirements for laying by 20 per cent. Others argued that it was more difficult to get the blocks out of hand-operated machines and that the blocks were often slightly damaged. It seems likely that laying time is greater with blocks produced by hand-operated machines than with vibrated blocks.

The initial assumption of homogeneity of product is thus wrong, since the hand-operated machine produces blocks of more variable and generally lower strength from the same mixture and the blocks are more uneven and more difficult to lay. Whether this is a serious objection to them depends on the requirements. The engineer of the Machakos County Council was perfectly satisfied with the results, as suitable for low-cost single-storey housing and other buildings (including schools). The quarrier in Nanyuki was also satisfied. Only in Nairobi, where multi-storey houses abound, are these characteristics a decisive disadvantage. In the statistics, therefore, no allowance has been made for the inferior quality of blocks produced by hand-operated machines, although these quality differences must be borne in mind in assessing the results.

The vibrating characteristic had other implications. Obviously such a machine had fuel costs which the hand-operated machine did not. [1] The vibratory action also increased repair and maintenance requirements since the machine was vigorously vibrated, with the block, hundreds of times a day. Thus these machines either have to be built of extremely tough material and well put together, or they need considerable repairs and have a short life. Estimates of durability are almost always largely guesses, but there was sufficient uniformity about these guesses to make them of interest for nine hand-operated machines (category 1)—five used intermittently and four continually. All the answering firms stated that the machines could be maintained indefinitely. Two similar replies were received for normal locally produced stationary vibrating machines (2a), the other seven replies for this category indicating a length of life of anything between two and ten years, although the lowest estimates were all made by organisations that either no longer operated such machines or made only occasional use of them. The only estimate for a specially adapted machine (2b) was four years. The four estimates for imported stationary vibrating machines (3) were very vague; viz. "indefinite", "no estimate, but 17 years old and still going well", "longer than local" and "over ten years". For laying machines there were two replies, one indicating a life of eight or nine years and the other, concerning a machine in only occasional use, 14 or 15 years. Taken together, the estimates suggest that whereas hand-operated

---

[1] Fuel costs are discussed in detail below. The minimum cost for operating the block-maker alone was about 10 cents a block.

machines lasted indefinitely, irrespective of whether they were operated inter-mittently or continuously, the vibrating machines were of limited durability. The durability of vibrating machines clearly depends on use: one firm suggested that the life of its machine would be halved if they operated a double shift. From the opinions expressed, it would also seem that the local machine generally has a shorter life than its imported counterpart.

Estimates of the durability of a machine should not be divorced from repair and maintenance costs, since its life can be prolonged by heavy maintenance. Estimates of repair costs are possibly the most unreliable of all for a number of reasons: since it was impossible to check independently, this figure provided an opportunity to exaggerate the costs of the business; it was also easy for some spares to be forgotten; and ambiguity about whether labour required was or was not included also reduces the comparability of the figures. [1] One year's repair costs may not be typical; and one would expect repair costs to increase with age. A further problem arose from the replacement of moulds. Mould replacement has not been included in the repair estimates, apart from the high figures for two of the laying machines and one of the others. The amount of replacement required was greatest for the stationary vibrating machines, less for the laying machines and nil for the hand machines. While the manufacturer of the imported machine estimated that replacement would be required once a year, one of the firms suggested replacement every 60,000 blocks, which is likely to be two or three times a year. The cost of mould re-placement is around 2,000 shillings.

Table 41 shows the repair cost estimates together with the age of the machine when the survey was made because that is likely to have influenced the estimates. Hand machines require virtually no repairs, whether used intermit-tently or continuously. For all other types repairs were substantial if the machines were used continuously, though they appear to have been less con-siderable if the machines were used only intermittently. The figures do not reveal any tendency for repairs to increase with age; however, no conclusion can be drawn from this fact because, where repair cost estimates were supplied, sometimes only spares were included, sometimes labour used was also included, and often it was unclear what the figure represented. The higher average age of the imported vibrating machines might be regarded as independent evidence for the view that they have longer lives. However, since local production of machines started only in 1965, none of these machines was more than four years old at the time of the survey in 1969. It can be concluded that the hand-operated

---

[1] In cases in which the owner of the machine repaired it himself he normally did not include his labour as a cost. To the extent that the opportunity cost of his labour was zero he was right, but the skilled labour involved is none the less in many cases a real cost of operating such a block-maker.

Table 41. Examples of annual repair costs in individual cases, by category and age of machine

Category of machine

1		2a		3a		3b		4	
Age (years)	Repair costs (EA shs.)	Age (years)	Repair costs (EA shs.)	Age (years)	Repair costs (EA shs.)	Age (years)	Repair costs (EA shs.)	Age (years)	Repair costs (EA shs.)
2	0	1	200[1]	3+	1 000	2	1 000[2]	3+	2 000[1]
4	0	1-1/2	0[3]	6	1 000-2 000	17	2 000[1]	3+	2 000[1]
5	200	1-1/2	2 400	6	3 000-4 000[4]			5	23 750[4,5]
7	25	2	2 000-3 000	8-9	300-400[3]			13	31 250[4,6]
12/13	100	2	3 000[1]					14	1 000
15	100[7]	3	1 000						
17	0	4	2 000						
50	25								

[1] Spares only. [2] Labour also included. [3] Machine only in intermittent use. [4] Includes pattern replacement. [5] Spares 12,500 shs., mechanic 5,000 shs., wear and tear of tools 6,250 shs. [6] Spares 20,000 shs., mechanic 5,000 shs., wear and tear of tools 6,250 shs. Daily cost supplied; assumed 250 repair days per annum. [7] After five years.

machines last almost indefinitely and require no repairs. In contrast, vibrating machines subject to continuous use are of limited durability and require regular and sizable repairs. Repair requirements for the local stationary vibrating machines are similar to those for the imported machines but the latter generally last a good deal longer.

The third characteristic of the use of vibrating machines, as compared with that of hand machines, is that it is to a greater extent machine paced. [1] In no case did the pace of a vibrating machine entirely determine the pace of work, since the machine could be slowed down, but the machine did impose a regular rhythm which was lacking in hand operation. This impression was suggested more by watching operations than by any statistics. In the case of block-making, with vibrating machines each person performed his function in regular time ready for the next step. In the case of hand operation the whole thing was more like making mud pies, filling the mould, hitting it and then waiting to see whether it collapsed or not. Several of those interviewed commented that a more regular (and faster) mode of operation was achieved by the mechanised version.

PERFORMANCE

The output obtainable from the different machines varied according to the speed of operation, the size of the blocks, whether they were solid or hollow, and the hours in operation. The speed of operation was in turn dependent on supporting machinery (such as mixers), the number of workers and efficiency of operation. Since the hand machines were less suitable for hollow blocks (though some organisations did make hollow blocks with them), and in the interests of comparability, only solid blocks are considered here. The figures should be reduced by about one-third for stationary machines producing hollow blocks. It is assumed that each machine is operating a full single shift of 8 or 9 hours, which is what the machines were working except for those in only intermittent use.

Table 42 shows the rate of output per shift reported by the operators. There were significant design differences among the laying machines, with three different manufacturers and each machine of a somewhat different design; these differences are reflected in the very wide variations in output rate reported, and it is therefore almost impossible to generalise about this group. Within the other categories all the machines were of a similar design. Among the hand-operated machines the manufacturing firm varied but the machines were similar.

---

[1] The concept of machine pacing was introduced by A. O. Hirschman, who suggested it might be associated with capital-intensive techniques. See his *The strategy of economic development* (New Haven, Conn., Yale University Press, 1958).

Table 42. Number of blocks produced per shift of 8 or 9 hours in individual cases, by category of machine and size of block
(averages in italics)

Category of machine: 1	2a	2a	2b	3a	3a	3b	3b	4	4
Size of block									
9×9 in.	6×9 in.	9×9 in.	6×9 in. only	6×9 in.	9×9 in.	6×9 in.	9×9 in.	6×9 in.	9×9 in.
60-70	180[1]	90[1]	1 800	—	800	2 000	2 000[2]	525	350
150 <	200	150		500	300	2 800	1 400	875	525
200	600	400		1 200	600-750	3 000	1 200	2 000	1 250
200	1 000	500				4 000	2 000	2 300[3]	.
200-300	1 200	600		1 200	800	4 000	2 000	3 400	.
230	1 200	700-800							
250	1 600	800							
300	1 800	1 000							
300-400									
> 350									
500[4]									
260	*972*	*536*	*1 800*	*967*	*644*	*3 160*	*1 720*	*1 820*	*.*

[1] Used a different local machine from all others; also few workers and no mixing machines. [2] Sometimes too fast for workers, then about 1,750 at slower speed. This is a different type of imported machine that makes the same number of 6×9 in. or 9×9 in. blocks. [3] Hollow, but makes little difference with laying machine. [4] Flat out.

Seven of the eight local stationary vibrating machines were identical, being produced by the same firm, while the eighth was of a similar scale and design. All the small imported machines had the same manufacturer, and variations of design were slight. Four of the five large machines came from the same firm. The fifth displayed some design differences. The locally manufactured machine was basically a copy of the small imported machine. Differences between these two categories lay largely in the method of manufacture and the materials used, not in basic design.

Apart from the laying machines (category 4), therefore, any variation in the rate of output within each category is not to be attributed to machine design. A maximum speed of operation is imposed by the machine design, but variations below this occur according to the number of workers, the organisation of work and other factors. Despite these variations table 42 permits some generalisations. On average the hand-operated machine produced a substantially lower rate of output than the vibrating machines, about half the 9×9 rate of production and a bit over a quarter of the 6×9 rate of the locally manufactured vibrating machine. The locally manufactured and small imported stationary machines had similar rates of output, as was to be expected given the basic similarity of design, while the rate of output of the large imported machine

Table 43. Average output and employment, by scale of production and category of machine

Group and category of machine	Size of block (in.)	Output (O)	Average employment (L)	Blocks per man-day (O/L)	Man-days per block (L/O)
Group I:					
1	6 x 9	260	6	43.3	.023
Group II:					
2a	6 x 9	948 ⎱	8.5	⎰ 111.5	.009
	9 x 9	536 ⎰		⎱ 63.1	.016
2b	6 x 9	1 800	19	94.7	.011
3a	6 x 9	967 ⎱	9	⎰ 107.4	.009
	9 x 9	644 ⎰		⎱ 71.6	.014
4a	6 x 9	700 ⎱	9	⎰ 77.8	.013
	9 x 9	438 ⎰		⎱ 48.7	.021
Group III:					
3b	6 x 9	3 160 ⎱	11.5	⎰ 274.8	.004
	9 x 9	1 720 ⎰		⎱ 149.6	.007
4b	6 x 9	2 567	15.5	165.6	.006

was over twice as great. The rate of output of three of the five laying machines was similar to that of the large stationary machines. The modified local machine's output was the same as the maximum rate reported among the unmodified machines.

In terms of scale of output the machines fall into three groups, as follows: Group I: the hand-operated (category 1); Group II: the small stationary vibrating, whether imported or locally produced (2a, 2b and 3a) and the two small laying machines (4a); Group III: the large stationary vibrating machine (3b) and the large laying machine (4b).

Tables 43 and 44 show output and employment associated with the machines according to these scale categories. Except for the local modified machine, average employment requirements rise with the scale of output but less than proportionately, so that labour productivity rises, and labour input per block manufactured falls, as the scale of output increases. As can be seen from this data, the performance of the locally manufactured and the imported small stationary machine is very similar. However, within each category there are divergences in performance according to the efficiency with which the machines are run, and the averages in table 44 conceal considerable variations in output and employment. Averages are used there because if other "typical" or "representative" figures are used it is very easy to bias the evidence in selection. However, particularly for categories 2a and 3a, the average rate of output does seem to diverge from what appeared to be typical of a fairly efficient operation (a rate of output of 1,800 blocks of 6×9 in. and 800 of 9×9 in.). In table 45 two sets of

Table 44. Average and "typical" performance of small stationary vibrating machines (categories 2a and 3a)

Item	Size of block (in.)	Average	"Typical" performance
Output	6 x 9	950	1 200
	9 x 9	570	800
Employment	6 x 9 ⎱ 9 x 9 ⎰	8.5	12
Blocks per man-day	6 x 9	111.8	100
	9 x 9	67.1	66.7
Man-days per block	6 x 9	.0089	.01
	9 x 9	.0149	.015

figures have been used to represent this category of machines—the average for the group as a whole, and what is taken to be a " typical " performance.

Table 45 compares the rate of output, employment and productivity of the power-driven machines with those of the hand-operated machine.

On the whole, labour requirements per block tend to fall with scale as labour productivity rises, so that the largest machines use about one-quarter [1] of the labour of the hand machines, while the small power-driven machines use about half as much labour per block; labour intensity defined in terms of man-days per block therefore falls considerably as the scale of production rises. The labour involved was generally unskilled. Training on the job normally takes

Table 45. Performance of various categories of machines in relation to that of hand-operated machines (hand-operated machines = 1)

Group and category of machine	Size of block (in.)	Output O	Average employment L	Blocks per man-day $\frac{O}{L}$	Man-days per block $\frac{L}{O}$
Scale II:					
"Typical" 2a and 3a	6 x 9	4.6 ⎱	2.0	2.3 ⎰	0.43
	9 x 9	3.1 ⎰		1.5 ⎱	0.65
Average 2a and 3a	6 x 9	3.7 ⎱	1.4	2.6 ⎰	0.39
	9 x 9	2.2 ⎰		1.5 ⎱	0.65
2b	6 x 9	6.9	3.2	2.2	0.46
4a	6 x 9	2.7 ⎱	1.5	1.8 ⎰	0.56
	9 x 9	1.7 ⎰		1.1 ⎱	0.89
Scale III:					
3b	6 x 9	12.1 ⎱	1.9	6.3 ⎰	0.16
	9 x 9	6.6 ⎰		3.5 ⎱	0.29
4b	9 x 9	9.9	2.6	3.8	0.26

[1] Depending on which block is produced.

less than a day. Some of the firms had a foreman whom they paid above the unskilled rate, but he too was trained on site in a short period. Skills were required for repair and maintenance; as repair and maintenance requirements rose with scale, so did the requirements for skill.

INVESTMENT AND RUNNING COSTS

With regard to capital costs, any definition of capital raises difficulties: and questions of the measurement of capital raise insuperable theoretical problems. [1] Some of the problems may be avoided here since we are concerned with the investment costs of different techniques, not an aggregate measure of capital.

The firms supplied figures for the cost of acquisition of the machines. Since they were bought at varying times in the past, with changing prices, the figures supplied may not be on a comparable basis; moreover, some of the machines were bought new, others second-hand; and some firms gave a gross figure only, inclusive of mixer machine and other capital costs. Table 46 shows the acquisition cost, as reported with no adjustments. Table 47 shows the investment per man employed and investment productivity, or daily output per shilling of acquisition cost. Again the figures used are the unadjusted acquisition cost as shown in table 46. The substantial variations between machines in each group make it difficult to generalise. On average, investment costs per worker rise as the scale of the machine increases. The locally produced vibrating machines have an acquisition cost per worker, on average, nearly three times that of the hand machines. The small imported machines involve nearly twice as much investment per worker as the locally produced machines, while the large imported stationary machines require eight times the investment per man of the hand machines and over one-and-a-half times that of the small imported machines. [2] The investment-output ratios in table 48 are for 6×9 in. blocks. If the figures for 9×9 in. blocks were used, the ratio for all categories except hand machines would roughly double, and hence the comparison would be substantially more favourable to the hand machines.

---

[1] Some difficulties are powerfully presented in Joan Robinson: "The production function: and the theory of capital—a reply", in *Review of Economic Studies*, Vol. XXIII, 1955, p. 247, and more recently further theoretical problems have been raised in the re-switching controversy; see articles by L. Pasinetti and P. Samuelson under the general heading "Paradoxes in capital theory: A symposium", in *Quarterly Journal of Economics*, Vol. LXXX, No. 4, Nov. 1966. For a review of the controversy over capital measurement and its relevance to the question of choice of technique see Amartya Sen: "On some debates in capital theory", in *Economica*, Vol. XLI, No. 163, Aug. 1974.

[2] Though this comparison should not be taken too seriously since one of the observations includes mixer, chute, etc.

Table 46. Acquisition cost of machines, whether bought new or second-hand, in individual cases
(East African shillings; second-hand purchases in italics)

Category of machine

1	2a	2b	3a	3b	4
0	*6 000*	24 000 [1]	*6 000*	*15 000-20 000*	*6 000*
500	7 000 [2]		23 000	35 000	*13 000*
1 000	10 000		*25 000-30 000*	35 000	30 000
1 800	10 000		28 000	60 000	400 000 [3]
1 800	10 700			*110 000* [1]	500 000 [3]
2 000	12 000				
2 800	12 000				
4 000	12 000				
5 000					
8 000					

[1] Includes mixer, chutes, etc. [2] Cost to make. [3] Estimate of capital cost of full plant if new 1969. Estimated cost of similar block-making machine only, 54,000 shs.

The ratios used are not comparable to capital-output ratios as normally defined—that is the ratios of capital stock to annual output in value terms: the ratios in table 48 are ratios of acquisition cost to daily output in volume terms. This does not invalidate the comparison between categories, but deliberately makes it difficult to compare with normal capital-output ratios. [1] The unadjusted ratios suggest that, on average, the investment-output ratio for hand machines is lower, and investment productivity higher, for hand than for mechanised machines. Thus the hand machines appear technically efficient, saving investible resources in relation to output, as well as employment. On the other hand, the ratios show greater investment costs in relation to output for the small vibrating machines (whether imported or locally produced) than for the large vibrating or laying machines. This suggests that the small vibrating machines may be inefficient at any factor prices, using more of both factors as compared with the large machines.

Historic acquisition costs may be of little relevance to current opportunities. For the economy as a whole current opportunities include making or importing new machines or importing second-hand machines. None of the data included imported second-hand machines. Surveys of producers and retailers of equip-

---

[1] The somewhat odd concept of the ratio of acquisition cost to daily output has been deliberately adopted here, to make it difficult to neglect the many reasons why capital-output ratios can be misleading. Some of these are discussed later in the present chapter. For a comprehensive discussion see G. Myrdal: *Asian drama* (London, Allen Lane The Penguin Press, 1968), Appendix III, and W. B. Reddaway: *The development of the Indian economy*, op. cit., Appendix C.

Table 47. Individual and average ratios of acquisition costs[1], employment[2] and output[3] (6 × 9 in. blocks)

Category of machine

1 $\frac{I}{L}$	1 $\frac{I}{O}$	2a $\frac{I}{L}$	2a $\frac{I}{O}$	2b $\frac{I}{L}$	2b $\frac{I}{O}$	3a $\frac{I}{L}$	3a $\frac{I}{O}$	3b $\frac{I}{L}$	3b $\frac{I}{O}$	4 $\frac{I}{L}$	4 $\frac{I}{O}$
100	2.5	333	3.3	1 263	13.3	1 833	34.4 [4]	2 188	5.8	667	11.4
200	15.3	892	7.6			2 000	12.0	2 917	8.75	1 444	14.9
286	5.7	1 273	43.7			2 333	23.3	3 273	12.5	2 143	15.0
300	7.2	1 412	10.0			3 833	19.2	5 435	30.0	3 000 [6]	23.5 [6]
300	7.8	1 600	12.0					6 875 [5]	27.5 [5]	3 600 [6]	15.9 [6]
467	11.2	1 667	8.3								
500	17.8	2 000	20.0								
1 000	16.7	2 000	50.0								
1 333	16.0										
498 [7]	*11.1* [8]	*1 397*	*19.4*	*1 263*	*13.3*	*2 500*	*22.2*	*4 137*	*16.9*	*2 171*	*16.1*

Figures in italics denote averages. Key: I = Investment; L = Labour; O = Output. [1] In East African shillings. [2] In man-days. [3] Number of blocks produced per day. [4] 9×9 in. [5] Includes mixer and chutes. [6] Using estimated cost for block-maker alone derived from costs for similar machine. [7] The averages are 598 for machines bought new and 150 for those bought second-hand. [8] The averages are 11.8 for machines bought new and 9.0 for those bought second-hand.

Table 48. Investment-labour, investment-output and output-labour ratios, by category
of machine (6 × 9 in. blocks)

Category of machine	Price (thousands of EA shs.)	Daily output	Employ- ment	$\dfrac{I}{L}$	$\dfrac{I}{O}$	$\dfrac{O}{L}$
1	2 [1]	260	6	333	7.7	43.3
2a "Typical"	12	1 200	12	1 000	10.0	100
Average		950	8.5	1 412	12.6	111.8
2b	24 [2]	1 800	19	1 263	13.3	94.7
3a "Typical"	23	1 200	12	1 917	19.2	100
Average		950	8.5	2 706	24.2	111.8
3b	35	3 160	11.5	3 043	11.1	274.8
4b	54 [3]	2 300	15	3 600	23.5	153.3

[1] Price of locally produced machine; import price 2,100 shs.   [2] Includes chutes, mixers, etc.   [3] Within the laying category (4), reliable data including acquisition cost for a new machine were available only for one laying machine, with a daily output rate of 2,300 6 × 9 in. blocks, so that only the data for this machine are included in the table.

ment, together with the information supplied by users who had recently (i.e. within a year or so of the survey) bought equipment new, suggested the prices indicated in table 48, which also shows the various ratios calculated on the basis of these prices.

Investment intensity, when defined as investment per man, is substantially higher for all the power-driven categories than for the hand-operated machine. Investment requirements per unit of output are also lowest for the hand-operated machines, while labour requirements per unit of output are highest for the hand machines. Table 49 shows the ratios in table 48 expressed as a proportion of the figures for the hand-operated machines.

The ratio of acquisition cost to output makes no allowance for different asset lives, and is therefore likely to be misleading if used to compare assets with substantially different lives. The acquisition cost may be converted to an annual investment cost: the conversion rate depends on the interest rate assumed. [1] Table 50 shows the annual capital cost of each asset on the assumption of 0, 10 and 20 per cent interest. Because of its very long life the hand machine appears substantially cheaper than before while the short life of the

---

[1] The formula for converting a constant annual cash flow, A, into present value is well known:

$$PV = \frac{A}{r} \left[1 - \frac{1}{(1 + r)^n} \right]$$

where r = interest rate and n = number of years.
In this case the inverse operation is performed, and an initial acquisition cost, which may be thought of as the present value of the capital cost, is converted into a constant annual

cash flow, A, by dividing the initial cost by $\dfrac{\left[1 - \dfrac{1}{(1 + r)^n} \right]}{r}$.

Table 49. Investment-labour, investment-output[1] and labour-output ratios of various categories of machine in relation to those of hand-operated machines (hand-operated machines = 1)

Category of machine	Cost of machine	$\dfrac{I}{L}$	$\dfrac{I}{O}$	$\dfrac{O}{L}$	$\dfrac{L}{O}$
1	1.0	1.0	1.0	1.0	1.0
2a "Typical"	6.0	3.0	1.3	2.3	0.43
Average		4.2	1.6	2.6	0.38
2b	12.0	3.8	1.7	2.2	0.45
3a "Typical"	11.5	5.8	2.5	2.3	0.43
Average		8.1	3.1	2.6	0.38
3b	17.5	9.1	1.4	6.4	0.16
4b	27.0	10.8	3.1	3.5	0.29

[1] Output in 6×9 in. blocks.

locally produced vibrating machine just offsets its lower acquisition cost as compared with the imported machine. At a discount rate of 10 per cent the annual investment costs of the two types of small stationary machine (locally produced and imported) become very similar. The extremely short life assumed for the specially adapted local machine substantially increases its relative investment costs, so that at low rates of discount it exceeds that of all other techniques.

Table 51 shows how the adjusted ratios (at 0 and 20 per cent discount) affect the relative investment costs in relation to employment (I/L) and output (I/O). The annual adjustment improves the relative performance of the hand machine for both ratios, and to an extent that increases as the interest rate falls. The ordering of the investment-labour ratio is not seriously affected by the annual adjustment, although the small stationary machines come much closer together, while the ratio of the adapted local machine increases substantially.

Table 50. Annual investment cost by category of machine (East African shillings)

Category of machine	Estimated life (in years)	Annual investment cost at percentage discount rate indicated		
		0	10	20
1	30	70	210	400
2a	5	2 400	3 170	4 010
2b	4	6 000	7 570	8 030
3a	15	1 530	3 030	4 920
3b	15	2 340	4 600	7 490
4b	12	4 500	7 930	12 170

Table 51. Investment-labour and investment-output ratios by category of machine, in relation to those for hand-operated machines, after adjustment for different assumptions concerning asset life and the rate of interest (hand-operated machine = 1)

Category of machine	Investment-labour ratio			Investment-output ratio		
	Unadjusted acquisition cost	Annual cost discounted at		Unadjusted acquisition cost	Annual cost discounted at	
		0 per cent	20 per cent		0 per cent	20 per cent
1	1.0	1.0	1.0	1.0	1.0	1.0
2a "Typical"	3.0	17.9	5.0	1.3	7.7	2.2
Average	4.2	25.3	7.0	1.6	9.7	2.7
2b	3.8	28.3	6.3	1.7	12.8	2.9
3a "Typical"	5.8	11.4	6.1	2.5	4.9	2.6
Average	8.1	16.2	8.6	3.1	6.2	3.3
3b	9.1	18.2	9.7	1.4	2.8	1.5
4b	10.8	26.8	12.1	3.1	7.5	3.4

It remains true, broadly [1], that the investment-labour ratio rises as the scale increases. All techniques use more investment in relation to output than the hand techniques: how much more depends on the exact basis of the calculations.

When the mechanically powered techniques are compared, the large imported stationary machine is found to have significantly the lowest investment-output ratio: it also (see table 49) has the highest output-labour ratio. In contrast to this, the laying machine and the small stationary vibrating machines appear to represent inferior techniques, since their investment and labour costs per unit of output are both higher than those of 3b. This is illustrated in figure 4. All the techniques are to the north-east of the line joining the hand technique to the large imported stationary machine. This indicates their inferiority. It is true both at 0 and at 20 per cent interest. The diagram does not take the scale of production into consideration. Since the techniques are designed for different scales the conclusions may be altered when scale is considered. There are also important dimensions of cost which have been omitted from the analysis. These are repair costs, mould replacement and fuel consumption.

As seen earlier, the different techniques are associated with different annual repair costs. The vibrating machines also require periodic replacement of the patterns or moulds in which the blocks are formed. This does not apply to the hand machine. Mould replacement costs between 2,000 and 4,000 shillings a year for the small stationary vibrating machines and somewhat more for the larger machines which require more expensive moulds. Repair costs for the

---

[1] The term "broadly" is used because the investment-labour ratio (adjusted) of the laying machine (and of the local adapted machine at low interest rates) are greater than that of the large stationary machine, although the scale of output is smaller in the former categories.

Figure 4. Relative efficiency of block-making techniques using different types of machine

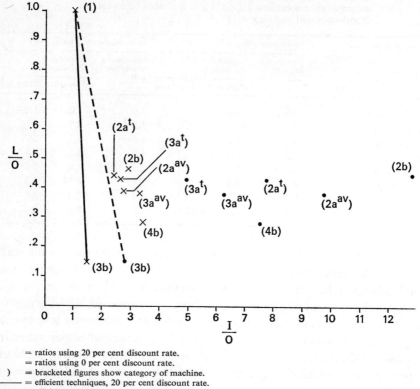

x   = ratios using 20 per cent discount rate.
.   = ratios using 0 per cent discount rate.
( )   = bracketed figures show category of machine.
———— = efficient techniques, 20 per cent discount rate.
– – – – = efficient techniques, 0 per cent discount rate.

machine varied, as can be seen in table 41. They were much smaller for the hand techniques than for the powered techniques, but there appeared to be little consistent difference between the repair requirements of the three stationary vibrating techniques: all were around 2,000 shillings a year. Thus allowing for pattern replacement and repairs would raise the annual costs of these techniques by between 4,000 and 8,000 shillings a year, which is as much as or more than the annual adjusted investment costs. The repair costs of the hand machines were much lower, about 50 shillings a year. Inclusion of repair costs and mould replacement reduces the relative costs of the hand machine, but does not significantly alter the comparison between the stationary vibrating machines.

The estimated annual repair costs of the laying machine (inclusive of pattern replacement) were higher than those of the stationary techniques. Repair costs for the locally adapted machine, which had only recently been installed, were not supplied. However, the intention was to use the equipment fully for four

Table 52.   Fuel costs by category of machine

Category of machine	Per block, in cents	Per annum, in shillings
1	0	0
2a ⎫ 3a ⎭	1-2.5	2 500-6 250
3b	0.7-1	5 250-7 500
4a	0.3-0.9	3 060-3 150
4b	0.3-0.9	5 175

years and then scrap it, and repair costs were expected to be correspondingly low, thus partially offsetting the high annual investment cost due to the short life of this equipment.

Estimates for fuel consumption were supplied on a number of different bases. For some no estimates were supplied; for others a total figure was given inclusive of mixer, and sometimes quarry works. Piecing the evidence together, and deducting for quarry and mixer, suggests the approximate estimate given in table 52.

Table 53 brings together the estimates of investment, repair and fuel costs. As the table shows, the non-labour costs per block are substantially lower for the hand technique than for the other techniques.

Table 54 shows the total costs per block when investment costs are valued using a 10 per cent discount rate, as labour costs vary between 2.5 shillings and 10 shillings a day.

If the two "efficient" techniques (those using the hand machine and the large stationary vibrating machine) are compared, it will be seen that the savings in non-labour costs on the hand machine are outweighed by the extra labour costs even at wage rates as low as 2.5 shillings a day. (The minimum urban wage at the time was 7 shillings a day; the estimated opportunity costs of rural labour around 2 or 3 shillings a day.) [1] With a 10 per cent discount rate the switching wage for these two techniques is 1.03 shillings—i.e. at wages above this the vibrating machine is cheapest, while at wages below it is the hand machine. Using discount rates below 10 per cent would raise the switching wage, while using discount rates above 10 per cent would reduce it. The costs per block of the other powered techniques are substantially higher than that of the large stationary vibrating machine, as shown in table 55. The small sta-

---

[1] Estimates produced by M. Scott and N. Stern in their cost-benefit analyses. See N. H. Stern: "Experience with the use of the Little-Mirrlees method for an appraisal of small-holder tea in Kenya", in *Bulletin of the Oxford University Institute of Economics and Statistics*, Vol. 34, No. 1, Feb. 1972, and M. Scott: "Estimates of shadow wages in Kenya" (1973, mimeographed).

Table 53. Annual non-labour costs by category of machine
(East African shillings)

Type of cost	Category of machine				
	1	2a[1] 3a	2b	3b	4b
Annual investment[2]	210	3 100	7 570	4 600	7 930
Repairs and moulds	50	5 000⎫	10 000[4]	⎧8 000	15 000
Fuel[3]	0	4 380⎭		⎩6 380	5 180
Total[5]:	260	12 480	17 570	18 980	28 110
Absolute figures	260	12 480	17 570	18 980	28 110
Cents per block	*0.4*	*5.3*	*3.9*	*2.4*	*4.9*
Ratio (hand-operated machine = 1)	*1*	*13.1*	*9.8*	*6.0*	*12.2*

[1] Estimates for the imported and locally produced stationary vibrating machine are so close that they are treated together here. [2] Assuming a discount rate of 10 per cent. [3] Taking mid-point of fuel estimates shown in table 52. [4] Estimate. [5] Assuming 250 working days a year.

tionary vibrating machine is the nearest to the hand machine in price and scale, and consequently in some sense its closest competitor. The switching wage between these two techniques, again using a 10 per cent discount rate, is 3.40 shillings, which is higher than the competitive rural wage rate, and probably higher than the urban opportunity cost of labour. Irrespective of whether it is locally manufactured or imported, the small stationary vibrating machine is a product of an "old" technology from an advanced country; and it is accordingly interesting to compare its costs with those of the technique as adapted to local conditions. In fact, it appears that total costs were very similar, with the adapted technique involving a somewhat higher element of labour costs and lower non-

Table 54. Costs per block by category of machine
(East African cents)

Type of cost	Wage rate (EA shs.)	Category of machine				
		1	2a, 3a	2b	3b	4b
Non-labour[1]		0.4	5.3	3.9	2.4	4.9
Labour, at wage rates indicated	2.50	5.8	2.2	2.7	0.9	1.6
	5.00	11.6	4.5	5.3	1.8	3.3
	7.00	16.2	6.2	7.4	2.5	4.6
	10.00	23.1	8.9	10.6	3.6	6.5
Total, at wage rates indicated	2.50	6.2	7.5	6.6	3.3	6.5
	5.00	12.0	9.8	9.2	4.2	8.2
	7.00	16.6	11.5	11.3	4.9	9.5
	10.00	23.5	14.2	14.5	6.0	11.4

[1] Assuming a discount rate of 10 per cent.

Table 55. Non-labour costs per block, at the output level of a hand-operated machine, on the assumption that investment costs do not vary and on different assumptions about other non-labour costs, by category of machine (East African cents)

Category of machine	Assuming that the other non-labour costs vary proportionately with output	Assuming that half of the other non-labour costs do not vary proportionately with output
1	0.4	0.4
2a } 3a }	8.8	14.0
2b	13.8	20.4
3b	8.9	19.0
4b	15.7	15.7

labour costs. Hence the adaptation, though not dramatic, was in the right direction.

The calculations so far have ignored scale. In fact each of the techniques is indivisible in the sense that investment costs cannot be saved by operating at a lower scale. If operations were on a smaller scale the other costs would be less, but not proportionately less. If we assume that investment costs are invariable with respect to output, but that all other costs vary proportionately with output (i.e. costs per block are constant for each technique), then if all techniques were utilised to the point at which output was equal to the 8-hour output of the hand techniques (working 250 days a year), the investment costs per block of all the techniques other than the hand technique would rise correspondingly. The switching wage between the hand machine and the large stationary vibrating machine would then be 4.35 shillings. In the rural areas where wages are generally less than this the hand machine would be the sensible choice. In the urban areas, the social cost of labour is probably lower than this figure.

These calculations assume that all non-acquisition costs are variable. In fact there are other costs that are unlikely to vary proportionately with output. Suppose we assume that labour costs vary proportionately with output but that half of the repair, maintenance and fuel costs do not vary with respect to output, then to use the powered techniques to produce the level of output of the hand machine would raise their costs significantly more. For example, the switching wage between hand machines and large stationary vibrating machines would rise to just over 9.00 shillings. In this situation the costs of the small stationary vibrating machine would be less than those of the large machine for wages of less than 9.43 shillings. [1] Table 55 shows the non-labour costs per block

---

[1] All these calculations assume a 10 per cent discount rate.

for each technique, for output levels of the hand machine, according to the two assumptions.

The figures are illustrative of the importance of scale and capacity utilisation in determining the relative costs of different techniques designed for different scales of output. Because of heavy transport costs (a solid $6 \times 9 \times 9$ in. block weighs 52 kg) the scale of production is determined by the market for blocks in the immediate vicinity. [1] In the rural areas the market is often smaller than the output obtained from 8-hour capacity use of a hand machine, and the machines lie idle much of the time. In such situations, while the costs of all the techniques would rise, those of the hand machine would rise least, since the greatest proportion of its costs are labour and therefore escapable; conversely, if each of the techniques were operated for more than the one shift a day all costs would fall, but those of the capital-intensive techniques would fall most.

The considerations discussed suggest why rural and urban choice technique may differ. First, there is the question of product standard: blocks in large cities may have to be strong enough for more than one-storey accommodation. Secondly, there is the question of scale. Thirdly, wage costs differ, wage rates being lower in rural areas. All these considerations indicate that more labour-intensive hand-operated machines are more suitable for rural use.

THE PROCESS APPROACH

The production of any good can be split into a number of activities. The analysis so far has been entirely concerned with examining the costs associated with different types of block-making machinery. But in the production of blocks the operation of the actual block-making machine is only one part of the production process. The production process as a whole can be split up as follows:

1. Production of raw materials.
2. *(a)* Transfer to site.
   (b) Transfer to mixing area.
3. Mixing raw materials.
4. Transfer of mixture to block-making machine.
5. Operation of block-maker.
6. Transfer of blocks from machine to drying area.
7. Watering blocks when drying.
8. Stacking blocks.
9. Transporting blocks from site to where needed.

---

[1] In addition to the other factors determining capacity utilisation.

The first phase itself covers a large number of activities. The divisions are to some extent arbitrary, and reflect the type of machines available. Thus, if a single machine were always used to mix and produce the blocks these two activities might be amalgamated into one. In each of the activities described some choice of technique is possible. Thus transport (of materials and then of blocks) may be done in a number of ways, depending on distance, quantities and costs. The technique used in the production of raw materials was not included in the survey. [1]

The following range of techniques was observed for different activities:

2a and 9 Transport to and from site: lorry.
2b Transfer to mixing area: spade, wheelbarrow or lorry, depending on scale and distance.
3 Mixing: spade or either local or imported mixing machine (diesel or electric).
4 Transfer of mixture to block-making machine: spade or automatic chute.
5 Block-maker: already covered.
6 Transport of blocks to drying area: manual or by wheeled trolley.
7 Watering: women with watering cans or hoses or automatic hoses.
8 Stacking: manual or by hand-operated trolley or electrically operated stacking machine.

There was no unbreakable connection between choice of technique at one stage and choice at another: in theory one could combine very labour-intensive mixing (on the ground) with a more mechanised block-maker; but in practice, choice of technique at one stage did partly determine choice at another. This was in part a question of speed, and in part of scale. The larger block-makers required a steady and rapid flow of mixture—which virtually ruled out mixing on the ground; generally the larger machines were combined with automatic mixers so that the mixture was automatically transferred to the machine. The quality of the blocks, in terms of strength and uniformity for a given ratio of materials, tended to be greater for automatic than for manual mixing, and also, it was claimed, if the mixture was transferred automatically from mixer to machine.

In general, therefore, the most labour-intensive methods of block making (the hand machines) were also combined with the most labour-intensive methods

---

[1] The choice of technique in the production of material used has been established elsewhere. For different methods in quarrying see E. J. Wells: *The production and marketing of ballast in Kenya*, op. cit.; for cement, see in particular Leonard A. Doyle: *Inter-economy comparisons: A case study* (Berkeley, University of California Press; London, Cambridge University Press, 1965), and Carlos F. Diaz Alejandro: *Labor productivity and other characteristics of cement plants*, Discussion Paper No. 105 (Yale University Economic Growth Center, 1971).

Table 56.   Typical investment costs of cement-mixing techniques, by type of equipment

Type of mixing equipment	Cost (EA shillings)	Suitable for use in association with
Spade	25	(1) hand-operated block-maker (2) possibly small vibrating machine
Small mixer	6 000-10 000	(1) hand-operated block-maker (2) small vibrating machine
Large mixer	40 000-50 000	(1) large stationary vibrating machines (2) laying machines (large)

of mixing, etc. while the converse applied for the more capital-intensive methods. This partly reflected different attitudes to mechanisation on the part of the entrepreneurs, which influenced all their decisions; but it also reflected the fact that requirements of scale and standard of block which led to the adoption of capital-intensive methods of block manufacture led to similar decisions in mixing. For stacking and watering the use of mechanised methods was confined to the larger mechanised block-makers, but many of them used the most labour-intensive methods for this part of the process.

Since different block-making machines may be used in combination with any mixing process, etc. the figures for relative costs of the different block-making machines do not need to be amended to allow for different costs of mixers. Hand mixing was observed for some of the smaller power-driven machines as well as the hand machines. However, as already stated, in practice the type of block-maker does partly determine the mixing method, and a comparison of costs of techniques should take this into account. In the typical case the hand machines were used with manual mixing, the small vibrating machines with small local or imported mixers and the larger vibrating machines with larger mixers combined with a system of automatic chutes. The typical investment costs might therefore be affected as shown in table 56. Inclusion of costs of mixing may thus make a substantial difference to the comparison. Very roughly the cost of the typical mixer is in line with the cost of the block-maker. Mixers also need repairs and use fuel. Generally, the repair cost seemed substantially less than that for the block-makers (about a third), but the fuel costs were of a similar magnitude. On the assumption that the hand-operated machines use no mixer and that their investment costs are unchanged but that the non-labour costs of the other categories increase by 50 per cent as compared with table 55, the costs per block become those listed in table 57. Inclusion of the mixer increases the costs of the mechanised machines in relation to the hand machines but does not alter the over-all results.

Other investment costs might also be included. The costs of erecting the necessary buildings and floor can be substantial, but since such costs are not

Table 57. Costs per block by category of machine, including provision for corresponding mixers
(East African cents)

Category of machine	Size of block (in.)	Non-labour costs [1]	Total costs at wage rate indicated (EA shillings)	
			5.00	10.00
1		0.4	12.00	23.50
2a and 3a	6×9	8.2	12.90	17.10
3b	9×9	3.6	5.40	7.20
4b	6×9	7.4	10.70	13.90

[1] Discounted at 10 per cent, and taken from table 54: category 1 same as in table 54; other categories one-and-a-half times the figures in table 54.

associated with particular block-making machines they have not been included. Generally, inclusion of these costs is likely to raise the costs of all the powered techniques, in relation to the use of the hand machine, since the latter is often used with a minimum of infrastructural and building investment.

PRODUCT CHOICE: OTHER BUILDING MATERIALS

The survey covered different methods of making cement blocks in Kenya. However, the labour intensity of material manufacture and construction may also be altered by selecting different building materials. To supplement the survey, therefore, some of the other materials available in Kenya are briefly described below with a report on the associated input requirements, based on interviews conducted in 1972.

Mud and wattle

Mud and wattle are the materials from which traditional housing is made; the vast majority of the rural population are housed in mud and wattle houses. The construction of these houses is almost exclusively labour-using, in the sense that they are made with local materials, local labour and no purchased equipment. [1] They are not suitable for much urban housing because of lack of materials within easy distance, and because they generally fall below the standards set by urban housing authorities. However, although they do not meet those requirements, they are often of far better quality than the houses

[1] See Jomo Kenyatta: *Facing Mount Kenya* (New York, Vintage Books, 1961), pp. 78-82 for a description of the way in which these houses are constructed by the Kikuyu.

made of cardboard boxes, newspapers, etc., which urban dwellers often use as building materials. At Thika, the City Council has produced some experimental houses with mud and wattle walls covered with various plaster and cement washes. The Council found these unsatisfactory, as they deteriorated rapidly and required more labour and materials than dried blocks. Mud and wattle is labour-intensive in terms of maintenance requirements as well as of initial construction. It is therefore most suitable, and most often to be found, as a building method where labour is costed at close to zero, as in the subsistence sector.

Sun-dried clay blocks

Sun-dried clay blocks are made entirely from local materials. The clay is mixed with *murram* [1] and a little grass, and moulded in wooden moulds; the blocks are then dried in the sun. It is estimated that one man makes up to 60 blocks in a day (not including the labour for digging up the materials). Block manufacture thus normally includes no purchased equipment, and in this sense is almost purely labour-intensive. [2] Labour used in constructing buildings made of these blocks is similar to that in concrete block building. The walls may be covered with cement or plaster to improve their resistance to the rain. The production of clay blocks is more labour-intensive than that of concrete blocks, both in the manufacture of materials to make the blocks and in the manufacture of the blocks themselves; but the quality of the blocks is inferior. This type of block can be produced in a block-maker similar to that used for concrete blocks. In one case where a machine was used, six untrained men produced 450 blocks in an hour, or nearly 80 per man per day. The machine cost 1,500 shillings and was similar to the hand-operated block-makers described above.

A stronger clay block was made by the prisoners at Thika. To form the blocks they used a machine (costing 50,000 shillings in 1956), which made 2,500 blocks a day and employed 100 people, including those digging and moving the blocks around. It is impossible to compare the (apparently low) labour productivity with that of the concrete block-makers because of the inclusion of labour for digging in the prison figures. The investment-output ratio (at 20) was similar to that of the mechanically operated block-makers and

---

[1] A local red earth.

[2] In all the methods which use only local labour and no machinery bought from abroad, or from the industrial sector of Kenya, some investment occurs in the sense of delay between the input of the labour and the final output. However, in so far as the scarce investment resources consist of imported equipment or equipment from the industrial sector, this investment does not involve the costs associated with capital-intensive techniques requiring purchased equipment.

above that of the hand machine (see table 48). The very low labour cost (the prisoners were paid 10-12 cents a day) was passed on in the block price, which as a result was also very low.

Murram-enforced blocks

Murram-enforced blocks are made from murram, cement and sand. On the Nairobi City Council estate they were made with a hand-operated block-maker, which cost 1,800 shillings when new. Four men could produce 300 blocks a day. The chief difference between this operation and that of hand-operated cement block manufacture was the use of murram instead of stones. This led to a considerable cost saving—a typical block cost 40 cents instead of 1.25 shillings—but the quality was substantially lower. The crushing strength of the murram blocks was found to be 150-220 lb/sq. in. on one testing, compared with the official requirement of 400 lb/sq. in. The labour requirements for building were slightly above those for concrete blocks because of the uneven quality of the blocks.

Black cotton bricks

Black cotton is the type of soil found in many areas where the richer murram is absent. It is mixed with water and put in moulds. After drying and removal from the moulds the bricks are stacked and dried, using a primitive oven consisting of a hole in the gound, as in traditional charcoal burning. Two people can produce 250 bricks a day. (The bricks are about half the size of a normal cement block.) Black cotton bricks are not as strong as murram but are cheaper in the areas where murram is not available.

Stone

The possibility of using stone depends on the availability of nearby quarries since stone is too heavy to transport far. The use of stone is highly labour intensive. It is estimated that one man can quarry six blocks of stone in one (intensive) hour. They then require a good deal of skilled or semi-skilled labour for building because of their uneven edges. The resulting building is of high quality.

Timber

Building with timber tends to be fairly skill-intensive, though little capital equipment is required for either production or building. In an experimental construction project [1], Nairobi City Council employed nearly as many carpenters (15) as unskilled labourers (20) on the site, and considerably more carpenters (8-10) than labourers (about 4) in the workshop where the panels were prepared.

A breakdown of the costs of a single-storey timber house (total costs per house EA£800-850) showed that labour costs varied between 30 and 47 per cent of total costs, material costs varied between 47 and 53 per cent and plant cost between 4 and 11 per cent. The proportion of labour costs was higher than that for the construction sector as a whole in 1962 [2]; labour costs accounted for 29.2 per cent of gross output, material inputs for 51.3 per cent and operating surplus for 19.5 per cent. Three different low-cost housing projects around Nairobi gave labour costs as 30 per cent of total costs, with materials accounting for between 50 and 60 per cent. [3] These three schemes used cement blocks or reinforced *murram* blocks. The relatively high proportion of labour costs in the timber project reflects high skill levels and wage rates rather than greater employment generation. Moreover, a greater proportion of labour employment in preparation of materials has been included, as compared with the other figures.

Pre-cast concrete panels

There are various methods of forming pre-cast concrete panels. Broadly speaking, their manufacture is similar to that of the manufacture of cement blocks, and involves mixing cement and other materials, vibrating to strengthen and then leaving to dry. The labour intensity of the method varies according to the machinery used, which tends to vary with scale. One pre-cast production unit visited had the following characteristics:

Cost of mixer	1,500 shillings
Cost of air vibrator	30,000 shillings
Cost of moulds	no estimate
Output	14 panels a day (8 ft. × 3 ft.)
	$\simeq$ 300 blocks per day
Cost of wire for reinforcement	5 shillings per panel
Employment	5 unskilled labourers

---

[1] *Timber housing pilot project at Kariobangi, Nairobi*, report by R. S. Ryatt, Resident Engineer (mimeographed, 1972).

[2] Figures from D. Turin: *Notes on housing and construction*, paper prepared for the ILO-organised comprehensive employment strategy mission to Kenya, 1972, para. 5.5.

[3] Ibid., para. 5.8.

Output-labour ratio (block equivalent)	60 blocks per man-day
Investment-labour ratio ex mixer and ex moulds	6,000 shillings per worker
Investment-output ratio ex mixer and ex moulds	100 shillings per block
Investment-labour ratio including mixer and allowing 2,000 shillings for moulds	6,700 shillings per worker
Investment-output ratio including mixer and ex moulds	112 shillings per block

Mixture used: 1 part cement, 2 of sand and 4 of ballast.

If this one example of a pre-cast production unit is compared with the earlier figures for cement block manufacture, it appears that the ratio of investment cost to output and investment cost per worker is higher in the case of the pre-cast unit than in any examples of block manufacture. The rate of labour productivity, moreover, appears to be somewhat lower than that of most of the block-makers. However, this might be offset by improved labour productivity in construction, since the pre-cast units avoid much of the labour involved in laying blocks. However, the general impression given was that such labour saving in construction was slight because the panels were so heavy that a considerable amount of manpower was needed to lift and place them in position. Hence the pre-cast technology observed would appear to be inferior to the block-making technology, being more capital-intensive with little if any consequent labour saving. [1] This conclusion was supported by the fact that demand for the existing units was extremely low, and the factory was producing them only intermittently.

Pre-cast concrete panels made with a foaming agent

The major problem with pre-cast concrete panels is their weight. This problem has been partially overcome by a new technology involving the addition of a foaming agent to the mixture, which by creating bubbles in the panel improves insulation and makes it lighter. One production unit using such an

---

[1] This is also the conclusion to be drawn from W. P. Strassmann's figures which show a fall in the labour share in construction costs, a rise in materials costs and a rise in absolute costs per square metre, which can of course be offset by reduced land area and costs, as the weight of buildings rises and a switch is made from blocks to pre-cast panels. See W. P. Strassmann: *Mass production of dwellings in Colombia: A case study* (Geneva, ILO, 1974; mimeographed).

agent was observed. It was pre-casting the panels on site for the building of 93 houses, and had the following characteristics:

Cost of mixer from the United Kingdom	29,000 shillings
Cost of motorised carrier	18,000 shillings
Cost of mould (100 at 500 shillings each)	50,000 shillings
Output	50 panels (of 2.3 square metres a day) $\simeq 1,000$ blocks a day
Employment	30 workers (of whom 2 skilled)
Output-labour ratio (block equivalent)	33.3 blocks per man-day
Investment-labour ratio ex moulds and ex carrier	967 shillings per worker
Investment-output ratio ex moulds and ex carrier	29 shillings per block
Investment-labour ratio including moulds and carrier	3,233 shillings per worker
Investment-output ratio including moulds and carrier	97 shillings per block

Mixture: 1 part cement and 3 of sand and foaming agent.

Labour productivity in the manufacture of these pre-cast panels appears to be somewhat lower than in the making of cement blocks. Investment per worker is higher than in the case of some of the cement-block-making techniques. Much depends on how many moulds are included, what allowance is made for other equipment such as metal pipes, and whether costs of the mixer are included. Investment costs per unit of output tend to be higher than in the case of the block-makers for cement. In general, it seems that this type of production of panels is somewhat more capital-intensive than block-making. The cost of the foaming agent, owned and licensed by a United Kingdom company, must also be added. It is in construction, however, that the chief differences arise. The use of these pre-cast panels yields a considerable saving, which the manager of this unit estimated at 80 per cent, in labour. It also requires more capital equipment in construction—particularly cranes. Consequently, over all, the use of the panels involves greater capital intensity and less labour use than concrete blocks. Although it would require much more detailed research to establish precise magnitudes, this conclusion emphasises the validity of a general principle in discussion of choice of techniques, namely that looking at one stage of a chain production process can be highly misleading.

Cement blocks made with a chemical additive

Recent technological developments include the addition of a foaming agent to the mixture for the manufacture of cement blocks, making them lighter and better insulated. The operation requires careful supervision by scientifically qualified personnel, and is for large-scale production. The Ministry of Works received proposals for the establishment of a factory making these blocks from a multinational enterprise. The estimated capital cost of the proposed factory was EA £1 million, and the estimated labour force 150-200 workers and supervisors—a capital cost per workplace of between EA £5,000 and EA £7,000 or between EA £100,000 and EA £140,000 all told—over ten times the capital cost per worker of any of the block-makers. Among those employed 19 were to be managerial, including a number of chemists. Total block production was estimated at 400 cubic metres per year, on a two-shift basis, which is enough to build 7,000 houses. Since the total number of houses built in Kenya in 1970-71 was 7,000, the factory would displace most of the existing block-makers. There is an estimated 10 per cent saving in construction labour because of the even quality of the blocks. The very high quality of the blocks—even, light and strong, meeting all international specifications—was stressed. The material requirements are 5-10 per cent cement, 70 per cent lime and the balance sand. Particular specifications of lime and sand are required, so that the materials would need to be either transported over considerable distances or imported; to meet the requirements locally might well involve mechanisation of the production of materials to ensure the uniform quality required. The proposed factory, which had not (in 1972) been accepted, illustrates many aspects of choice of technique. First, the relationship of choice of technique to product specifications; given sufficiently detailed product specifications to meet international requirements [1], the proposed factory might seem the only possibility available, despite the capital intensity of the technique. Secondly, technological developments in the advanced countries threaten to overtake older and more labour-intensive techniques. While in fact the proposal would probably have involved a higher-cost block, with greater productivity associated with further technical advances, it might have produced cheaper blocks than the block-makers for cement, and consequently have displaced 2,000 or so workers in block manufacture and as many more in construction. Thirdly, indivisibilities in modern techniques are such that marginal additions to productive capacity, which may meet the additional demand without creating excess capacity, are often not possible.

---

[1] In block making, transport costs effectively prevent international trade and hence remove the need to compete as a reason for meeting international requirements; but in traded goods this becomes an important consideration. See F. Stewart: "Trade and technology", in P. P. Streeten (ed.): *Trade strategies for development* (London, Macmillan, 1973).

CONCLUSIONS

If a choice of building materials is included in choice of techniques, the choice is widened well beyond that which appears if only one building material is taken into consideration. Once a range of products is considered, two aspects of choice of technique that are easy to ignore in looking at a single material become of key importance: one is the question of product quality, and hence of consumer requirements, which are closely related to levels of income distribution; the other is that discussed above, namely the need to look at the productive process as a chain, in which choice of technique at one stage helps to determine choice of technique at other stages. In relation to building materials the following steps in the chain need detailed examination:

(1) Production (excavation and processing) of raw materials.
(2) Transport of raw materials to next stage of processing.
(3) Processing of raw materials—i.e. block manufacture, preparation of timber panels, etc.
(4) Transport to the building site.
(5) Building techniques.

The use of a particular material may appear labour-intensive at one stage, but impose different requirements at subsequent stages. Investigation of technical choice at each stage is needed.

These points are illustrated by a comparison of the characteristics of different material technological requirements, as shown in table 58. The table sheds some light on the question, which has exercised a number of people [1], whether there is any necessary association between the labour intensity of a technique and product characteristics, or in other words whether products intended for the lower income groups are necessarily more labour-intensive. Common sense suggests that there need be no such association, but the table suggests that there is in fact quite a close association in this particular case: the higher-income products are associated with more capital-intensive techniques, with the exception of stone production, which is a high-income product produced in a labour-intensive way. Some such association between product characteristics and labour intensity of techniques is likely in practice for historical reasons. The bulk of the population in traditional societies tend to be low-income, and to use labour-intensive techniques because they lack the knowledge and money required for the use of capital-intensive techniques. Hence, traditional methods

---

[1] The association was discussed by F. Stewart in "Choice of technique in developing countries", op. cit.; and questioned by D. Morawetz: "Employment implications of industrialisation in developing countries: A survey", op. cit., and G. K. Helleiner: *Technology and the international firm* (mimeographed, 1974).

Table 58. Choice of building materials in relation to quality of building, consumer income, source of basic material and labour intensity

Material	Quality of building	Income class [1] of consumer	Source of basic material [2]		Labour intensity [3]		
			Place	Economic sector	Material production	Processing	Building
Mud and wattle	low	subsistence	local	informal	high	high	high
Sun-dried clay blocks	low to medium	slightly above subsistence (rural)	local	informal	high	high	high
Murram-enforced blocks	medium	low (urban)	local	informal / formal [4]	high	high-medium	high
Black cotton bricks	high	mainly high	local	informal	high	high	high
Stones	high		local	informal / formal	high	high	high
Timber	medium	medium and high	local	informal / formal	.[5]	.[5]	.[6]
Pre-cast concrete panels	medium	medium and high	local	formal	medium [7]	medium [7]	medium
Pre-cast concrete panels made with a foaming agent	medium	medium and high	local and imported	formal	medium [7]	medium [7]	low to medium
Cement blocks made with a chemical additive	high	medium and high	local and imported	formal	low	low	low to medium
Cement blocks	medium	medium and high	local	formal	medium [7]	medium [7]	medium

[1] Low: rural subsistence and most small-scale market production, urban unemployed, under-employed—including most informal-sector activities. Medium: unskilled and semi-skilled formal-sector (mostly urban) workers. High: skilled and professional. The use of this classification means that the majority of the population should be classified as low-income. [2] This refers both to geographical location (local or imported) and to sector of the local economy, which is regarded as being divided into an informal and a formal sector. Broadly, the informal sector is more labour-using, while formal-sector production involves more equipment, often imported. [3] To classify production methods according to labour intensity implies that each material may be identified with a single technique; as the survey of cement block making showed, that is not the case, and considerable variation may be possible. [4] Cement. [5] Intensive use of skilled labour and natural resources. [6] Cement. [6] Intensive use of skilled labour and some use of machinery. [7] Varies (see footnote 3).

—mud and wattle, sun-dried blocks, etc.—tend both to be labour-intensive and to be directed to production for a low-income market. In contrast, technology recently developed in the advanced countries tends to be suitable for their income levels (i.e. high-income consumers by comparison with those in developing countries) and their factor availability (i.e. suited to factory methods, skill-intensive and capital-intensive). This is illustrated in the case of the pre-cast units with foaming additives. Techniques such as concrete block manufacture, which originated some time ago in developed countries, tend to be intermediate (by modern standards) in terms both of product characteristics and of technology used, reflecting the intermediate factor availability and income levels of advanced countries some decades ago, when the techniques were developed. But of course, there is not a necessary association. Thus craft production, as often suitable for high-income consumers, and modern technology can be harnessed to the production of low-income goods.

This brief look at different building materials has thus illustrated some of the complex relationships between choice of product and choice of technique and between income levels and distribution. For high-quality construction meeting international specifications and demanded by high-income consumers, there are broadly two possibilities, namely the craft-intensive high-quality construction in timber and stone or the mechanised methods of manufacturing pre-cast concrete panels or blocks. Low-income consumers, on the other hand, provide the market for hand-operated cement block manufacture and the various other blocks and bricks formed with primitive technology from local materials. Thus choice of product and technique cannot be divorced from income levels among consumers. Choice of technique also influences income distribution: the mechanised large-scale techniques tend to be associated with relatively high wages, and with profits of medium-scale to large-scale entrepreneurs. In contrast, the low-income techniques are often used outside the formal sector, and generate low incomes among those they employ.

CAPACITY OF THE ENGINEERING INDUSTRY IN COLOMBIA

9

by the Institute of Technological Research [1], Bogotá

The need for specially designed equipment appropriate for use in the developing countries with factor endowments differing from those of the developed ones has been voiced frequently at both national and international levels. [2] An indigenous capacity for machine fabrication and some degree of development of the engineering industry as a whole are virtually essential if labour-intensive technical change is to be generated in developing countries. No attempt to increase labour intensity is likely to be successful unless it is accompanied by technical change in the labour-intensive direction. This is important, since most of the technical progress that takes place in the developed countries is inevitably capital-using and tends to dictate the nature of technical change in the developing countries.

The development of a local capital goods sector is required, first, in order to supply light machines for the small-scale "informal" sector as well as for the modern sector. Secondly, it facilitates the introduction of new machine designs and of labour-intensive technologies and processes that are more suited to the local factor endowments of the developing countries. In the absence of industries producing capital goods, the range of technical choice will tend to remain narrow, guided only by the history of technological progress in the developed countries. [3]

---

[1] Instituto de Investigaciones Technológicas (IIT). This chapter is a summary by the editor of a much fuller report which the Institute has prepared for the ILO under contract.

[2] See United Nations: *World Plan of Action for the Application of Science and Technology to Development* (New York, 1971), pp. 18-19. More recently, various reports of the United Nations Advisory Committee on the Application of Science and Technology to Development (ACAST) have made this point. In particular see *Report of the Ad Hoc Working Group on Appropriate Technology*, Advisory Committee on the Application of Science and Technology to Development, 20th Session, Geneva, 21 Oct. to 1 Nov. 1974 (United Nations document E/AC.52/XX/CRP.1, 6 Aug. 1974; mimeographed), and Annex II: "Draft statement by ACAST to the Second General Conference of UNIDO".

[3] Howard Pack and Michael Todaro: "Technological transfer, labour absorption and economic development", in *Oxford Economic Papers*, Vol. 21, No. 3, Nov. 1969.

In addition, and contrary to what might at first glance be suspected, the experience of several countries, such as the United States and Japan, suggests that the manufacture of machinery is one of the more labour-intensive branches of industrial production and can be carried on efficiently on a small scale.

HISTORY OF THE CAPITAL GOODS INDUSTRY IN COLOMBIA

It is useful first to consider the evolution of the capital goods industry in Colombia in the course of the industrialisation of the economy.

The first undertakings of the metalworking industry were concerned with the maintenance of imported machinery and equipment, including motor cars and railway equipment; this led to an expansion of metalworking shops. With the installation of chemical, petroleum refining and steel plants some of these workshops developed into manufacturers of equipment.

By 1956 import substitution was almost complete with regard to products for which substitution was easy. [1] In order to ensure industrial expansion at any cost through import substitution, regardless of the suitability of the industry [2], the Government proceeded to reinforce protective tariffs. Unfortunately, however, the domestic market was not big enough, especially for machinery and equipment produced within the country, to allow production on an economic scale. Either industries established to produce capital goods were inefficient, or they did not obtain the support needed to look for markets abroad. For these reasons the indigenous design and planning of equipment did not develop properly; foreign equipment was often copied poorly, with technical imperfections.

Some industries resorted to the purchase of technology to avoid the cost and risk involved in the development and design of equipment. The experience of these industries demonstrates that not all imported technology was suitable for Colombia: differences in the availabilities of raw materials and labour skills meant that the imported designs and technologies could not be applied without modification, especially in view of the much smaller volume of production. In other words, to acquire technology it was necessary to develop engineering skills as well as physical production capacity within the country.

On the other hand the ease with which foreign credit could be obtained for imports of capital goods contributed to a reduction in the demand for locally produced equipment.

---

[1] República de Colombia, Departamento Nacional de Planeación: *Plan de Desarrollo. Industria* (1971), Part 3, p. 20.

[2] S. Macario: "Protectionism and industrialization in Latin America", in *Economic Bulletin for Latin America* (New York, United Nations), Vol. IX, No. 1, Mar. 1964.

As a result of the unfavourable effect of this general situation on the balance of payments as well as on employment, an export promotion policy began to be favoured in the early 1960s, and capital goods industries were encouraged to seek foreign markets for their products. International competition made it necessary to improve the designs and the quality of products offered; this requirement called for more attention in organising engineering departments, especially in the larger firms. This process is still continuing, and was stimulated further by the organisation of the Andean Common Market in 1960 and by the over-all industrial planning of the participating countries, which began in the metalworking sector, especially in relation to capital goods.

Furthermore, in the early 1960s certain organisations such as the Institute of Technological Research (IIT) and the Institute of Agriculture and Animal Husbandry—Instituto Colombiano Agropecuario (ICA)—began to realise the necessity of developing equipment and implements to meet the particular needs of Colombia and to make the best use of available resources without causing social dislocation and unemployment. It was this recognition that led to designs for the subterranean potato storage silos (completed in 1966), and the development of agricultural implements. However, up to the present great difficulty has been encountered in the adoption of these designs and equipment in commerce and agriculture, for a variety of reasons which are explained below in the case study examples of equipment designs.

RECENT EQUIPMENT DESIGNS

Agricultural equipment

Agricultural mechanisation through the use of power-driven machinery and implements is doubtless one of the most important factors in bringing improved technology to agriculture and the basic method of increasing production and productivity in cultivated areas. In Colombia the process of agricultural mechanisation began during the first few years after the Second World War, when machinery was imported on a large scale. Often this machinery was found unsuitable for Colombian conditions. In some cases it also created serious social problems: for example, cotton-picking machines each displaced about 600 man-days, and in the region where they were introduced this caused redundancy.

The present Colombian agricultural equipment and implement industry is composed of various private enterprises. The industry produces tractor-drawn implements such as cultivators, ploughs, rakes, weed cutters, trailers, planters and spreaders. The raw materials used are mainly domestic, with imported raw materials constituting only about 6 per cent of the cost of the products. Some firms have engineering staff who work on the development of new products or on design modifications, to reduce costs or to adapt the equipment produced to markets and production processes. However, most of the designs in production have been acquired from foreign firms with some local adaptations.

About 39 per cent of the Colombian population lives in rural areas. However, the use and marketing of the implements made by Colombian industry has been limited to medium-sized and large farms on flat or gently sloping land, whereas the rural population lives for the most part on the hillsides, and it is in those areas that "mini-farms" tend to be concentrated; hence the great interest shown in the development of simpler agricultural implements suitable for small farms on hillsides. The ICA has launched an agricultural machinery programme for the manufacture of suitable low-cost agricultural equipment. Since 1969 attempts have been made to design a small two-wheel tractor, an agricultural grader, and animal-drawn ploughs.

The first mini-tractor prototype, which was made by a commercial firm, had a 7 h.p. engine, rubber tyres, adjustable plough attachments and variable speed transmission. Testing in the field led to the substitution of a 10 h.p. engine, and of metallic tyres for rubber ones to reduce slipping. This two-wheel tractor was found more suitable than the large four-wheel tractor, especially for small farmers, as a substitute for animal traction or manual work wherever appropriate. Similarly, a farm grader designed and field-tested by the ICA has many advantages over an imported leveller: first, it reduces power requirements by 40 per cent, which is important, especially in the light of the current energy crisis and shortage of fuel in many developing countries; secondly, simple attachments permit its use with any type of agricultural tractor; and thirdly, it costs only a quarter as much as imported levellers.

Unfortunately there is a lack of communication between manufacturers of agricultural equipment and institutions devoted to research, design and development connected with such equipment. Of all the agricultural implements designed and manufactured by the ICA, none is now being produced on a commercial scale. This is due to various reasons worth analysing for the development of future programmes of required designs: in the first place, there has been too little continuity in the testing and development of prototypes to obtain a totally satisfactory design; secondly, the work carried out by the ICA in designing and developing agricultural machinery has not been widely publicised; and thirdly, there has been no promotional campaign among farmers and manufacturers, so that the advantages and benefits of locally designed tools and equipment are not known to them. This explains the lack of interest in the manufacture of potentially viable prototypes and their eventual application by the end users.

Methods of commercialising prototypes require special study. The existence of extension services to farmers, and the availability of sufficient credit to allow them to acquire implements without too much outlay, are two essential pre-requisites for the creation of a market for the products of the engineering industry. Co-operatives and trade unions of farmers cultivating similar agricultural products might also help to overcome problems relating to the introduction of machinery for the cultivation of crops. The advantages of such organisations might be multiplied if they were closely linked to universities and other research institutions within the country.

Potato silo

A potato silo has also been developed by the IIT for specific use under Colombian conditions. Potatoes are of vital importance in Colombia: production in 1972 reached 800,000 tons. A silo adapted to national conditions was needed because in the past the prices of potatoes followed a pronounced cycle owing to various factors such as the lack of effective organisation for planning Colombian agriculture, severe variations in climatic conditions and the lack of adequate systems for regulating the marketing for agricultural produce. Sometimes the prices reached did not even cover production costs, and in consequence production in subsequent periods was not enough to meet the demand. The IIT accordingly undertook the development of a silo in which potatoes could be kept for longer periods, up to seven months, so as to allow the market to be regulated.

On examining the problem it was found that a high proportion of potatoes is grown on widely dispersed small farms at a distance from electrical distribution networks, without mechanical power and without workers capable of maintaining electromechanical equipment.

Broadly, two technological alternatives were evident, namely—

(a) installations based on advanced technology (ground-level construction, thermal insulation of synthetic material, forced air circulation, mechanised conveyors and automated control of storage conditions); and

(b) installations based on alternative technology (subterranean construction, thermal insulation based on materials available in the country, manual handling and manual control of storage conditions).

The second alternative was selected. In the subsequent development work it was thought that use could be made of the natural environmental conditions (relatively low temperature and high moisture during the night) in order to avoid cooling or ventilation systems which consume energy. Meteorological data showed that in the plain of Bogotá the temperature and relative humidity would enable the contents of the silo to be cooled by natural ventilation. This region presented the additional advantage of being near the main potato-consuming centre. The most adequate type of silo appeared to be an underground silo which would facilitate thermal insulation. In its construction cheap materials which farmers could afford were used, namely asbestos-cement tiles, hay, polyethylene, wood, earth, cement and sand. The dome of the silo was insulated with tar and earth.

With the first prototype (capacity 12 tons) it was proved that by opening the doors at night and closing them early in the day it was possible to maintain temperature and relative moisture preserving the potato in good condition for 10 months. These results were carried over into a commercial silo with a design capacity of 100 tons.

When a comparison was made between 50 100-ton silos of this type with commercial storage in more developed countries (a 5,000-ton warehouse equipped with forced air ventilation and conveyor handling), it was found that the Colombian silos were less capital-intensive: fixed investment per ton stored amounted to 428.40 pesos in the Colombian silos and 620.10 pesos in the warehouse, the corresponding figures for storage and management costs being 182.10 and 228.10 pesos respectively.

Later the IIT participated in the installation of 128 silos, each of 100-ton capacity, for a government agency that began operations in 1967. Besides the silos, the plant has equipment to clean, select and handle potatoes.

Initially it was envisaged that 54 wage earners and 10 salaried employees could operate this plant, but in practice many difficulties arose, related mainly to the techniques of handling the potatoes. For these and other reasons the plant did not operate for two years, but afterwards it began to operate on an intermittent basis, and recently with the current surplus production of potatoes it has been necessary to use it at full capacity. The personnel has increased to 120 people.

Direct employment generated in the construction of a silo amounts to 290 man-days. According to data on the volume of production 200,000 tons must be stored annually in years of higher production to damp down the price cycle. At present there are 130 silos where 13,000 tons can be stored, which is only 6.5 per cent of the total needed. Consequently, at least 1,870 additional silos should be constructed in the various producing areas; at first sight it would appear preferable that silos should be widely distributed over producing regions, rather than in large complexes, so as to reduce the cost of transport and of the labour required for the cleaning, classification and sorting of potatoes. However, silo complexes operated by government agencies for storage facilitate massive intervention on the potato market and the rediscounting of the working capital of farmers.

The construction of the silos required would generate about 542,300 mandays of temporary work. However, the benefit obtained with silos as a price-regulating instrument is much more important because it is advantageous to both farmers and consumers. Once the proposed number of silos have been built, about 3,000 jobs would be generated. In addition, the support for a definite sowing programme would provide security for farmers, and possible losses would be reduced by approximately 400 million pesos.

Portable sawmill

For forestry operations a portable sawmill was designed by a foreign engineer for use in the tropical forests which cover a large proportion of Colombia's land surface. The portable sawmill has some advantages over the more traditional stationary sawmill, because the poor roads and transport facilities render the movement of felled round wood difficult. Basically the sawmill consists of an air-cooled Volkswagen four-cylinder engine driving two circular saws that cut simultaneously in the horizontal and vertical planes. The sawmill can be erected and operated by four men. It is now being manufactured domestically, 40 per cent of the output being exported. Table 59 indicates some basic features of the economics of the portable sawmill as compared with those of stationary models.

This table reveals that the choice of technology as between the portable sawmill and the stationary mills is by no means a clear one. With a given amount of investment resources it appears that more employment can be created with stationary sawmills (except the very large ones) than with the portable type. On the other hand, the table does not include the cost savings associated with portability, especially for the transport of the raw material (round wood); and the portable sawmill does not require investment in land and buildings. These cost savings may make the cost per unit of volume less for the portable sawmill, at least in some regions. Clearly, the major advantage of portability is that it permits the processing of timber in areas too remote for the transport of round wood to be contemplated. In these circumstances, the technology of the portable sawmill would seem to be both appropriate and employment-generating.

Woodworking equipment

In the woodworking industry the larger firms are still using imported machinery, and this situation cannot easily be altered because the domestic market is too small to support the domestic production of this machinery. The small and medium-sized firms are, however, supported by well established enterprises in Colombia which manufacture an extensive variety of wood-

Table 59. Economic data on types of sawmill commonly used in Colombia

Item	Type of sawmill					
	Portable[1]	Stationary				
		Very small [2]	Small [2]	Medium size [2]	Large	Very large
Production capacity per sawmill (m³/year)	400	425	1 400	2 500	4 200 [3]	13 000 [3]
Average investment [4] in machinery per sawmill	120.0	140.0	204.0	366.0	468.0	4 017.0
Total investment [4] per sawmill	120.0	225.0	260.0	443.0	605.0	4 545.0
Number of workers per sawmill	4	7	11	19	23	83
Average investment [4] per worker	30.0	32.1	23.6	23.3	26.3	54.8

[1] One 8-hour shift, 250 days a year. [2] One 8-hour shift, 190 days a year. [3] In two-shift working it is estimated that the capacity of the second shift is 70 per cent of that of the first. The capacity with such a schedule rises to 7,150 and 22,100 m³/year respectively. [4] Thousands of pesos.

working equipment that can be produced much more cheaply than the imported product. Moreover, the pieces of equipment produced locally can also be maintained and serviced with local spare parts, so reducing running costs.

In view of the employment situation in woodworking it is interesting to note that between 1958 and 1970 production increased by 65 per cent whereas employment increased by only 30 per cent (in the industrial group "woodworking other than furniture"). This may well be a good example of productivity-augmenting appropriate technology, in which the efficiency of a traditional industry is increased by simple machinery so as to render the industry competitive with its very modern counterpart. A similar development occurred in the complementary industrial group, "wood furniture and accessories", in which production increased by 40 per cent between 1958 and 1969 although the level of employment remained stationary. In this industrial group there are reported to be 400 traditional enterprises and 7 modern plants.

Food processing equipment

The food products industry is largely composed of small firms. Most of its equipment is domestically produced, being essentially copied from equipment imported in the past, but with a less elaborate finish and with modifications to adapt it to local needs and to permit the use of domestic raw materials. The domestic equipment is less costly by a very substantial margin, owing to import taxes and transport charges that raise the price of imported goods and owing to the use of second-hand spare parts in the fabrication of domestic equipment.

The fact that such industry groups as dairy products and preserved food and vegetables account for a very large proportion of private foreign investment may partly account for the lack of adaptation in machine fabrication for these industries by comparison with others financed largely by domestic capital.

Petrochemical equipment

The story of the chemical industry in Colombia is encouraging to other developing countries. This industry comprises advanced technology plants such as petroleum refineries, but also related sub-sectors such as fertilisers, paint and varnish. Domestic production of petrochemical equipment can save transport and assembly costs in comparison with imported machines because of the large size of tanks, boilers and other such vessels and the lower cost of locally available labour. Initially, the lack of skilled personnel was a major problem, but this was gradually overcome by grants for training abroad and training within the framework of the National Apprenticeship Service (SENA). Although at first the technicians thus trained served mainly the chemical engineering industry, their activities are increasingly becoming directly and indirectly useful to other sectors of the economy. The progress of the chemical industry itself is evidenced by the fact that in the decade 1970–80 a total investment of 1,200 million pesos is to be made, of which 300 million will be spent on domestically produced equipment.

As is to be expected, foreign technology has played an important role in the development of the engineering capacity of these industries. In that initial technology Colombian engineers have found a basis for elaborating their own designs, adapted to local needs and conditions. However, equipment designed to date has generally been based on engineering standards previously established and used internationally; the chemical processes are also more or less the same as those used in other countries. Nevertheless, local manufacture of this equipment has served to create a domestic demand for steel and iron products and for quality steel, which has promoted techniques for the development of the steel and iron industry in the country.

It is expected that considerable advances will take place in the coming years in the field of chemical process design. The IIT, for example, has begun working in this field with the design of a process to defluorinate phosphate rock.

Building materials

Adaptation and innovation have occurred in the building materials industry, partly in response to a tremendous boom in house building in recent years, in the course of which the prices of both bricks and cement blocks rose considerably.

The construction of machines for the production of cement blocks has been successful, and they are even being exported. Their price is lower than that of similar but more automated machines. Machines of different degrees of automation are offered for the manufacture of cement blocks and are finding a ready market since they are well adapted to the needs of different users; exports have begun to neighbouring countries. Machines for moulding clay bricks produced in the country have not found a good market; they do not produce bricks of optimum quality because they lack the vacuum-processing features needed to increase the brick density. The market for brick-making machines was apparently limited; consequently, their designs have not been developed sufficiently to compete with imported equipment. Even if heavy import duties were imposed it is believed that the domestic demand for these machines would be insufficient because only a few new brick factories are opened each year.

Another machine of an innovatory nature, for making soil-cement blocks, was promoted by the Inter-American Housing Center. Although largely successful in a technical sense it has not penetrated the materials market as deeply as might have been expected owing to the traditional preferences for mud plaster, wood, bamboo and other housing construction methods. This is indeed surprising because soil-cement blocks cost little more than one-fifth of the cost of ceramic or cement blocks.

CONCLUSIONS

Foreign and domestic technology

An examination and review of the capacity of the engineering industry in Colombia and the existing machine fabrication facilities in the country shows that much of the equipment and many of the designs are imported from abroad. There are no doubt minor modifications and adaptations in addition to copying of imported designs: for example, it was noticed that the investment requirement in the food products and woodworking industries has probably been lowered by modification of imported designs. Machinery for making blocks for construction work has also been copied with some modification to take into account the desired scale of production and the raw materials available domestically. The cement block machines perform well and provide a base for small production units requiring little capital; on the other hand, in many regions the machines have not been used because of consumer preferences for more traditional materials. The design of equipment in the chemical industry, although based on imported technology, represents a more advanced level of modification of existing technology, the objective being essentially to save foreign exchange rather than to create additional employment directly.

Several examples of innovation have also been noticed. First, the identification of the need and justification for farm machinery and implements has provided experience of some value for the production of indigenous designs. In the initial phases of the programme the body concerned with the project was markedly influenced by technical assistance from abroad; in that first phase the identification and evaluation of the projects lacked the objectivity needed to avoid failure. In the second phase, greater importance was attached to the evaluation of projects and testing of prototypes. During that phase of the programme attention was concentrated on equipment that would raise the productivity and incomes of small farmers in the hills where the farm population is more dense. Unfortunately, a sufficient outlay of funds to develop a more thorough campaign which would carry the demand for the designs to a level sufficient for commercial production was not contemplated.

The second example of innovation is the design of the portable sawmill. This arose out of the observations and experience of the designer in tropical forests (such as those found in Colombia), without any prior scientific investigation, so that this was a case of a response to a particular situation which led to the commercial manufacture of new machinery.

Notwithstanding these examples of modifications of imported designs and of innovations appropriate to Colombian conditions, the development of the engineering industry has been considerably limited by the lack of adequately trained manpower, inadequate planning of technological innovations and field testing of prototypes, and lack of adequate links between the producers of equipment and its potential users.

Training of engineers and technicians

To promote and improve the capacity of the engineering industry it is necessary in the first place to provide special training for Colombian engineers and technicians in particular aspects of machine building. The training and qualification of engineers does not end with graduation or on the completion of postgraduate work. Such graduates are qualified to learn the intricacies only of the particular field of specialisation they have chosen. However, the requirements of industry are manifold, and each often involves a multiplicity of kinds of engineering from metallurgical to chemical, as well as quality and cost controls.

The Government can subsidise engineer training within industry by supporting graduates for a year or two in plants where the equipment they design can actually be built and tested. If part of the cost were borne by a government subsidy, manufacturers would have an incentive to train such staff.

It is also possible to provide engineering training by financing trips for skilled personnel to observe production systems and existing designs abroad. Knowledge of the way in which things are done abroad has proved to be very effective in generating positive ideas, to an extent that depends on the preparation and motivation of the technical personnel who make the visits, as well as on the nature and content of their previous training.

It is also advisable to have the active participation of the country's engineers under contract in projects concerning the design and construction of industrial plants and equipment by foreign firms, because in this way national engineering capacity can be effectively increased. Such participation was achieved in the construction of equipment for the petrochemical industry. Governments should therefore require that the nation's own engineers be enabled to participate with those of foreign contractors on industrial projects, especially during the design stages and in all tasks that fall within the field of activity of local engineering firms. It would be generally advisable to avoid "turn-key" contracts by which the supplier commits himself to turn over a complete plant in operation, supplying and erecting all that is required. In certain cases contracts should provide for the employment of the country's engineers in key positions, starting from the design stage. In any case, whenever equipment can be domestically produced at a favourable cost it should not be imported. If the State is the main buyer, it should give preference to domestic manufacturers wherever feasible.

Organisation of research and development

It is necessary to carry out sectoral and, above all, sub-sectoral studies in order to discover what technology is being used and what machinery and equipment would be needed to improve productivity. Without such studies, which are of fundamental importance for the design of equipment, it is unlikely that the engineering and design work would fulfil its function in the development of a country. The studies should cover economic conditions and education for the potential users of recommended equipment; they should assess the social and economic benefits to be derived from its use, and should consider the possible adaptation of the equipment to the local environment. As a preliminary step it is also necessary to facilitate the gathering of data on equipment and process designs and blueprints through industrial information centres. Such information is essential to avoid the copying of obsolete machinery or equipment which would retard technological development, to facilitate design work on ideas which could be useful, and to assist in the selection of the most appropriate manufacturing technologies according to production volume and market size.

Once the need for certain equipment is identified, the costs and benefits of its manufacture need to be carefully evaluated. Often this becomes difficult because of uncertainty about mechanical characteristics and associated risks in estimating market potential. Financial assistance might be provided for feasibility studies of new machinery and equipment. Such assistance could be given, for example, through supervised loans, the reimbursement of which would depend on the achievement of positive results in a project, or on its favourable evaluation.

After feasibility studies have been carried out, it is necessary to decide what institution or industry should be responsible for manufacturing prototypes and undertaking pilot tests. Since such an undertaking can be expensive in terms of both time and money, positive government support is highly desirable at this stage too.

Subsequently, once the prototypes are constructed and their usefulness proved, it is almost always necessary to promote their use and manufacture on an industrial scale, especially when dealing with original designs to meet the specific needs of each industry. This promotional effort may be even more expensive than the design and production themselves. Without such promotion, the engineering and design efforts of specialised institutions make no sense.

Basic and applied research at universities and specialised centres and the theses of students in such subjects as mechanical engineering are of great value for increasing engineering capacity. However, without due co-operation from manufacturers and investors these studies will be of no practical effect. To achieve the necessary co-ordination the industrialist must be convinced of the benefit he can obtain from co-operation with universities and technology centres, and those institutions must be organised in such a way that they can provide the technical support required by industry. A good starting point for co-operation between universities, technology centres and the technical personnel of industry would be a quality control service that the institutions may be able to provide.

TECHNOLOGICAL CHOICE IN METALWORKING, WITH SPECIAL REFERENCE TO MEXICO [1]

10

by G. K. Boon [2]

Current concern about the unemployment problem in developing countries has generated renewed interest in the employment implications of production techniques. The theory of production in economics normally presupposes that a choice of technique exists. In economic terms such a choice takes the form of various combinations of capital and labour to produce a well defined output in a specified quantity. The theory also allows for situations where no choice of technique exists, while economies of scale in production, implying that factor input requirements are increasing less than proportionately to output increases, are also distinguished.

Although the theory allows for this diversity of production situations, the question of factor substitution has long been, and for some economists still is, a controversial one. The reason for this controversy is that economists have until recently done very little micro-research on the question of factor substitution. Much more attention has been devoted to empirical research, at various levels of aggregation, on the theory of consumption. Presumably the fundamental explanation for the neglect of micro-research in production theory is that, unlike consumption behaviour, the investigation of production behaviour touches on another discipline besides economics, namely engineering. [3]

---

[1] This chapter summarises parts of the author's forthcoming larger study to be entitled *Optimal technology behaviour: Possibilities and limitations*.

[2] The author is a senior member of the research staff and Professor at the Centro de Estudios Económicos y Demográficos, Colegio de México. He is grateful to A. S. Bhalla for editing a fuller earlier version of this chapter and reducing it to its present form.

[3] Empirical work on choice of technique is not a completely recent development. In the 1950s such work was started at the Netherlands Economic Institute in Rotterdam. The author undertook a number of micro-empirical case studies on woodworking, metalworking and earth-moving operations, which are included in his earlier book: see G. K. Boon: *Economic choice of human and physical factors in production* (Amsterdam, North-Holland

(footnote concluded overleaf)

A TASK-LEVEL ANALYSIS

In the present chapter, technological choice is discussed on the basis of a task-level analysis, i.e. an analysis of the efficiency of performance (in terms of handling and machining times) of different types of machines that can be used for certain basic operations. The degree of disaggregation is unusual, and the reader may wonder whether conclusions reached at this level of disaggregation are meaningful and whether these results can be generalised at higher levels of aggregation. A task analysis about choice among alternative technologies does not provide the same information about substitution possibilities as a product or plant analysis or an analysis at a higher level of aggregation. The information which an analysis at the task level provides is in fact unique and has an independent validity, so that it is not a substitute for but is complementary to results obtained by more aggregated analyses.

By conducting an analysis on a man-machine level, one obtains a clear insight at this level, which at any higher level of aggregation remains concealed. This is a fact of importance for a number of reasons. From a theoretical point of view, it is of interest to follow more closely the process by which the optimum technology is determined. By varying not only economic parameters but also key physical parameters, the choice problem can be seen in an additional dimension. Secondly, decisions on specialisation or geographical location of production are increasingly made on the basis of process analysis, and not on the basis of final product analysis or partial or sub-assembly analyses.

Technological choice at the micro-economic level requires the highest degree of disaggregation. Each decision maker at that level will translate man-machine alternatives into a product-specific choice, weighing-machine-balancing considerations and specific production implications. If the analysis is conducted at a higher level of aggregation, for example for a whole group of machines or for some hypothetical machine plants by alternative technologies and alternative capacities, the results are less meaningful to the decision maker at the micro-economic level, unless the analysis relates to his particular product.

The universality of the results obtained provides a solid basis for generalisations, of importance not only to decision makers at the micro-economic level but also to economic planners and others interested in technological behaviour in its theoretical as well as in its practical implications. One may wonder how it is possible to arrive at wide-ranging generalisations on the basis of such a

Publishing Co., 1964); idem: *Optimal capital-intensity in metal-chipping processes*, Progress Reports 1 and 2, Oct. 1965 and May 1966 (Stanford University, Institute of Engineering-Economic Systems); and idem: "Factor intensities in Mexico with special reference to manufacturing", in H. C. Bos (ed.): *Towards balanced international growth* (Amsterdam, North-Holland Publishing Co., 1969).

disaggregated analysis; yet the fact is that results of aggregated analyses have their policy implications not only at the level of aggregation conducted but often also at higher aggregation levels, although not at levels of aggregation lower than the one used.

THE DOS METHODOLOGY

The method of analysis used in this chapter is a further extension of the method used in an earlier work [1], namely that of break-even cost analysis, which works with total or unit cost functions per unit of time. This type of analysis is familiar to engineers or to managerial economists, whereas professional economists usually work with isocost curves, which apply only to homogeneous quantities. Break-even analysis explicitly introduces the time element, and therefore makes it easier to analyse output as a function of time as well as of primary and intermediate inputs. Moreover, it facilitates the introduction of product differentiation and indivisibilities in equipment. These features of break-even cost analysis enable allowance to be made for machine set-up or machine preparation times, and for lot sizes or production runs. If some of the additional aspects mentioned, such as indivisibilities and lot sizes, are taken into account, the isoquants from the empirical results turn out to be quite different from those suggested by theory.

The methodology used in this case is known as the DOS methodology, and consists of three elements, namely—

(a) a decomposition procedure (D);

(b) an analysis of optimality (O) using a minimum cost criterion; and

(c) a sensitivity procedure (S).

Actually in this particular case no break-even points between total cost functions are derived. Unit costs are calculated for certain economic parameters such as wage and interest rates, lot sizes and utilisation levels of equipment expressed in hours. The technology with the lowest unit cost is regarded as being the optimal one for a fixed and given task.

The utilisation of equipment is a variable. One can make the analysis for one-shift operation (2,000 hours of machine running time) or for two-shift operation (4,000 hours). It is customary to make calculations on the basis of machine utilisation expressed in production hours and not in physical output quantities. The reason is that the production tasks to be carried out show wide variations in physical characteristics, in quality and in type of raw materials: for example, it is impossible to say whether the capacity of a drilling machine is

---

[1] G. K. Boon: *Economic choice of human and physical factors in production*, op. cit.

50,000 holes a year, because unit machine time depends on the type of metal, its thickness, the diameter of the hole and so on. In contrast to flow process industries, for which annual capacities of installations can usually be expressed in physical terms, in discrete process industries this is normally not possible. Unit costs are not compared for different physical quantities but for identical numbers of machine-working hours per technology. This fact is not a disadvantage but rather a point in favour of the methodology, since this is the way in which decisions on technology are made in practice. The physical capacities of different technologies for given machine utilisation hours vary widely, because they are a function of the lot size and the latter influences total unit production times. By dividing these annual utilisation levels of equipment expressed in time units (hours) by the unit production times, also expressed in hours, the physical capacity is obtained.

In the analysis the lot size, like the wage and interest rates, is an exogenous variable: in other words it is assumed that the decision maker on technological choice has no influence over these parameters. Although the model of completely free competition is not necessarily the basis of the present analysis, it is implicit that the price of the product and the production quantity are given, and that the prices of the primary inputs are also given. The lot sizes are dictated by market conditions, but the enterprises may have some possibility of influencing their market share mainly by competing in quality and fashion and, therefore, indirectly may have some choice in relation to lot size. Such a market situation is rather common in developing countries where governments interfere with the market mechanism.

By taking as a base a given number of machine-working hours, the essential aspect of the choice of technology—that is the difference in the degree of mechanisation and possible automation among alternative technologies—is taken into account. An increasing degree of mechanisation usually implies that the unit production time declines and the annual capacity of the machine increases: the productivity of both labour and capital increases. However, highly mechanised and productive equipment becomes mechanically more complex and therefore the time to prepare such equipment for operation increases; whereas a simple machine may require only one to three hours to set up, no less than three days may be needed to prepare complex machinery for a production lot. This fact clearly has an influence on the lot size; economical production is very unlikely if three days are needed to prepare a machine for work while the lot size time is only one hour. Consequently the economical application of highly productive equipment tends to call for large lot sizes. [1]

---

[1] This issue is discussed in depth, with numerical examples, in G. K. Boon: *Economic choice of human and physical factors in production*, op. cit., pp. 20-23.

This factor of lot size is highly important in the choice of discrete technology, and the methodology fully covers this aspect of the choice. The choice among various capital intensities for discrete process technology (or in other words the choice among various degrees of mechanisation) is determined at least as much by the size of production lots as by the prices of capital and labour. In order to bring lot size into the analysis one does not have to start by determining the number of hours of utilisation of different machines, which implies varying physical outputs; one can also fix a physical output target which itself implies a varying number of hours of utilisation according to the particular technology. However, the difference between the two starting points is not a matter of principle; one may start from either of them since the methodology is not affected. Taking a fixed number of hours of utilisation as a starting point makes the analysis slightly easier from a computational point of view, because in that case only one unit of machinery is considered for each alternative. In the case of a fixed output target, more than one unit of machinery may have to enter into the calculations for one of the alternatives. Instead of full linearity one has stepwise linearity in the functions, except in the case of full utilisation, for which one may assume linear homogeneous functions. The fact remains, however, that in practice decisions on this type of technology are made on the basis of the number of hours of utilisation.

The sensitivity aspect of the analysis is introduced by applying a number of discrete variations in the main parameters. Two classes of parameters are distinguished—economic parameters and physical parameters. Parameters in the first group include lot sizes and wage and interest rates. The latter two are used as a set together, and are differentiated according to level of economic development: a high wage rate is associated with a relatively low interest rate and a low wage rate with a high interest rate. A total of seven lot size values are distinguished, and four capital and labour price sets, which can be roughly identified with highly industrialised, industrialised, semi-industrialised and under-industrialised countries. Other economic parameters considered are prices of factory floor space, utilisation levels of equipment, prices of new equipment and the allowance to be made for the proportion of nominal working time that is not actually worked, or not efficiently worked. This last economic parameter varies with the type of technology: it is higher for less mechanised equipment than for the more mechanised; it also reflects climatological and working conditions in general. Principally, the allowance reflects time for rest and personal care, and therefore the time by which the number of productive hours of work is reduced, as is a worker's annual output. By doubling these allowances, which are based on working conditions in the United States, more unfavourable climatological and general working conditions are simulated.

However, one can also regard this allowance as reflecting skill differences among individual operatives, since less well trained workers can achieve only a certain fraction of the output of a fully experienced worker in the same productive time. It is clearly important to analyse the effect on the choice of technology of a reduced amount of efficient working time per worker per year, for whatever reason.

In order to acquire a deeper insight into what determines the optimality of a technology, one needs to know not only how optimality is influenced by the variation of the parameters but also how each parameter independently influences optimality. This is necessary because our purpose is not just to establish the optimal technology given certain numerical values of the parameters but to establish the influence of each parameter on technological optimality separately. For this purpose the methodology is decomposed to allow the influence of the variation of each parameter on a technology to be separated. Essentially, the method followed in that respect is one of comparative statics: by changing a parameter from one discrete value to another and keeping all other parameters constant, a change in the optimality of technology can be explained by the variation of a particular parameter.

INITIAL ANALYSIS OF MACHINE-TOOL SUBSTITUTION POSSIBILITIES

The samples of engineering data to which the DOS methodology is applied consist of handling and machining times for carrying out certain production tasks on different models of machine. The machines correspond to different levels of mechanisation, and therefore to different investment requirements and different inputs of labour and capital. The sample of metalworking machines covers virtually the whole range of basic metal-chipping and cutting operations. These data were established by Markowitz and Rowe in the mid-1950s for the Rand Corporation. [1] Their economic and technological relevance to current conditions were checked before they were processed for the present analysis, and they were found sufficiently up to date, except that the latest and most sophisticated technology—for example numerically controlled machine tools and electrical erosion technology—is not covered.

The data used in the basic analysis cover 88 metalworking tasks. A task is an elementary machining operation defined, for a particular type of metal, in terms of precision and the shape and size of the workpiece. The production of most metal parts involves a combination of tasks. It may be possible with

---

[1] A. J. Rowe and H. Markowitz: *An analysis of machine tool substitution possibilities*, RM 1512 DDC Document No. AD 87449 (Rand Corporation, 1955).

specialised machinery to carry out a number of essentially different tasks simultaneously. Such equipment can be specialised by process or by product. The present analysis covers equipment of the former type but not equipment for the manufacture of special products because that would have reduced the generality of the information considerably. The data are said to be " process-orientated", not "product-orientated". This implies that analysing alternative technologies for a particular product would require a different problem formulation: in such a formulation, product-specific technology might enter, but it is not included in the sample. Especially in car and motor cycle production, product-specific equipment is used in connection with such parts as engines and gearboxes and the like. However, such equipment is often specific not only to the manufacture of a particular product but also to a particular make and firm, and is either built to order or made by the company manufacturing the product itself. However, for the majority of metal operations an analysis based on technological processes is sufficient. Plant managers are usually quite knowledgeable about the allowances to be made for the peculiarities of their own production, and are more interested in trends and tendencies with regard to alternative technologies than in a special analysis of their particular line of production, which would require a product analysis.

In the sample of 88 tasks it was possible to distinguish between two groups of tasks, namely the one in which tasks are suitable for machine optimality analysis since technical choice existed, and the other consisting of tasks for which no technical alternatives are available. The first group consists of 51 tasks, plus 4 special tasks with numbers 52 to 55. The second group contains the remaining 33 tasks. The application of the DOS methodology is confined to the first group of 55 tasks, which imply 55 case studies. In each of these case studies, seven discrete variations are introduced in the lot size, as well as four in capital and labour prices, two in factory floor space prices, two in the utilisation level of machines, two in the prices of machine tools and two in the allowances made for the proportion of nominal working time that is not actually or efficiently worked.

Suppose we have four factor price sets, I, II, III and IV, two lot size values Z_1 and Z_2 and two tasks which have the same shape and size classification, but task 1 has a lower precision requirement (say 1) than task 2 which has a tolerance classified as 2. The optimality results are illustrated in table 60. For example, for task 1, we observe for given factor price sets I, II and III, a shift to higher capital intensity. This change in optimality is caused exclusively by increase in the lot size from Z_1 to Z_2. By comparing the optimality positions for a given lot size and varying factor price sets, that is, within each block, influence of the variation in factor price sets on machine optimality is established. Comparing the optimal positions for given lot sizes and given factor prices

Table 60. Illustration of optimality of technology choice (X markings indicate which machine is optimal)

Production task no.	Shape	Size	Precision	Model of machine	Z_1 I	II	III	IV	Z_2 I	II	III	IV
1				A	–	–	X	X	–	–	–	X
	1	1	1	B	–	X	–	–	–	–	–	–
				C	X	–	–	–	–	X	–	X
				D	–	–	–	–	X	–	–	–
2				A	–	–	–	X	–	–	–	–
	1	1	2	B	–	–	X	–	–	–	–	X
				C	–	X	–	–	–	–	X	–
				D	X	–	–	–	X	X	–	–

[1] A, B, C, D in ascending value of capital-labour ratio.

(that is, in terms of table 60, a vertical comparison or a comparison between different tasks which differ only in one physical characteristic, e.g. the precision requirement), it is seen that a change in optimality is caused by a change in the quality output specification.

The optimality results are further processed by means of a sensitivity ranking. In this ranking, in the first instance only the factor price variable and the lot size variable are explicitly distinguished. Five sensitivity classes are established, the criterion of sensitivity being the number of different optimality positions that occur each time the value of a variable is changed. The choice of a criterion of sensitivity is always more or less arbitrary, and another criterion such as the number of changes in optimality positions could have been chosen equally well, but a different criterion would not lead to different conclusions.

The 51 tasks in the sample for which there appear to be substitution possibilities have been classified into four categories on the basis of the criterion of the influence of lot size and factor price on the machine optimality (table 61). That table and more detailed task-level studies [1] indicate that the variations in lot size and factor prices can allow the tasks in the sample to be classified in the following four sub-groups:

(a) for a number of tasks no choice of technique exists;

(b) the machine optimality in one group of tasks appears to be predominantly sensitive to variations in the size of the lots;

---

[1] For further details see G. K. Boon: *Optimal technology behaviour: Possibilities and limitations*, op. cit., Ch. 3.

Table 61.   Breakdown of tasks according to the influence of lot size and factor price on
optimality

Categories of variables	Number of tasks	
	Non-adjusted [1] classification	Adjusted [1] classification
I Lot size	11	14
II Factor price	9	4
III Both lot size and factor price	24	24
IV Neither of the two	7	9

[1] The adjustments are made not only for biases in the data but also for minor inconsistencies in the basic data. These adjustments were made in consultation with engineering experts.

(c) for another group of tasks, machine optimality is predominantly influenced by variations in the factor prices; and

(d) in the biggest group optimality is influenced by variations in both lot size and factor prices.

A somewhat different type of empirical analysis was also undertaken using the same data and methodology. The objective of this analysis was to establish sensitivity to earlier results on optimality if relative factor prices and lot sizes were kept constant but if the following additional variables were changed one by one:

(a) cost of factory floor space;

(b) price of equipment;

(c) existence of multiple shift working; and

(d) the allowance made for the proportion of nominal working time that is not actually or efficiently worked.

Taking cost of factory floor space first, it was noticed that a change of respectively $15 and 30 per square foot, keeping all other factors constant, causes little change in optimality. If the cost of space is increased by $15 per square foot while not varying other factors, 17 changes in optimality occur, the majority of them implying a shift to higher capital intensity of machinery. When the variation in the cost of space is doubled, the number of changes in optimality also doubles, to 35. Clearly, if costs of space increase, the machines that minimise the costs of machine and operator space per unit of output (by implication more automated machines) come closer to optimality. From this point of view, therefore, the concentration of industrial production in or very near urban centres tends to favour capital-intensive equipment.

What happens to machine optimality when the equipment price is doubled? If there is one man per machine, twice as much capital per machine operator is required, and that implies a shift in optimality in 154 cases; all these changes are shifts to less automated machines. The change in equipment price by 100 per cent gives an effect identical to that of a change in utilisation from 100 to 50 per cent, provided that the assumed lifetime of equipment is kept constant. Other things being equal, increased equipment prices due to such factors as rises in transport costs or import duties favour the choice of less automated machines, which are cheaper.

Thirdly, multiple-shift work can be particularly helpful in allocating capital and labour economically. If the lifetime of equipment is assumed to be given, in the present analysis at 10 years, if a second shift is assumed to reduce the capital requirement per unit of output as well as per worker to half, and if the number of machine-working hours is more than $2 \times 2,500 \times 10$, or 50,000, or in case of three shifts $3 \times 2,500 \times 10$, or 75,000 hours, then operating the equipment for 50,000 or 75,000 hours instead of 25,000 hours ensures a proportional saving in capital per worker and also per unit of output. However, these savings are achieved only if the capital equipment would indeed be abandoned after 10 years; if not, the conclusions obtained here do not hold. Suppose the technical lifetime of equipment is 150,000 machine-running hours before a major overhaul is needed. Assuming 2,500 hours per shift per year, and assuming further minor technical progress, the machine could serve 60 years on a one-shift basis, 30 years on a two-shift basis and 20 years on a three-shift basis. In this case no major savings in capital can be obtained by operating the machine on multiple shifts, but if the economic lifetime is assumed to be 20 years, multiple-shift production does yield savings in investment requirements per worker and per unit of output. [1]

In our analysis only two-shift operation is considered. This is often more realistic for most countries, where working a third shift reduces the workers' efficiency. It is further assumed that demand conditions justify two-shift operation more or less permanently. Given these assumptions, the results show that introducing a second shift makes the use of more capital-intensive machinery optimal; in fact, this is so in 138 cases. This result is understandable, given the assumptions, since 16-hour operation reduces capital requirements per worker by half.

In principle, any firm has an option to start two-shift operation right from the start. It may, therefore, choose less capital-intensive equipment, of lower capacity but operated on two shifts, rather than more capital-intensive equipment with a higher output capacity operated on one shift. However,

---

[1] In the case analysed in this study, a 10-year economic life is assumed, which in practice is extremely low for the technology under discussion. However, this case is put forward only as an interesting example from an analytical point of view. Also, in the version of the DOS methodology applied in the study summarised in this chapter, no discounting procedure is applied. It is further assumed that annual costs and benefits are constant over the lifetime of alternative techniques.

no such choice is normally made in practice. Unless it is necessary for techno-logical reasons to operate the equipment on a three-shift basis, as may happen with continuous-flow processes, equipment is normally chosen to operate on a one-shift basis, especially in production involving discrete processes such as metalworking. One of the reasons for this is that in general a new enterprise has complex organisational problems, particularly in developing countries; to organise production on a double-shift basis from the start is even more complex. Secondly, no management likes to plan capacity so narrowly that it buys lower-capacity equipment, which immediately has to be operated on double shifts; rather it plans initial capacity in the light of expected future demand patterns. Since it is difficult for any enterprise to estimate its future market share, in practice most enterprises tend to introduce second shifts for individual machines, instead of buying machines with larger capacities. By the time all machines are operated on double shifts, the machines with which double-shift working was started have to be operated on three shifts, or if this is legally or otherwise not possible they have to be disposed of and replaced by machines of larger capacity. Thus, in practice, double-shift working in discrete processes is introduced only gradually, and very seldom from the start—an outcome which is at odds with economic theory. This means that multiple-shift working is normally not a factor which plays a role at the time of making a decision on choice of technology.

The last exercise with our sample consists of a variation in the allowance made for the proportion of nominal working time that is not actually or efficiently worked. The basic time data are adjusted to make allowances for fatigue, personal care and work delay. These allowances vary with the task, the type of machine and the size of the lots. The allowances used in the basic sample refer to working conditions in the United States at the time when the sample was established. The present variation consists of doubling these allowances, introducing the consequence of differences in skill and general efficiency. This variation results in 112 changes in optimality; in practically all the cases this change makes more capital-intensive machinery optimal. This is because machines that are more automated depend less on the human factor and therefore maintain their efficiency level better under these conditions; and it is precisely for this reason that there is a tendency among entrepreneurs to favour higher levels of capital intensity.

CORRELATION WITH ANSWERS GIVEN BY MANAGERS IN A RECENT MEXICAN SURVEY

The accuracy of the results of the initial optimality analysis for the 88 metal-working tasks was also tested in actual production conditions in Mexico, through interviews with the managers of a number of factories. After the plant manager had been informed of the aims of the study, he was shown a small technical drawing of a task, with specifications of the dimensions and the material. Next, a number of machines were mentioned with an indication also of their technical capabilities, and the plant manager was asked which of those machines he would choose to carry out the task if the production lot size was either 5,100 or infinite. In cases where, among those mentioned, no machine of the plant manager's choice existed, he was requested to specify which other machine he would prefer. After this preliminary interview, the plant was visited to examine actual production. During this tour of the plant practical questions were raised, for example, about factor proportions, adaptation and other utilisation and capital-stretching policies. The question of capital and labour price variation was not raised; only the variation in lot sizes was mentioned in the interview. Implicitly the capital-labour price ratio plays a role in the optimal choice as expressed by the plant manager, since he is asked for his choice in Mexico, naturally for the prevailing market factor prices.

The findings on optimal machine choice so obtained were compared with the optimal choice which the analysis based on the Rowe-Markowitz data indicates for semi-industrialised countries, i.e. for a wage rate corresponding to $0.45 an hour and an annual interest rate of 10 per cent. In 1973 in Mexico a more relevant price set would be a labour cost, including all social security charges, of about $1.00, while the market interest rate would vary from 12 to 15 per cent. These values differ rather substantially from the above ones. Nevertheless, the testing of the sample is meaningful, especially considering that the shadow rate of semi-skilled labour, including social security charges, would be closer to $0.45 an hour anyway. The shadow rate for capital would be close to the actual market price, at about 15 per cent.

Optimality of machines does not only depend on the relative prices of capital and labour, and even for these variables optimality may change only for significant jumps in the price sets. The alternative machine choices introduced in the study mean discrete choices, often with wide discontinuities between their capital and labour requirements.

Seven factories were visited in and around the Federal District of Mexico. Not all these factories produce metal products exclusively. For this reason the breakdown of size class by number of workers in table 62 refers in each case to the metalworking section and not to the whole enterprise.

There were 55 tasks in respect of which the results of the basic analysis of machine-tool substitution possibilities were to be tested against the answers given by managers in the Mexican survey. In several cases, however, testing in respect of one task yielded an outcome valid for a number of related tasks. For

Table 62. Breakdown of factories in the Mexican sample by size class

Number of factories	Size class	Number of workers
2	I	$\leqslant 15$
4	II	15-75
1	III	$\geqslant 75$

tasks in which the shape of the workpiece is the same but its size and the tolerance both differ, the technological alternatives and the economically optimal machines may turn out to be the same. Therefore, although on the average the results relating to only 54 per cent of the tasks were tested, the outcome represents 92 per cent of all the listed machine choices, so that the coverage of the sample is almost complete.

Of the seven lot sizes on which the analysis is based, correlation was tested in respect of only three. These three, however, are representative of small, medium and large lots, that is the two extreme values and an intermediate value. The intermediate value poses a problem, because although plant managers may be certain about their choice for extreme values of the lot size, for an intermediate value they have to guess, since a definite answer can be given only after a precise calculation. Presumably a part of the non-conformity between the choice of the manager and the results of the basic optimality analysis can be attributed to this factor.

The results of the tests are summarised in table 63 according to both the non-adjusted and the adjusted classification of tasks. The seven enterprises visited are ordered according to the size of their metalworking plant measured by number of workers employed. The first observation on the results is that the adjusted classification gives a better correlation with the initial optimality analysis than the non-adjusted classification. This is understandable since the adjusted classification allows, somewhat more realistically than the original Rowe-Markowitz data, for the effect of large lot sizes, mainly the infinite value, on machine optimality. Secondly, the answers of the largest firm show the least agreement with the results of the initial optimality analysis: in this case the percentage of correlation is considerably below the average and the choice of technology conforms to that made in more industrialised countries like the United States and the countries of Western Europe. The two next largest firms also show a correspondingly low degree of correlation, although they do not show a strong tendency to choose a technology more advanced than that covered in the initial analysis, and consequently they show the highest degree of negative correlation.

Table 63. Correlation between the results of the initial optimality analysis and the choices of Mexican managers

Factory size class	Number of workers employed		Correlation (percentages)					
	In metal machining only	By the firm as a whole	Positive		Negative		Not comparable [1]	
			n.a.	a.	n.a.	a.	n.a.	a.
I	5	40	60	73	35	27	5	0
	6	63	58	65	27	27	15	8
II	40	140	64	69	28	28	8	3
	50	500	62	67	21	21	17	12
	58	68	56	61	39	39	5	0
	68	80	58	64	35	35	6	1
III	.	.	48	54	33	33	19	13
All sizes	.	.	58	65	31	30	11	5

n.a. = non-adjusted classification of tasks (see table 61).
a. = adjusted classification of tasks (see table 61).
[1] Because for the tasks in question the technology chosen by the Mexican managers was ahead of that considered in the Rowe-Markowitz survey.

One explanation for the above may be the difficulty for the plant manager of estimating the optimal machine for the intermediate lot size value. This difficulty may be greater for the plant managers of the smaller firms than for those of the larger firms, who are generally more experienced. Therefore we may presume that the negative correlation estimated for the three firms that have the largest metalworking section and also have the highest degree of negative correlation suggests a bias towards more capital-intensive technology. Negative correlation can be distinguished from cases in which the technology preferred by the Mexican manager lay outside the terms of the comparison; in the latter cases optimal technologies were singled out for different relative factor prices which prevailed in Western Europe and the United States, whereas in cases of negative correlation the preferred technology, although slightly less advanced, nevertheless had in general a higher capital intensity than the technology indicated by the initial analysis. The explanation might be that relative factor prices are somewhat different for the larger and the smaller firms respectively. Capital may be cheaper for the larger firms since they have easier access to bank credits, or, in the case of a subsidiary of a multinational firm, capital may be obtained from the parent company, while on the other hand larger firms have a better management and are more careful in their compliance with government regulations and legislation concerning labour. This attitude tends to make labour costs higher for them than for smaller firms, which do not always apply labour regulations.

Other differences in optimality as determined by the results of the interviews and those obtained by our earlier analysis might be explained by the following factors:

(a) the differences between actual market factor prices in Mexico and those used in the earlier analysis;

(b) ignorance on the part of plant managers about existing alternatives; and

(c) inexperience of the plant manager in the matter of machine choice, and prevalence of non-economic considerations such as:

(i) prestige value attached to the use of more sophisticated machinery;

(ii) non-economic personal preferences; for example, a plant manager may be biased towards technological sophistication, which he likes as an engineer;

(iii) a preference for higher mechanisation in order to reduce human and legal problems (labour laws are such in Mexico that labour can be dismissed only with great difficulty).

To obtain a better understanding of the cases where deviations occur, the sample test outcome obtained from the most experienced plant manager was analysed in greater depth. For the three lot sizes mentioned above, 26 tasks were tested, covering 144 machine choices. For 11 tasks, covering 48 machine choices, the plant manager expressed an optimality choice different from what the sample indicated. For these 11 tasks the main causes of difference in optimality choice were:

(a) in one case, a definite error, of a technological nature, in the plant manager's choice;

(b) in three cases an arbitrary difference (for example, the initial optimality analysis pointed to a vertical milling machine whereas the plant manager chose a horizontal one; the two machines have the same technical capabilities for the task concerned and their prices, when purchased new, are very close);

(c) in six cases the plant manager selected a machine that, without doubt, is the optimal one, whereas the majority of the sample attributed optimality to another machine which was theoretically a possible choice but could not be used in practice; and

(d) in one case the plant manager pointed to a machine that would no doubt be economically optimal but was not included in the sample.

Specific plant situations

It is worth while to report in more detail on the experience of an engine plant which belongs to a multinational enterprise. At the start of the interview the management stated that the Mexican plant used exactly the same technology as the parent plant, whereas during the interview it became clear that the technology used by the Mexican subsidiary departed significantly from that of the parent plant.

In the engine plant, no major differences could be observed in the special-purpose machinery used in Mexico as compared with that used in the parent plant. In certain cases, however, the machinery installed was second-hand, since it had previously been part of the parent plant, and, being technically completely operational, had been shipped to Mexico for further use. The new machines installed in place of the old ones in the parent plant were essentially the same, with minor changes in technical capability. This is a typical situation for metal-chipping machinery, in relation to which technical progress is slow; in other production processes, however, such a substitution may entail significant changes in the degree of mechanisation or automation in the parent plant, and the second-hand, less capital-intensive machines are transferred to a subsidiary plant in a developing country on purpose, because there the plant is optimal at the prevailing factor prices.

The reason for shifting used machinery from the mother plant to the subsidiary, even if the technology is essentially identical, might be that utilisation levels, and therefore wear and tear, in the subsidiary plant are lower, and that the used machinery can therefore serve longer at lower depreciation rates. Another reason might be great differences in import taxes on new and used equipment in developing countries; this does not seem to be the case for Mexico.

In the Mexican plant under discussion, only for one particular operation was the level of mechanisation lower than in the parent plant. Although the machine used was optimal for Mexican wage and interest rates, the plant manager nevertheless wanted to introduce the more automated machine for quality reasons.

In the Mexican engine plant no overhead conveyors were used, whereas such an installation existed in all the engine plants in the home country of the concern. Fork-lift trucks were used in both countries, while in Mexico an overhead crane was also available in some parts of the factory. Internal transport in the Mexican plant was therefore more labour-intensive. In addition, the materials handling in the production line was non-mechanised, whereas in the parent plant machine-powered transfer mechanisms were used. In Mexico tables with free-moving steel rollers were used between machines;

the worker pushed the heavy workpieces by hand over the rollers from one machine to another. The human energy required for the moving of the workpiece was relatively low.

In Mexico some machines in the engine plant were operated on a two-shift basis. Such machines were those which could not be used to produce the required volume in eight hours. Instead of disposing of these machines and substituting others of larger capacity, which usually means a higher level of capital intensity, they were operated on a second shift.

As already pointed out, the level of automation of the machine tools in the engine plant was basically the same in Mexico as in the home country. Use of equipment that was less mechanised would imply either doubling lines of equipment or introducing a second or third shift for all machines. In the first case, more factory floor space would have been required since more workers would have needed more working space. Floor space, generally, is expensive in Mexico. Although the introduction of a full second shift and a partial third shift would not have called for more factory floor space, it would have called for more skilled labour and more supervisory personnel. However, skilled and experienced labour is less abundant than one may think, which made recruitment for a second and third shift difficult. The firm therefore preferred the existing level of mechanisation in its main production processes, and carried out auxiliary processes in a more labour-intensive way. Quality was always mentioned as an additional argument for not going back to less capital-intensive techniques. On the other hand, these considerations did not play a role in internal transport, for which more labour-intensive alternatives were chosen; more labour-intensive methods were also applied in other auxiliary operations like packaging, handling and storing of spare parts.

The multinational firm visited seemed responsive to relative factor prices in choosing factor proportions, and adjustments were made whenever it could be done without affecting quality. However, in one respect the subsidiary firm might have been hampered in achieving optimal resource allocation. From the headquarters of international firms all kinds of control guidelines were issued, e.g. organisation schemes, and guidelines stipulating the lot sizes and capacity at which a higher degree of mechanisation is desirable. The local management did not always realise that such guidelines often do not fit Mexican conditions.

Machine adaptations

The machine adaptations that are most important in metal-chipping for Mexico consist in the transformation of universal machines into special-purpose machines or the adaptation of special-purpose machinery in such a way that it becomes somewhat more universal. Evidence of adaptations was found during the factory visits. In particular, interesting machine adaptations were noticed in one of the factories visited. They mainly involved the adaptation of a basic universal machine tool such as a common lathe into a special machine. The common lathe (bench lathe) was often bought second-hand and u ed as a stand. With a small additional investment a special-purpose machine was obtained which would have required a higher investment if bought new. A case in point was a common lathe transformed into a boring machine: a similar boring machine bought new would have cost 2.5 times more than the adapted lathe that was in fact used. This is a typical case of capital-saving innovation. [1] Another example is the conversion of a common bench lathe into a special production machine capable of carrying out several machining operations simultaneously. However, such examples are very rare.

Over-all results

The study of the Mexican metalworking industry suggests that Mexican entrepreneurs respond to local factor prices, as well as to lot size and product characteristics. Although there are very few cases of capital-saving machine adaptations, auxiliary operations are often undertaken with labour-intensive methods.

The test has enabled us to check the technological appropriateness of the various possible means of performing the tasks, and consequently also the extent to which technological progress has made certain machines obsolete. It may then be concluded that the sample of tasks used is still relevant for basic metal-chipping and that technological progress has had only a minor influence on the relevance of the sample. On the other hand, it has been discovered that in a small number of cases the sample lists machine options that are no longer realistic, or are otherwise doubtful. This can be explained by the fact that the original data were compiled by Rowe and Markowitz in the 1950s and had, presumably, been established from technical manuals on time data for machining operations; they were not based on actual plant observations.

---

[1] In the lathe, the workpiece rotates; in the boring machine the tool rotates. The lathe is adapted by mounting a shaft combined with the tool between the centre point and the claw; the workpiece is moved from left to right on the carriage.

CONCLUSIONS

The foregoing study has important implications for technological choice and employment generation. In the first place, the empirical disaggregated analyses on metalworking confirm the existence of capital-labour substitution possibilities even though at a higher level of aggregation the technologically determined view of fixed proportions may seem to prevail. On the basis of an analysis at the micro-economic level, it can be seen that the tasks that are not sensitive to economies of scale can be performed in decentralised units while tasks sensitive to capital-labour price variations can be performed in low-wage labour-surplus countries, an argument relevant to the international division of labour.

Some generalisations can also be ventured on the basis of empirical findings of our analysis of variations in economic and physical parameters on the choice of technology. Taking economic parameters first, an increase in the price of labour in relation to capital, and in that of factory floor space, tends to favour more capital-intensive operations. A high price of floor space in particular favours capital-intensive machines because they reduce the ratio of building costs to output. The variations in the physical parameters show that the availability of a choice of technique can in several cases be explained by the physical characteristics of the product. Generally speaking, when physical characteristics are very particularised, the choice of technology narrows down and the coefficients of production tend to become fixed: when shapes move from regular to irregular and complex, when sizes move from small to large and extremely large, when precision moves from normal to high, the range of possible technologies narrows down considerably. On the other hand, very common tasks, regular shapes, small and normal precision requirements are well suited to mass production. The demand for products such as screws, nuts and bolts and the like, for either final or intermediate use, is enormous. For such products a range of possible technologies does also exist, but special machines are developed which have such a high productivity that lot size tends to outweigh factor price considerations when the choice is to be made. Furthermore, for extremely common tasks only one machine becomes optimal.

PRODUCTIVITY AND EMPLOYMENT IN THE COPPER AND ALUMINIUM INDUSTRIES

11

by P. Della Valle [1]

This chapter examines the relationship between productivity and employment in the world's copper and aluminium industries, and analyses the reasons why labour productivity and utilisation vary among mineral producing developed and underdeveloped countries. In discussing employment generation, it is argued that the indirect employment aspects discussed earlier, in Chapter 3, may be more important in mining than the direct employment aspects, so that in order to appreciate fully the total employment associated with extractive industries, indirect employment in the regions in which the principal ore producing areas are located should also be taken into consideration.

In the light of the technology in use in the extractive industries (see Appendix IV), the capital-labour substitution possibilities in copper, bauxite and aluminium production in selected developed and developing countries are examined by using the CES production function analysis extensively reviewed in Chapter 2. Since a great proportion of mineral ores are produced by international corporations, an appendix is also devoted to the effect of foreign investment and technology in the developing countries.

DIRECT AND INDIRECT EMPLOYMENT

The importance of the mineral industries to developing countries can be seen from the following statistics: the United Nations Conference on Trade and Development (UNCTAD) has projected that by 1975, exports of minerals, metals and fuels will account for 50 per cent of the total exports from developing countries. This will represent a significant rise from the 1964 value of

---

[1] Associate Professor of Economics, University of Windsor, Ontario.

42 per cent. [1] On the other hand, it is estimated that the value of agricultural production will fall from 40 per cent of the total value of exports in 1964 to 37 per cent in 1975. From 1960 to 1968 the value of output grew approximately 64 per cent faster in the mining sector than in the manufacturing or construction sectors in developing countries. It is interesting to note that much of this growth has been occurring in Africa, where the mining industry had a later start and is now catching up. The African nations are typical of mineral-dependent less developed countries: copper accounts for over 40 per cent of the total exports of Zaire [2], and mining and quarrying for 50 per cent of the gross domestic product of Zambia.

It is easily established that as a country develops, the concentration of employment moves from the primary sector, including mining, to the manufacturing and service sectors. For example, the proportion of the labour force employed in mining amounts to less than 1 per cent in the United States, but to no less than 10 per cent in Zaire and 16 per cent in Zambia. It should be also noted that capital is increasingly being substituted for labour in the world's mining industries. For example, it has been stated that "labour requirements for bauxite mining have been greatly reduced in recent years by extensive use of modern mechanical equipment such as bulldozers, draglines, large trucks and in particular enormous rotary bucket excavators which have been in use for some years now in Surinam and Guyana". [3]

It might be argued, therefore, that the expansion of the mineral industries in developing countries will not significantly increase the long-run employment opportunities in those countries. This argument would be misleading for the following reasons. In the first place, any investment in a new mineral industry offers a developing country an opportunity for an expanded export base. Not only does the new industry provide more jobs and export earnings but the new taxes generated also enable the government to re-invest the monetary returns of the minerals sector in other parts of the economy. Secondly, in terms of total employment generated it would be erroneous to consider solely the employment involved in the extraction and processing of ore. The exploitation of a new ore body usually involves the establishment of a new town, or at any rate the constitution of some new zone of comparatively advanced economic activity. Mining towns are often in isolated areas of the country, and provide

---

[1] United Nations Conference on Trade and Development: *The longer-term prospects for commodity exports from developing countries*, doc. TD/9/Supp. 2 (Geneva, United Nations, 1967; mimeographed).

[2] During the period considered in this inquiry, the Republic of Zaire was known by various names: Congo (Léopoldville), Congo (Kinshasa) and the Democratic Republic of the Congo. Throughout this chapter the country is referred to as Zaire.

[3] United Nations Industrial Development Organisation: *Aluminium products from various ores*, doc. ID/WG. 11/10 (Vienna, 1967; mimeographed), p. 19.

ample new employment opportunities for all types of labour from construction workers to merchants and grocers' assistants. Thus by generating indirect employment the mining industry makes a significant contribution to the solution of the unemployment problem in less developed countries.

The dearth of statistical information makes it all but impossible to quantify the full indirect employment effect of developing the mineral industry. However, we can establish the patterns of the growth of this employment by observing some simple trends. For example, from 1963 to 1969 the population of Zambia rose by about 16.2 per cent but the population in the Copperbelt, which is the main mining area, increased by about 50 per cent. [1] Since the total direct employment in the mining industry grew by less than 10 per cent, we can only assume that, given the birth rate, indirect employment accounted for a large part of the increase in population in the area, i.e. there must have been an influx of traders and of skilled and semi-skilled workers who were attracted by the employment opportunities created by the copper mining industry. Indirect employment often occurs in independent communities which derive their livelihood from the extraction of ores elsewhere. In Chile's Gran Mineria, for example, one finds the main copper producing centre of Chuquicamata which has a population of approximately 25,000 people. A few miles away lies Calama, a town half the size of Chuquicamata. Calama's main source of income lies in the provision of goods and services for the larger copper town. Further examples of such indirect employment can be given for other mineral producing developing countries: for instance in Guyana's bauxite producing region the number of workers employed in bauxite production multiplied by the average family size in the region equals approximately 77 per cent of the local population, and it can be assumed that most of the remaining 23 per cent is indirectly supported by the bauxite mines.

PERFORMANCE AND LABOUR UTILISATION IN THE UNITED STATES AND SELECTED DEVELOPING COUNTRIES

The main purpose of the following empirical examinations is to analyse and compare the performances of the copper and aluminium industries in the United States and in the developing countries. Emphasis will be placed on the copper producing countries since the data for those countries is more homogeneous. However, comparisons will be made for the developed and less developed countries in each industry and between industries.

---

[1] *Monthly Digest of Statistics* (Lusaka, Central Statistical Office).

Copper

An examination of tables 64 and 65 will enable us to compare the levels and rates of change of the relative production variables in copper mining in Chile, the United States and Zambia. All variables are indexed on the United States data for 1960.[1] No comparable variables are available for Zaire but it was none the less felt that useful conclusions could be drawn from an examination of the rate of change of the Zairian variables. It must also be noted that the absolute values of the data used must be considered in the light of the slightly differing definitions of the relevant variables in each country. In this regard, it was therefore important to compare the time trends of those variables when significant trends did exist.

Since the periods under consideration differed somewhat between the countries studied, the relevant variables were examined in two steps. First, a comparison was made of the average of the actual observed variables in the original periods; and secondly, those variables were examined in the light of a common 1975 projected value. The projected values will give a common yet hypothetical comparative year which incorporates the rate of change of the variables for each country through time.

An examination of tables 64 and 65 indicates that the observed output level for Chile was 48 per cent that of the United States while the output of Zambia

Table 64. Variations in over-all output, employment, wages and output per head in Chile, Zambia and the United States

Item	Chile		Zambia		United States
	Absolute figures	Ratio to US data	Absolute figures	Ratio to US data	
Over-all output:					
Average observed value	45.8	.48	53.2	.55	96.4
Projected 1975 value	60.1	.47	77.5	.60	129.0
Employment:					
Average observed value	53.4	.51	201.2	1.93	104.5
Projected 1975 value	58.7	.	226.8	.	—[1]
Wages:					
Average observed value	30.1	.31	29.8	.31	96.8
Projected 1975 value	61.4	.39	52.6	.34	156.8
Output per head:					
Average observed value	.86	.92	.26	.28	.93
Projected 1975 value	1.04	.83	.35	.28	1.25

[1] Not significant.

[1] See Appendix I for the data used in the empirical tests.

Table 65. Average percentage changes in over-all output, employment, wages and output per head in Chile, Zaire, Zambia and the United States

Item	Chile	Zaire	Zambia	United States
Over-all output	2.2	4.6	3.7	2.2
Employment	0.7	— [1]	1.0	— [1]
Wages	7.2	6	6.1	4.0
Output per head	1.5	4.6	2.7	2.3

[1] Not significant.

was 55 per cent that of the United States. Since output in Zambia was rising at an annual rate of 3.7 per cent compared to one of 2.2 per cent for the United States and Chile, the projected value for Zambia output should rise to 60 per cent of the United States value by 1975, assuming of course that the observed trends continue. In addition, the expected rise of 4.6 per cent in the annual output rate for Zaire indicates that copper production in the main African producing nations will account for a larger percentage of the future world output of the metal.

The labour market statistics lead to the following observations. Employment is not changing significantly in the United States or Zaire and is increasing at a very small average rate in Chile and in Zambia. It must be recalled, however, that the effect of mining on employment in a country cannot be measured solely by the number of people actually involved in the production of ore: the earlier discussion on the subject indicated the importance of indirect employment as a result of the development of the mining sector.

Over the period studied, the average wage of the copper miners in both Chile and Zambia was about $30 a week, while the average United States copper miner received $97. Since the periods covered are not quite the same, it is most useful to compare this average wage to the projected 1975 levels. [1] If that is done, it can be seen that Chilean wages should rise to 39 per cent of the United States equivalent while Zambian miners' wages should increase to 34 per cent. It should also be noted that mining wages in Zaire were increasing by about 6 per cent per annum, which is comparable to the 6.1 per cent rise for Zambia, while being slightly less than Chile's 7.2 per cent increase. In all the developing countries, however, wages in the industry have been rising at a faster rate than in the United States. This may be explained in part by the same reasons as account for the fact that skilled-unskilled wage differentials have been falling over time in developed countries: as the mining industry in the developing

---

[1] It should be noted that the wage rates here do not include fringe benefits, which are more significant in the developing countries than in the United States.

countries evolves, the average skill level of the workers will increase, so that a rise in their wages in relation to those of their Northern American counterparts might be expected. In addition, a more rapid growth of trade unionism in developing countries may have accelerated the upward pressure on mining wages in those countries. [1]

In terms of productivity comparisons, we find the output per head figure for the United States to be greater than those of the developing countries. Chile's average output per head for the observed years was 92 per cent of the United States value and would seem to have been decreasing as a percentage of the United States figure. The relative labour productivity of the Zambian mines seemed to be remaining constant at approximately 28 per cent of the United States level. The considerably lower value for Zambia can be explained by the nature of the underground mining process in that country, which is less productive than open-pit mining. [2]

In terms of capital expansion, capital in the United States has been growing at the rate of 6.4 per cent per annum, while in Chile it has averaged only 3.5 per cent. The expansion in the United States seems to have been accomplished at the expense of labour, since the capital-labour ratio for the United States has been rising at 7.5 per cent while that for Chile was increasing by 2.9 per cent. A higher rate of substitution of capital for labour is not confined to developed countries, since capital has increased by 6.1 per cent in Zaire and the average rise in the capital-labour ratio has been a substantial 6.2 per cent in that country.

Aluminium (bauxite)

International comparisons of the main economic variables cannot be made as directly for the aluminium industry as for copper, the reason being that less developed countries do not produce substantial amounts of aluminium as such but produce bauxite or alumina, unlike developed countries such as the United States. Because of variations in the product, precise inter-country comparisons in terms of output per head and the capital-output ratio would be of limited value. It is still interesting, however, to examine some production variables if we

---

[1] For a discussion of organised labour in the minerals industries of developing countries, see Phillip E. Church: "Labour relations in mineral and petroleum resource development", in R. F. Mikesell (ed.): *Foreign investment in the petroleum and mineral industries* (Baltimore, Johns Hopkins Press, 1971), Ch. 4.

[2] Referring to the United States industry, McMahon states: "The largest increases in production and productivity rates have occurred in open-pit mines. Because these mines have contributed most to domestic production, they tend to lift the productivity average for the whole industry". A. D. McMahon: *Copper: A materials survey* (Washington, DC, United States Department of the Interior, Bureau of Mines, 1965), p. 302.

remember that the differences in those variables are often to be explained by differences in the final product. [1]

Since the United States industry is engaged in producing refined aluminium, which is a more capital-intensive process than the earlier stages of production, we find the capital-labour ratio to be higher for the United States than for the less developed countries. For example, the capital-labour index for the United States was 1.00 whereas for Guyana it was 0.23 during the period studied.

Weekly wages received by labour in the less developed countries were substantially lower than in the United States—$106.10 in the United States as compared with $31.10 for Jamaica and $25.90 for Guyana. In addition, the ratio of wages to product selling price was lower in the United States than in either Jamaica or Guyana (1.11, as against 15.0 and 13.3 respectively). The differences in the ratio can in part be explained by the differences in the final product sold by each country, i.e. an average index price of 96.2 for aluminium in the United States as compared with the lower-priced bauxite and alumina for Jamaica and Guyana (2.08 and 1.95 respectively).

In terms of labour utilisation the United States index was 110.0 while that for Jamaica stood at 31.9 and for Guyana at 21.1 The United States labour force was also growing at a faster rate, and that growth probably reflects an increased use of labour in the final production stage of aluminium goods such as cans and cables. Labour demand at the strictly mining end of the operation is not growing as fast as the labour used in the production of manufactured aluminium products.

ESTIMATION OF CES PRODUCTION FUNCTION

Chapter 2 in this volume discussed at length the conceptual and measurement aspects of CES production functions. Below we estimate this function with the data on copper and aluminium.

The CES production function yields the following reduced form equation in terms of output per head:

$$\frac{Q}{L} = m \left(\frac{W}{P}\right)^{\sigma} \qquad \ldots(1)$$

In our empirical tests for the copper and aluminium industries Q = tons of output produced, L = the number of workers employed, W = the weekly earnings in US dollars, P = the average price of the product sold, and m is a parameter. The second parameter, σ, stands for the elasticity of substitution between capital and labour.

---

[1] As was the case for the copper data, all variables are given in index form (United States data for 1960 = 100). See Appendix I for the description and tables of data used.

Table 66.   Regression estimates

Country	Equation nos.	σ	η	R^2
Zaire	1	.54		.63
		(.13)		
	2	.58	1.35 [1]	.74
		(.12)	(.68)	
United States	1	.83		.52
		(.19)		
	2	.77	.52	.70
		(.15)	(.16)	
Zambia	1	.52		.37
		(.18)		
	2	.08 [1]	1.31	.68
		(.18)	(.37)	
Chile	1	.20		.36
		(.08)		
	2	.15	1.01	.71
		(.06)	(.31)	

Key: σ = factor elasticity of substitution.
η = coefficient that measures the effect of capacity utilisation on output per head.
Standard errors are given in brackets.
[1] Not significant at 95 per cent level.

Copper

In table 66 we have the results of the estimates of equation (1) for each of our copper producing countries. It will be observed that for each of our developing countries the factor elasticity of substitution (σ) is significantly less than 1. For the United States, on the other hand, σ is much higher than the estimates for the developing countries, and, indeed, is not significantly different from unity. [1] The implication is that producers in the United States responded to relatively higher wages by substituting capital for labour at a faster rate than the producers in the developing countries. Since we are dealing with firms which are multinational, or at least are assumed to act like multinational firms, we might ask why they did not respond to relatively higher wages in the same way in the developed and underdeveloped countries. There are two possible answers to this question. The first possibility is that wages in the

[1] It should be stated here that the author had previously undertaken this test for the United States and Chile using different time periods and slightly different data. The results in that initial empirical work were somewhat different in the magnitudes of the estimates given for σ, but the ranking of the values of σ for the two countries was the same as that given here. P. Della Valle: "The elasticity of substitution and changes in capacity utilization in United States and Chilean copper mining", in *Applied Economics* (Oxford, Pergamon), Vol. 2, No. 4, 1970.

developing countries were quite low to start with, so that any increase in earnings did not represent a significant increase in cost to the companies involved. In such a case, relatively small increases in labour costs will not lead to substantial substitution of capital for labour. Secondly, we might consider the political and economic environment in those developing countries. Large, foreign-based corporations have always been suspiciously regarded by the governments and unions in developing countries, and any attempt rapidly to reduce the labour force has often been met with strong opposition. Such opposition could hinder the employment reducing response of mining companies to higher wages, and hence contribute to a lower elasticity of substitution of capital for labour.

In all cases the estimate given for σ is statistically significant. What one might say is that in both the developed and the underdeveloped countries, considerable wage changes lead to some substitution of capital for labour. In other words, in so far as the copper mining industry is concerned, market forces do not play an important role in the allocation of factor inputs in all countries. Two additional observations can be made on the average value of σ in the countries observed. The first is that σ is approximately the same for both of the copper producing countries in Africa. This would imply a relatively similar responsiveness of factor utilisation to factor cost in the African mines. Secondly, since $\sigma < 1$ for all the developing countries, we may say that labour's share of the value of production in those countries has been increasing through time. In fact, labour's share is rising at a faster rate in the copper mines of Chile, Zaire and Zambia than in those of the United States. Changes in relative shares through time depend both on the elasticity of substitution and on the nature of technological progress. Ferguson has pointed out that labour-using technological progress will increase labour's share, and offers some evidence that this has been the case for the United States economy since the Second World War. [1] It is therefore possible that technological progress has been more labour-using in the developing countries than in the United States industry.

Since the production function which yields equation (1) assumes the existence of full capacity, variations in capacity utilisation may present a serious problem. For example, suppose the world market demand for copper falls in any particular year. In such a case employers may decide to cut back on the use of capital and labour. The reported number of labourers used for the year would be lower than in the preceding year, but the reported amount of capital used may not be diminished. The company would simply report the amount of capital available in any particular year and not the amount of

---

[1] C. Ferguson: "Substitution, technological progress and returns to scale", in *American Economic Review, Papers and Proceedings,* Vol. LV, 1965, pp. 296-305.

capital used. The fall in output from one year to another would therefore increase the capital-output ratio as well as the ratio of capital to labour. The reported increase in the capital-labour ratio would not be a consequence of any change in the factor prices, and as a result σ would be erroneously estimated.

It was therefore decided to adjust equation (1) for variations in capacity utilisation. The testable hypothesis then becomes

$$\left(\frac{Q}{L}\right) = m \left(\frac{W}{P}\right)^{\sigma} \cdot (U)^{\eta} \qquad \dots(2)$$

where U = index of capacity, and η = coefficient that measures the effect of capacity utilisation [1] on output per head. Referring again to table 66 we can see the results of the tests for equation (2) in each of our countries. The introduction of capacity utilisation did not significantly affect the value of σ in Chile, Zaire or the United States. In Zambia, however, σ falls dramatically and indeed is not significantly different from zero.

It is most noticeable that η is much higher for the developing countries than for the United States: that is, capacity utilisation had a much greater effect on labour productivity in the developing countries. An explanation of the differing effects of capacity utilisation on productivity may lie in the concept of under-employment. Because of political and social pressures it may well be that there was an over-supply of labour in the mines in the developing countries. Hence, the labour force was at least partially under-employed or under-utilised. In such a situation, increases in market demand, which lead to an increase in capacity utilisation, would be met in part by increasing the utilisation of existing labour. Therefore output would increase while the number of workers remained much the same, and output per head would rise. Since we may safely assume that under-employment of labour was more prevalent in the developing countries than in the United States, we would expect η to be larger for the developing countries than in United States copper mining.

The under-employment hypothesis, together with a glance at table 67, may offer an explanation of why σ for Zambia falls to zero. We find that the average value of our index of utilisation was the same for both the United States and Zambia and was lower than the average U for Chile and Zaire. The coefficients of variation are significantly higher for the United States and Zambia than for the other two countries. It can be argued that the instability of capacity utilisation in Zambia, coupled with political and social pressures to

---

[1] The index of capacity utilisation which was constructed for each country was arrived at by comparing the deviations of the actual values of each year's output from some expected full capacity trend value of that output. The full capacity trend was established by the observation which represented the largest positive output deviation through time. See Appendix III for a detailed description of how the index of capacity utilisation was constructed.

Table 67. Average values and coefficients of variation of capacity utilisation

Country	Average values	Coefficients of variation [1]
Congo	94.8	4.4
Chile	92.7	5.0
Zambia	86.4	10.5
United States	86.4	12.8

[1] The coefficient of variation (CV) is defined as: $CV = S/\bar{u}$ (100) where $\bar{u}$ is the average value of capacity utilisation and S is its standard deviation.

preserve employment in the mines, led to the marked effect of U on Q/L in that country. These pressures would seem to have been great enough to offset the economically rational behaviour of substituting capital for labour as the factor-price ratio changes. The low productivity levels in Zambia can be accounted for by a combination of the above pressures as well as by the nature of the underground mining techniques. [1]

Since it is possible that technological change might influence Q/L in our time series observations, it was decided to introduce technological change into our tests in the following way:

$$\left(\frac{Q}{L}\right) = m \left(\frac{W}{P}\right)^{\sigma} \cdot (U)^{\eta} \cdot (e)^{\lambda t} \qquad \ldots(3)$$

where λ is the technological change parameter and t is the time variable. In only two cases, those of the United States and Zambia, did λ prove to be significant. Those equations (LN being Napierian logarithm) are given as:

$$LN\left(\frac{Q}{L}\right)_{US} = -.15 + .43\ LN\left(\frac{W}{P}\right) + .50\ LN(U) + .015t; \ R^2 = 0.88 \qquad \ldots(3a)$$
$$\underset{(.12)}{} \qquad \underset{(.11)}{} \qquad \underset{(.003)}{}$$

and

$$LN\left(\frac{Q}{L}\right)_{Zambia} = -1.33 + .03\ LN\left(\frac{W}{P}\right) + 1.07\ LN(U) + 0.021t; \ R^2 = 0.95 \ldots(3b)$$
$$\underset{(.08)}{} \qquad \underset{(.15)}{} \qquad \underset{(.003)}{}$$

It seems likely that technological change did not have a marked effect in the cases of Chile and Zaire because of the shortness of the periods studied for those countries. In equation (3a) we find that the value of σ for the United States has been substantially reduced by the introduction of technological change. The value of σ for Zambia in equation (3b) is not significant for reasons which were previously mentioned.

---

[1] For a discussion of the mining techniques for copper and aluminium, see Appendix IV.

The most interesting comparison given by equations (3a) and (3b) is the value of λ for the United States and Zambia. We find the rate of technological change to be greater in the Zambian mines. This is probably due to the more recent expansion of mining in Zambia, and hence the adaptation of the most modern techniques.

Aluminium (bauxite)

Empirical tests similar to those already described for copper were conducted on the aluminium industry of the United States and the bauxite industries of Jamaica and Guyana with the following interesting results.

For the United States aluminium industry equation (1) yields

$$LN\left(\frac{Q}{L}\right) = -0.32 + 1.08\ LN\left(\frac{W}{P}\right); R^2 = .88.$$
$$\phantom{LN\left(\frac{Q}{L}\right) = -0.32 + 1.08\ }{(.13)}$$

Equation (2) yields

$$LN\left(\frac{Q}{L}\right) = .009 + 1.08\ LN\left(\frac{W}{P}\right) + .47\ LN\ (U); R^2 = .93.$$
$$\phantom{LN\left(\frac{Q}{L}\right) = .009 + 1.08\ }{(.11)}\phantom{LN\left(\frac{W}{P}\right)}{(.21)}$$

From these results we can conclude that the elasticity of substitution was significant in the United States aluminium industry and did not differ significantly from unity, i.e. that changes in relative factor prices affected the substitution of capital for labour and that the change in the capital-labour ratio was roughly equal to the change in the ratio of wage rates to the selling price of the product.

When the foregoing tests were conducted on the bauxite industries of Jamaica and Guyana the elasticity of substitution was not significant. Therefore it would seem that the firms involved in the production of bauxite in those countries did not substitute capital for labour as wages rose with respect to profits. This result is not surprising if one compares the relative cost involved in bauxite production to the total cost of making aluminium. Girvan has estimated that bauxite labour costs for 1 ton of aluminium represent only J£1.58 out of total gross value of J£159.78. [1]

Therefore, it seems reasonable to assume that relative increases in labour costs would not significantly influence the substitution of capital for labour in that industry in the less developed countries. When we move into the final aluminium fabrication stage of production in the developed country, labour costs and σ play an important role in the choice of a production process.

---

[1] Labour costs at the earlier stages of production account for about 1 per cent of the total value added of aluminium production. See Norman Girvan: "The Caribbean bauxite industry", in *Studies in Regional Economic Integration*, Vol. 2, No. 4, p. 3.

Finally, capacity utilisation had a significant effect on labour productivity in Jamaica but not in Guyana. This may be due to the fact that Jamaican production is substantially larger than Guyanese production. Hence, Jamaican producers adjusted output and capacity to changes in market demand. Guyanese production, on the other hand, did not substantially influence the world market and was not as readily altered to demand fluctuations.

EMPLOYMENT EFFECT OF QUALITY OF ORE MINED

In the case of copper the types of technology applied in our examples are largely determined by the variations in the quality of the ore that is mined. It will be shown that this variation had a significant effect on the substitutability of capital for labour.

The mining firms used in the present comparison were chosen because they were both controlled by the same interest and have the following other similarities: they are both located in Zambia; the statistics reported for each are quite detailed and comparable; and they both operate underground mines, which are of approximately the same size in terms of blister copper produced in the period under study. We shall refer to the mines as Zambia Copper A and Zambia Copper B respectively.

Since both mines were quite similar in the above respects, we would expect their management to use similarly economical means of ore extraction and treatment in each case, i.e. to adopt the best possible mix of capital and labour which is technologically possible in each mine. In both cases the capital available for possible use must have been the same, as was the number of labourers, since labour was drawn from a common population. The main difference between mine A and mine B, however, was the quality of the ore extracted in each case. The copper content of the ore mined in A averaged only 2.14 per cent in the period studied, while mine B's copper content averaged 3.08 per cent. As a consequence of this fact, mine A had to process more ore in order to extract the same amount of copper output as mine B, and the techniques used to process the less concentrated ore were more capital-intensive.

The results of the varying methods of production in terms of the capital and labour used in each case can be examined by using Weitzman's technique to compare the isoquants derived from the CES production function which was defined earlier. [1]

---

[1] M. L. Weitzman: "Soviet postwar economic growth and capital-labor substitution", in *American Economic Review*, Vol. LX, No. 4, Sep. 1970. The production function used to derive the isoquants in figures 5 and 6 contains the parameter λ which represents Hicks-neutral disembodied exponential technical change. In addition, in order to reduce the

(footnote concluded overleaf)

Table 68. Detrended capital-output and labour-output ratios for Zambia Copper Mines A and B[1]

Observation	Mine A		Mine B	
	K^*	L^*	K^*	L^*
1	100.0	100.0	100.0	100.0
2	105.5	89.9	103.3	104.6
3	108.9	85.5	87.5	99.7
4	116.2	81.2	126.3	99.9
5	119.6	77.8	146.3	92.3
6	126.6	77.1	158.8	89.7
7	132.1	69.5	181.2	88.0
8	135.2	60.4	199.7	79.2
9	138.1	50.9	211.9	78.2
10	140.6	50.2	228.4	66.6
11	138.2	47.2	239.3	71.0
12	142.0	45.0	249.9	69.9

[1] See footnote 1, page 287, for a description of the techniques used to arrive at the values presented here.

An examination of table 68 shows that in both mine A and mine B capital was substituted for labour over time. However, when we estimate the elasticity of labour with respect to capital for the isoquants given in figures 5 and 6 we find it to be − 2.1 for mine A and − .4 for mine B. What this means is that in our standardised example a 1 per cent increase in capital in mine B leads to a .4 per cent decrease in labour in order to maintain a given level of output. In mine A, however, a 1 per cent increase in capital would lead to a 2.1 per cent decrease in labour. We therefore can say that, because of the lower quality of copper ore, labour will be replaced five times more quickly in mine A than in mine B. Furthermore, average capital productivity was found to be 12 per cent greater in mine A than in mine B. Although we are dealing with hypothetical conditions, and perhaps our results should be viewed in terms of ordinal rather than cardinal measurements, our example indicates the influence of ore quality on factor usage and substitution.

---

number of assumptions made about the production process, we have adjusted the total outputs in each mine for the previously mentioned index of capacity utilisation. Output adjusted for capacity now becomes Q^*, and the production function is given as:

$$Q^* = \gamma e^{\lambda t} (\delta K^{-\rho} + (1 - \delta) L^{-\rho})^{1/\rho}$$

The isoquants were then constructed using detrended values of capital and labour which were also adjusted for the scale of production in each mine. These variables are defined as follows:

$$L^* = \frac{L}{Q^* e^{-\lambda t}} \text{ and } K^* = \frac{K}{Q^* e^{-\lambda t}}$$

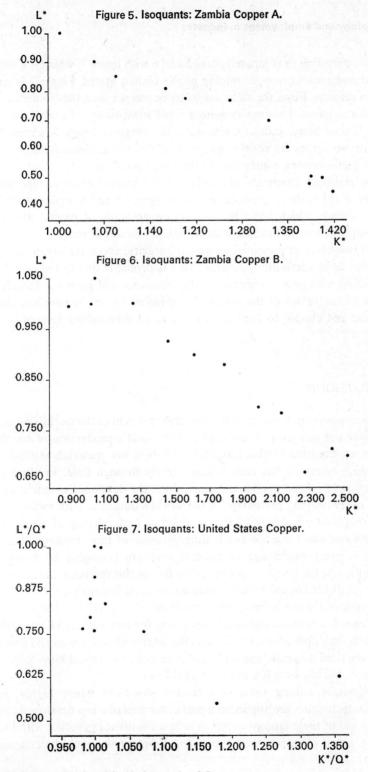

Figure 5. Isoquants: Zambia Copper A.

Figure 6. Isoquants: Zambia Copper B.

Figure 7. Isoquants: United States Copper.

The meaning of the symbols is explained in footnote 1, p. 287.
Source: table 68.

It is interesting to compare figures 5 and 6 with figure 7, which is a similarly derived isoquant for copper mining in the United States. Figure 7 is striking for two reasons. First, the slope seems to be steeper than the isoquants of the two African mines. This implies a more rapid substitution of capital for labour in the United States industry. Secondly, the shape of figure 7 is more traditional in so far as its convex nature indicates some diminishing returns in United States copper mining during the time period studied.

The resulting isoquants are adjusted for technological change and for changes in the scale of production. Hence figures 5 and 6 represent textbook-type isoquants which indicate the varying amounts of capital and labour that could be used to produce a standardised amount of output.

The isoquants as presented approach linearity, and therefore do not reveal the presence of diminishing returns. This is probably due to the fact that we are dealing with only a segment of the isoquants and have not included the more extreme values of the inputs. In addition, we might conclude that the firms did not choose to operate in the area of diminishing returns to either factor.

CONCLUSIONS

Developing countries account for over 40 per cent of the world's production of copper ore and over 50 per cent of the world's production of bauxite. In both cases, the total market production of these raw materials attributable to developing countries has been falling slightly through time. In terms of the production of refined copper and aluminium, the developed countries account for an overwhelming percentage of the world's output in both metals.

Throughout all stages of the production and refining of these mineral products one can trace the dominating presence of large multinational enterprises. A great percentage of mineral ores are produced by foreign corporations, and the single most impressive fact in this respect is the amount and growth of direct United States investment in these industries.

In view of the depleting resource position of developed countries, multinational enterprises are continuously searching for new sources of raw materials to expand their operations. In addition, the large-scale extraction and processing of ore required adequate financing and a degree of technical knowledge which is possessed solely by a few multinational firms.

Large-scale mining ventures represent situations where highly capital-intensive techniques are applied in a particular area of a less developed country, and the use of these labour-saving techniques is often in conflict with the host country's goal of employment expansion. Employment could be expanded by extending the vertical structure of production and marketing controls in the developing countries. In this regard, however, it is important to consider the international aspects of the supply and demand for minerals, since the marketing of copper and aluminium is dominated by a few international firms. The production of aluminium, for example, is almost completely integrated on a vertical basis, so that the main producers control all the steps in the production and marketing processes. Although there is an independent futures market for copper, its financial control lies outside the developing world, and large fluctuations in the market price of copper have made the metal quite unreliable as a foundation for stable economic growth in the producing countries.

At present countries expanding their minerals production to the final product stage may run into severe difficulties at the marketing end of their operations. Marketing knowledge must be increased in the developing countries before a more vertical structure of production can be used to create more employment.

The empirical tests yielded several interesting conclusions. Disguised unemployment went a long way towards explaining the lower levels of labour productivity in copper mining in the less developed countries, and in accounting for the fact that changes in capacity utilisation had a more significant influence on labour productivity in those countries than in the United States. In so far as world copper mining was concerned, it was generally noted that in all countries capital was substituted for labour when wages rose with respect to profits, i.e. when the elasticity of substitution of capital for labour was significant. However, our results showed that although elasticity of substitution has a noticeable effect on labour productivity in the United States aluminium industry, the results were not positive for the bauxite producing developing countries. The negative results were attributed to the fact that labour costs in the bauxite industry account for only a minor part of the total final cost of aluminium.

There is also some evidence that the substitution of capital for labour was influenced more by changes in relative factor costs in the United States industry than in the copper producing countries of the developing world. In addition, it was discovered that variations in the quality of copper ore significantly affected the capital-intensity of production, and the ease with which capital could be substituted for labour in two major mining firms in Zambia.

In terms of relative remuneration, in copper mines wage levels in the less developed countries were considerably lower, but increasing, by comparison with wages in the United States. Such was not the case in the world's aluminium industry, where wages in the less developed countries were increasing at a slower rate than in the developed country.

Although the mineral industry contributes greatly to the exports and gross national product of less developed countries, it is not a big direct employer of labour in those countries. In fact, employment in that sector seems to be inversely related to the level of economic development. However, the employment possibilities created by a mining industry are not limited to the direct employment of labour in the extraction and processing of ores: the development of new ore bodies often involves the establishment of a new town or other economic centre, and thus provides opportunities for the indirect expansion of employment. Indirect employment includes local traders as well as skilled and unskilled workers who earn their livelihood by virtue of the existence of an industry, but do not work directly for that industry.

Since indirect employment provides potentially large scope for employment expansion in the mineral producing, less developed countries, governments might attempt to induce multinational enterprises to invest in job creating industries which are outside the activities directly related to mining. Perhaps tax concessions could be granted to foreign corporations that re-invest part of their profits in such industries as housing construction. Employment in communities surrounding mining areas could also be stimulated by the grant of a larger number of loans to small businesses.

APPENDIX I

DATA USED

The data used in the case study were taken from the following sources.

United States
Output and prices from United States Bureau of Mines: *Minerals yearbook*.
Employment and wages from United States Bureau of Labor Statistics: *Employment and earnings*, SIC 102 (copper ores).
Since capital data were not available for SIC 102, a capital index was constructed using United States Department of Commerce: *Industry profiles*, SIC 331 (copper smelting and refining), 1950-69. For SIC 3334 (primary aluminium), all data were obtained from idem: *Industry profiles*, SIC 3334, 1958-68.

Chile
Output, prices, wages and employment for Chile's Gran Minería were provided by the Braden Copper Company.
A capital index was constructed from data originating in J. Grunwald and P. Musgrave: *Natural resources in Latin American development* and United Nations: *Economic survey of Latin America*, 1955-66.

Zaire
All data pertained to the Union Minière du Haut-Katanga and were taken in index form from J. Gouverneur: *Productivity and factor proportions in less developed countries*, op. cit.

Zambia
All data for Zambia Copperbelt Province were taken from Northern Rhodesia Chamber of Mines: *Yearbook*, 1955-63, and *Copperbelt of Zambia industry yearbook*, 1964-70. Data for the two Zambian copper firms came from private sources.

Guyana
All data for 1956-69 were obtained from D. F. Barnett: *A production function for the Guyanese bauxite industry*, report submitted to the Ministry of Economic Development (Georgetown, 1971).

Jamaica
Output, employment and wages from Central Planning Unit: *Economic survey of Jamaica*, 1959-68.
Value of output from United States Bureau of Mines: *Minerals yearbook*.

Indices of production variables

Table 69.   Zaire: copper

Year	Q	L	W	P	K	U
1950	100.0	100.0	100.0	100.0	100.0	89
1951	109.0	102.0	107.0	123.3	112.0	92
1952	119.0	111.0	125.0	144.9	124.0	95
1953	130.0	116.0	140.0	140.3	133.0	99
1954	134.0	119.0	150.0	139.2	141.0	97
1955	139.0	125.0	148.0	196.0	156.0	96
1956	147.0	132.0	162.0	184.1	170.0	98
1957	148.0	132.0	165.0	122.7	184.0	94
1958	141.0	104.0	168.0	110.2	186.0	86
1959	165.0	106.0	186.0	133.6	190.0	97
1960	176.0	107.0	194.0	137.5	196.0	100
1961	172.0	108.0	211.0	128.4	204.0	94

Key: Q = output
L = labour
W = wages
P = product price
K = capital
U = capacity utilisation

Table 70.   United States: copper

Year	Q	L	W	P	K	U
1950	79.7	101.3	61.0	64.8	.	87
1951	81.4	100.4	66.4	75.4	.	87
1952	81.1	101.8	72.8	75.4	.	85
1953	82.5	108.8	77.9	89.4	.	84
1954	72.9	105.8	74.2	91.9	.	73
1955	88.1	108.4	81.4	117.8	.	86
1956	97.8	125.7	84.9	132.1	.	94
1957	94.6	119.9	82.9	93.8	.	89
1958	86.9	100.4	80.6	81.9	.	80
1959	69.9	81.9	90.6	95.6	.	63
1960	100.0	100.0	100.0	100.0	.	89
1961	101.7	105.3	101.9	93.1	.	89
1962	122.2	103.5	103.4	95.9	.	96
1963	110.1	100.4	106.8	95.9	.	93
1964	113.8	97.8	111.7	101.6	.	94
1965	122.7	109.2	117.1	110.3	.	100
1966	125.0	115.9	119.9	112.8	.	100
1967	73.6	89.6	120.0	119.0	.	58
1968	108.1	94.2	138.5	130.2	.	84
1969	125.4	119.0	144.7	147.9	.	96

Key: as for table 69.

Table 71.   Zambia: copper

Year	Q	L	W	P	K	U
1955	33.4	184.0	18.8	134.9	.	72
1956	37.6	196.4	22.6	131.1	.	78
1957	40.5	203.4	20.9	89.5	.	80
1958	36.6	182.5	17.1	76.6	.	70
1959	51.9	174.8	22.9	95.9	.	95
1960	54.7	192.8	26.7	100.0	.	97
1961	55.0	205.9	25.4	92.9	.	94
1962	52.7	198.4	26.3	94.4	.	87
1963	55.6	199.9	26.6	94.4	.	89
1964	62.0	205.9	28.9	97.2	.	97
1965	66.1	209.3	31.4	110.2	.	100
1966	56.7	213.5	37.4	147.8	.	83
1967	59.6	212.6	40.9	157.3	.	85
1968	63.6	211.7	42.9	173.3	.	88
1969	65.9	213.4	42.9	243.2	.	89
1970	59.1	214.5	44.7	224.6	.	78

Key: as for table 69.

Table 72.   Chile: copper

Year	Q	L	W	P	K	U
1955	37.8	48.8	23.1	108.9	160.7	86
1956	42.8	52.2	20.1	105.9	147.8	95
1957	41.9	54.0	18.6	84.4	189.6	91
1958	40.4	53.1	20.9	78.7	197.3	86
1959	47.9	52.2	26.1	89.8	206.2	100
1960	46.2	52.7	30.2	98.9	210.9	95
1961	46.4	55.2	35.9	88.4	219.3	93
1962	49.2	53.7	33.2	89.9	222.1	97
1963	48.9	58.0	30.1	91.1	225.4	94
1964	50.9	51.9	38.4	97.3	228.1	96
1965	46.2	55.8	41.2	117.6	232.7	86
1966	50.8	53.6	42.8	141.4	237.3	93

Key: as for table 69.

Table 73.   Zambia: Copper Mine A

Year	Q	L	W	P	K	U
1950	8.0	54.3	7.9	48.5	10.4	75
1951	7.3	51.0	10.3	67.2	11.4	87
1952	7.9	52.4	12.1	78.6	12.3	93
1953	8.5	50.0	15.6	95.0	13.7	99
1954	8.7	50.0	15.8	89.5	14.7	100
1955	8.1	51.7	18.8	103.4	16.2	92
1956	8.7	48.3	22.6	128.6	17.6	98
1957	8.4	43.9	20.9	95.8	18.8	93
1958	7.8	38.8	17.1	67.5	20.1	85
1959	7.9	39.5	22.9	83.9	21.1	86
1960	9.0	39.0	26.7	93.5	21.8	96
1961	8.0	38.5	25.4	88.2	23.2	85

Key: as for table 69.

Table 74.   Zambia: Copper Mine B

Year	Q	L	W	P	K	U
1949	6.9	37.7	6.5	47.4	6.7	87
1950	7.5	41.6	7.9	49.4	7.3	92
1951	8.4	41.9	10.3	68.6	6.5	100
1952	7.4	41.7	12.1	80.7	9.9	86
1953	7.4	43.4	15.6	97.6	12.2	83
1954	8.6	44.5	15.8	90.2	13.9	94
1955	8.4	45.7	18.8	107.4	16.6	90
1956	9.5	43.4	22.6	133.7	19.3	99
1957	9.7	49.9	20.9	101.5	21.5	99
1958	9.1	40.1	17.1	73.7	24.3	91
1959	8.6	44.8	22.9	112.9	26.7	84
1960	10.1	46.6	26.7	97.6	29.4	96

Key: as for table 69.

Table 75.   United States: aluminium

Year	Q	L	W	P	K	U
1958	77.8	97.6	92.1	103.5	84.9	88
1959	97.0	101.1	97.8	103.5	96.8	100
1960	100.0	100.0	100.0	100.0	100.0	95
1961	94.5	91.0	103.9	98.1	102.2	83
1962	105.2	96.6	104.8	91.9	103.8	86
1963	114.9	101.7	106.4	86.9	104.7	88
1964	126.8	114.0	103.8	91.2	109.9	91
1965	136.7	116.3	108.6	94.2	114.9	93
1966	147.4	118.5	114.4	94.2	120.4	94
1967	162.3	133.7	115.1	96.2	132.9	99
1968	161.6	139.9	120.0	98.5	142.9	93

Key: as for table 69.

Table 76.   Jamaica: bauxite

Year	Q	L	W	P	K	U
1959	254.3	33.1	19.8	1.7	.	82
1960	285.5	26.3	28.3	1.6	.	86
1961	324.2	26.2	27.4	1.6	.	92
1962	372.7	30.6	30.3	2.1	.	100
1963	342.1	32.4	26.0	2.1	.	82
1964	387.1	28.1	31.3	2.3	.	94
1965	422.0	29.5	35.9	2.4	.	97
1966	442.0	33.8	36.8	2.1	.	97
1967	450.3	41.0	39.6	2.5	.	95
1968	415.8	38.1	35.2	2.4	.	84

Key: as for table 69.

Table 77.   Guyana: bauxite

Year	Q	L	W	P	K	U
1956	110.0	15.5	16.3	1.6	1.5	100
1957	97.7	14.7	19.0	1.7	2.4	84
1958	70.3	12.7	19.3	1.7	2.8	58
1959	74.3	10.1	22.4	1.8	3.3	58
1960	109.6	13.4	25.9	1.6	4.5	82
1961	105.0	18.6	24.4	1.8	4.8	75
1962	120.6	20.2	26.5	1.8	4.8	83
1963	103.9	20.8	24.5	1.8	4.9	69
1964	109.9	22.9	26.5	1.9	5.3	70
1965	127.4	26.8	30.5	1.9	5.8	78
1966	146.6	28.8	31.9	2.1	6.7	87
1967	147.5	29.7	33.6	2.2	7.7	84
1968	154.8	29.9	31.3	2.6	6.6	85
1969	170.3	31.1	30.5	2.8	6.7	91

Key: as for table 69.

ROLE OF THE INTERNATIONAL FIRM

A large proportion of mineral ores are extracted by foreign corporations. Firms incorporated in Belgium, France, the United Kingdom and the United States control the production of the world's principal mineral exports—iron, manganese, copper, bauxite, tin lead and zinc. [1]

During the period covered by our case study, Anaconda and Kennecott produced 90 per cent of Chile's copper output; Anglo-American and Rhodesian Selection Trust dominated Zambia's copper mining; and copper production in the Congo was owned by the Belgian firm Union Minière du Haut-Katanga. In more recent years those companies, as well as other international copper producing firms, have faced the problem of direct nationalisation or increased government pressure for local participation, but by and large the world's minerals market is still dominated by foreign concerns. For example, United States Steel and Bethlehem Steel produce much of the iron ore in Latin America, while the Aluminum Company of America and the Aluminum Company of Canada dominate the world production of bauxite and aluminium.

The introduction of direct foreign investment and technology into the developing countries is usually explained by the simple economics of demand and supply. On the demand side, large multinational firms hope to capture new markets for their products. As countries develop and incomes per head rise, large international industries expand into those markets to vie for a slice of the newly increased incomes. The expansion of the American take-out food industry in Europe is a simple example of this demand-oriented foreign investment. In the world's minerals industry, however, the following considerations can be put forward to explain the large amounts of direct foreign investment. First, multinational enterprises are continuously searching for new sources of raw materials to expand their operations. This is especially important in the light of the increasing depletion of the resources of developed countries and the vertical structure of production of many multinational firms. Secondly, the large-scale extraction and processing of ore requires a degree of technological knowledge which is possessed solely by a few multinational mining firms. The possession of this knowledge enables the multinational firm to exploit mineral deposits which smaller cor-

---

[1] Raymond Mikesell: "The contribution of petroleum and mineral resources to economic development", in R. F. Mikesell (ed.): *Foreign investment in the petroleum and mineral industries*, op. cit.

porations in host countries could not. In addition, large multinational firms are often the only source of the enormous amounts of venture capital necessary to discover and recover new ore bodies.

The above reasons would suggest why host countries have difficulty in entering into large-scale mining ventures. In this regard it is also important to mention the international aspects of the supply and demand for minerals: the marketing of the world's major minerals is dominated by a few international firms. The production of aluminium, for example, is completely integrated vertically, and the main producers control all the steps in the production process. The establishment of this vertical structure is quite beyond the financial and technical abilities of most developing countries. In the cases of other metals such as copper, tin and zinc, marketing channels are established through futures markets that are in the hands of financial interests in the developed countries. The nature of the financial control of the marketing of most metals makes it very difficult for developing countries to be completely independent in the production and distribution of their mineral resources.

In addition to the international aspects of the financial control of the metals market, a less developed country may encounter difficulty through its reliance on a single metal for a substantial part of its export earnings. The market price of a particular metal is usually unstable and hence quite unreliable as a foundation for a sustained growth in export earnings. In the case of copper, for example, the principal reasons for the large variation in copper prices from year to year can be listed as showu below. [1]

Copper

Markets

The markets for copper are multiple. One set of price quotations comes from Kennecott, Anaconda, and Phelps-Dodge. Another source of price quotations comes from custom smelters who use the output from small mines and large amounts of scrap ore. A third source of prices is the London Metal Exchange which is particularly important to producers who sell their output on the basis of the London price. Finally, wholesale metal dealers may have a slightly different price. Because of the profusion of these different markets and the multiple prices which arise in these markets, prices tend to fluctuate considerably more than they would if copper supply and demand were concentrated in a single market.

Demand

Since over 70 per cent of the output goes into the production of new capital goods and consumer durables [2], the copper industry is susceptible to extreme cyclical variations. Economic history tells us that the purchases of both capital goods and consumer durables can be postponed during a recession; consequently the purchases of the copper needed for the production of these goods can be postponed. During a down-swing, the highly mechanised large-scale producers must therefore lower their prices and find their profits reduced. Conversely, in a boom, these producers work at full capacity and prices are bid up in response to the increased demand for both producers' and consumers' durables.

---

[1] Olin T. Mouzon: *Resources and industries of the United States* (New York, Appleton-Century-Crofts, 1967), pp. 408 ff.

[2] Stanley Vance: *Industrial structure and policy* (Englewood Cliffs, NJ, Prentice-Hall, 1961), p. 106.

Inventory policies of consumers

Consumers of copper tend to over-buy in anticipation of shortages; therefore prices rise. On the other hand, these consumers will under-buy when the supply is adequate because they will have accumulated inventories on which to draw; in this case, therefore, prices will fall.

Supply

Since copper ore occurs only in a limited number of major deposits, when production in one of these areas is somehow adversely affected a large amount of copper does not reach the market and the world prices change. Also, since large-scale mechanised mining is the most efficient method of extracting copper, and since these mines require great capital investment, they will continue to operate during a recession in order to cover, at least partially, the high fixed costs. With present-day open-pit mining techniques, variable costs are a relatively small part of total operational costs, whereas the older type of underground mining techniques involved more variable and less fixed costs (i.e. more labour and less overhead capital).

Smelter policy

The nature of the smelting process requires continuous production, 24 hours a day and seven days a week. Therefore when sales begin to decline the smelter will cut his selling price to attract buyers rather than stop his furnaces from operating.

Interdependence of world producers

Because the producers and the governments of the various countries which supply the world with copper are extremely aware of each other's actions, any factor which interrupts the production of one country's output will affect the supply and the price of copper in the rest of the world. Government policies cause buying and selling waves in the world market and are a source of price disturbances, as may be seen when there is any indication that the Government of the United States may be altering its policy with regard to its stockpile of minerals.

ESTIMATING CAPACITY UTILISATION

It is assumed that the productive capacity of an industry grows at some linear rate through time, and that shortfalls in capacity utilisation are reflected in deviations from a trend in production through time. In other words, if we have some values of output, Q, through time we can estimate capacity utilisation by observing and quantifying the deviations of the actual values of each year's output from some expected full capacity trend value of that output. The methodology employed in creating an index of each year's capacity utilisation is given below.

Let Q be a measure of output for years $1 \ldots N$, where Q^* is the observation with the largest deviation of Q in that time period.

Step 1

Regress $Q = f(t)$ in the linear form. We then have:

$$\hat{Q} = a + b\,t \qquad \ldots(1)$$

where: $\hat{Q}_1 \ldots \hat{Q}_N$ are the estimated values of Q for years 1 to N.

We then define the full capacity estimates of Q from:

$$\hat{\hat{Q}} = a' + b\,t \qquad \ldots(2)$$

where $\hat{\hat{Q}}_1 \ldots \hat{\hat{Q}}_N$ are the full capacity estimates values of Q for years 1 to N and a' is the intercept value of a line passing through Q^* and having the slope b of equation (1).

The diagram on p. 302 illustrates the foregoing relationships.

Note that $a' = a + a_0$, and that equation (2) is simply equation (1) shifted upward by a_0 so that the estimated full capacity regression line passes through the year with the maximum output (Q^*).

Step 2

Given that $Q^* = a' + bt^*$

$$\therefore a' = Q^* - bt^* \qquad \ldots(3)$$

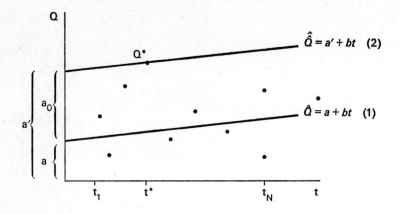

We are now able to calculate a full capacity estimate of $Q_1 \ldots \ldots Q_N$ as follows:

$$\hat{\hat{Q}}_1 = a' + bt_1$$

$$\vdots \qquad \vdots \qquad \vdots$$
$$\vdots \qquad \vdots \qquad \vdots$$
$$\vdots \qquad \vdots \qquad \vdots$$
$$\vdots \qquad \vdots \qquad \vdots$$

$$\hat{\hat{Q}}_N = a' + bt_N$$

where a' is estimated from (3).

It therefore follows that

$$\hat{\hat{Q}}_N = a + a_0 + bt_N$$
$$= a + bt_N + a_0$$
$$= \hat{Q}_N + a_0$$

Step 3

We are now able to create the index of capacity estimates for each observed Q.

It is given as $\dfrac{Q_1}{\hat{\hat{Q}}_1}, \dfrac{Q_2}{\hat{\hat{Q}}_2}, \dfrac{Q_3}{\hat{\hat{Q}}_3}, \dfrac{Q^*}{\hat{\hat{Q}}^*} \ldots, \ldots \dfrac{Q_N}{\hat{\hat{Q}}_N}$

It can be seen that $Q^*/\hat{\hat{Q}}^* = 1$ and that all other values of the index will be less than 1 and hence will be some proportion of full capacity.

Step 4

For empirical simplicity it was decided to calculate the index of capacity estimates as follows.

Let r_N be the residual of the original observation of Q_N from the estimated value of Q_N over time.

For example:

$$r_N = Q_N - \hat{Q}_N$$
$$r_N = Q_N - a - bt_N$$

subtracting a_0 from both sides

$$r_N - a_0 = q_N - a - a_0 - bt_N$$
$$a_0 - r_N = a + a_0 + bt_N - Q_N$$
$$a_0 - r_N = \hat{\hat{Q}}_N - Q_N$$

multiplying both sides by Q_N^{-1}

$$\frac{a_0 - r_N}{Q_N} = \frac{\hat{\hat{Q}}_N - Q_N}{Q_N}$$

$\therefore$ our index of capacity for observation N as given in step 3 is:

$$U_N = \frac{Q_N}{\hat{\hat{Q}}_N} = \frac{Q_N}{Q_N + a_0 - r_N} \quad \text{and is lower than } 1 \dots (4)$$

it can also be seen that in the case of Q^*, since $a_0 = r_N$

$$U^* = \frac{Q^*}{\hat{\hat{Q}}^*} = 1$$

Equation (4) will be used as an index of capacity utilisation for observations $Q_1 \dots Q_N$.

The estimate of capacity utilisation as defined above was constructed in a simplified way for usage on several industries with slightly differing and limited data. A. S. Bhalla and J. Gaude have correctly pointed out that such an index assumes a unique observed maximum of output whereas for longer time periods it would be more realistic to use more that one maximum from which to define the full capacity growth levels.

TECHNOLOGY IN EXTRACTIVE INDUSTRIES

Copper

Mining

In most recent years open-pit mining has accounted for the major part of the ore mined in Chile and the United States. In Zambia, on the other hand, a large proportion of the ore is extracted from underground mines. The choice of whether a particular deposit should be exploited by open-pit or underground methods is determined by several factors, such as the depth of ore body, surface conditions and topography, concentration of ore deposits, and availability of capital and labour. These factors will affect the relative costs of the two main extraction processes. Surface mining is employed in the case of large porphyry deposits which lie near the surface. Veins that contain a large percentage of ore are mined underground.

In an open-pit operation the ore is broken up by blasting and then large mechanical shovels load the pieces into railroad dump cars which transport the ore to the concentrator or smelter. This process of ore extraction first requires the removal of large quantities of overburden from the earth's surface. As the shovels work their way deeper into the deposit, the crater takes on the appearance of a huge amphitheatre with terraces about 70 ft. high and from 70-250 ft. wide. These terraces take the form of concentric ellipses, becoming narrower towards the bottom of the pit.

In underground mines the ore is extracted by miners and loaded into wagons drawn by electric locomotives which haul it to the shaft. It is then dumped into skips and hoisted to the surface. A new and very effective method of underground mining is called "caving". In the block caving method, the veins of ore are undercut and fall by gravity to a predetermined level, from which the ore is lifted to the surface.

Concentration, smelting and refining

Once the ore is extracted from the ground it undergoes three processes—concentration, smelting and refining. The concentrated ore is often smelted near the mine, whereas the production of refined copper generally takes place in industrialised countries. In recent years there has been some attempt to conduct both smelting and refining operations close to the mines, especially in the countries where government intervention has influenced the industry. Such attempts may do much to increase total employment in copper mining in the developing countries, providing that the relocation of the refining process does not drastically alter the economics of producing in those countries.

Native copper, copper oxides and copper sulfides are processed in three different ways. Native copper is concentrated mainly by the use of flotation machines. Copper concentrate is therefore formed by the mixing of crushed ores with water, oil, and reagents. Air bubbles collect the coated particles and float the concentrate to the surface. This concentrate is melted in reverberating furnaces, and air is introduced to oxidise the impurities which are slagged off. After the molten ore has undergone reduction, it is cast into ingots, billets, cakes and wire bars.

Oxidised copper ores are treated by leaching. In this process the crushed ore is placed in large vats, and leaching solutions are allowed to percolate through it. The sulfuric acid in the solution leaches out the copper in the form of copper sulfate. This copper sulfate solution is purified in an electrolytic tank, and metallic copper is formed on cathodes. The cathodes are then melted and cast into the usual commercial forms.

Sulfide ores, which represent the bulk of copper produced in the United States, are less than 1 per cent copper. The flotation process, "one of the greatest technological advancements in the recovery of copper from low-grade sulfide ores" [1], allows the economic extraction of the low-grade sulfide ores. From every 100 tons of ore treated, approximately 96 tons of material are discarded as tailings. Besides containing 30 per cent copper, the dried concentrate includes significant quantities of gold, silver, and other metals.

The smelting of sulfide copper ores [2] begins with roasting, which removes volatile components like sulfur and arsenic, and reverberatory furnacing, in the course of which the copper concentrate is heated to a molten state and the iron floats to the top while the matte of copper and sulfur settles to the bottom. After the periodic removal of the slag, the matte is placed in converters. In the converting process, air is forced through the molten mass of copper so that the sulfur is burned off. Next, the oxygen is removed in the form of carbon dioxide and steam through treatment in an anode furnace. The resulting copper is called "blister" copper, and is 99.6 per cent pure, though still too impure to be used for electrical purposes.

Refining blister copper is done through electrolysis, which removes the usual gold, silver, and platinum impurities. The process of producing pure "cathode" copper is thus completed. Cathode (or electrolytic) copper has a minimum purity of 99.9 per cent.

Fabrication and uses

Electrolytic copper is cast into various forms. Ingots are 10 in. in length and weigh approximately 20 lb. Wire bars, which weigh 250 lb, are designed for drawing into wire. Sheets and flatware are made from copper cakes, and billets are formed into copper tubing.

Half of copper consumption takes the form of electrical applications which include power transmission, electronics, home and office power supply, and wiring. Telephone and telegraph cables and wires are the two largest users of copper in the communications field, but numerous components in radios and television sets are also made of copper and its alloys.

Because of the non-corrosive quality of copper and its alloys, its construction uses include roofing, plumbing goods, and builders' hardware. "The automotive

---

[1] Stanley Vance: *Industrial structure and policy*, op. cit., p. 100. Vance points out that before the flotation process it was economical to mine only the high-grade ores, selectively.

[2] Olin T. Mouzon: *Resources and industries of the United States*, op. cit., pp. 390-391.

industry accounts for about 9 per cent of copper consumed in the United States, using about 30 to 40 lb of copper per vehicle." [1] Copper also plays an important part in the manufacture of aircraft, missiles, locomotives, cargo and passenger ships, submarines, home appliances, turbines, watches and clocks. The above-mentioned uses for copper and copper products give only a small indication of the versatility and importance of the product. [1]

Aluminium

Bauxite

Bauxite is mostly mined through open-pit operations. Underground mining, however, is common in European countries, including France and the USSR where mining costs are substantially greater than in open-pit mines. Bauxite deposits are often covered with up to 50 ft. of over-burden which must be stripped as the mining is carried out. Stripping ratios of as much as 10 ft. of over-burden to 1 ft. of ore are minable, and a ratio of 15 to 1 is considered feasible.

Although the amount of over-burden varies greatly from operation to operation, this alone is not the main determinant of the final cost of production. Transport cost, the amount of pre-drying and beneficiation necessary, as well as the required infrastructural investment, all tend to cause wide fluctuations in the final cost of bauxite. In addition, the scale of the operation is of great importance in achieving competitive costs since a significant investment in power, water, transport and port facilities is normally required. The minimum size of operation is estimated to be about 300,000 tons per annum if large amounts of infrastructural investment are not required. New projects now usually provide for an output of about 1 million tons.

As far as known deposits and future sources of supply are concerned, it would appear that, in the medium term, increases in production will increasingly tend to come from Australia and Guinea, where deposits three to four times the size of Jamaica's are reported; previously Jamaica's 500 million tons were considered the world's largest deposits. In general, known reserves of bauxite have more than doubled in the last decade and continued exploration will probably lead to still more discoveries: it is probable that known reserves of bauxite will again double in the next decade, Africa being the most likely source of new discoveries. [2]

Alumina

It is usually assumed that 2 tons of bauxite are required for the production of 1 ton of alumina. Approximately 90 per cent of the world's alumina is produced on the basis of the Bayer process. Although the basic principle underlying this process has remained unaltered, considerable technological change has occurred in the production of alumina, and this stage has reached a rather high technical level. Essentially, this process involves the mixing of bauxite with a solution of hot caustic soda in which the alumina is dissolved and the impurities eliminated through filtration. The resulting sodium aluminate solution is then hydrated and finally calcined to produce alumina. This process is often modified to suit the nature of the particular ores. A number of other processes for the recovery of aluminium clays, alumite and lower-grade bauxite ores, as well as the elimination of the alumina stage, are currently being attempted on an experimental basis. Once these techniques

---

[1] Albert D. McMahon: *Copper: A materials survey*, op. cit., p. 23.

[2] UNIDO: *World reserves and requirements for alumina raw materials* (Vienna, 1967).

are used on a commercial scale, the distribution of ore and alumina facilities could be altered considerably. However, wholesale innovations seem unlikely in the near future since all recent investment has been undertaken on the basis of the traditional bauxite-aluminium process.

Traditionally, the conversion of bauxite to alumina has occurred near the aluminium reduction facilities in developed countries. However, the recent trend has been towards increased production of alumina nearer the ore fields. Since the ratio of bauxite to alumina is approximately 2 to 1, significant savings in transport costs can be made by producing alumina at the mine site. Alumina plant size varies from 660,000 to 150,000 tons per annum, depending on whether the alumina plant is connected to the bauxite mining operation or to the aluminium reduction plant. Major aluminium companies usually think in terms of a 360,000-ton alumina plant adjacent to a 100,000-ton aluminium reduction facility. The minimum plant size for a self-contained alumina plant in North America is 250,000 tons per annum while in Japan and Europe it is 150,000 tons. Alumina operations connected with the mining of bauxite, as in Australia, Jamaica, Surinam or West Africa, usually run at about 200,000 tons per annum. UNIDO estimates the capital requirements for alumina plants to be about $200 to $250 per ton of annual output, and also states that the unit cost for a plant producing 100,000 tons annually is about 50 per cent greater than for plants producing 300,000 tons per annum. [1]

Aluminium proper

The industrial production of aluminium is based on the electrolysis of an alumina solution in molten cryolite. The reduction is undertaken in a series of electrolytic cells called potlines. Carbon anodes extend into the potlines, which are carbon-lined and act as cathodes. The direct current passing between the anodes and cathodes reduces the aluminium oxide, and the metal sinks to the bottom of the cell. Although other techniques for this reduction process are currently being studied, none has so far been proved to be economically feasible.

Inputs per ton of aluminium are as follows [2]:

Alumina	3,802 lb.	Anode carbon	1,100 lb.
Cryolite	40-60 lb.	Cathode carbon	40 lb.
Aluminium fluoride	40-60 lb.	Electricity	14-16 mW/minute.
Fluorspar	6 lb.	Labour	13-14 man-hours.

Labour inputs may vary considerably more than the above table indicates. According to UNIDO, labour use may vary between 10 and 120 man-hours per ton of metal, depending upon the degree of mechanisation, which in turn may depend on local labour market conditions. [3] Electric power is the most important economic factor in aluminium production: the electrolytic reduction of aluminium on an economic scale requires large quantities of cheap power.

---

[1] UNIDO: *Non-ferrous metals industry* (New York, United Nations, 1969), pp. 36-37.

[2] United States Bureau of Mines: *Mineral facts and problems* (Washington, DC, 1965), p. 31.

[3] Ibid., p. 31.

LESSONS FROM THE CASE STUDIES

by A. S. Bhalla [1]

12

One must be cautious in making generalisations from a heterogeneous sample of industries such as the ones included in this volume. However, most of these studies point in one important direction, namely that the determinist view that choice of techniques does not exist, especially in manufacturing, is incorrect. Whether one considers the production processes themselves, e.g. in metalworking and cotton textiles, or ancillary operations such as materials handling, transport and packaging, the empirical evidence supports the existence of possibilities of substitution between capital, labour and materials. This conclusion seems valid irrespective of the geographical region and manufacturing industry to which the studies relate.

It can be argued that unemployment and poverty are more severe in Asia and Africa than in Latin America, so that there might be a lesser tendency for countries in the latter region either to look for more labour-intensive methods or to choose from among those that are already known. In other words, the choice of appropriate technologies as instruments of employment and income distribution policies may not carry equal weight in different sets of economic, social and political environments.

SOME KEY ISSUES

Factor pricing and technological choice

In the neo-classical tradition, many writers on technological choice blame factor price distortions for choices in the direction of capital intensity. Their analysis is based on simplified substitution possibilities between two factors,

---

[1] International Labour Office.

namely capital and labour. It is assumed that if wages rise faster than the price of capital, factor proportions will tilt in favour of a relatively greater input of capital. Sure enough, as the empirical studies in this volume show, relative factor prices have an important role to play in the choice of techniques. Della Valle, in the study on the minerals industry, demonstrates that producers in the United States were much more responsive to higher wages than those in the developing countries. This is shown by the fact that the elasticity of substitution of capital for labour is higher in the United States than in developing countries such as Chile, Zaire and Zambia.

There is one interesting factor here, however. Most of the firms in the extractive industries in these developing countries are subsidiaries of multinational firms. Why then do they behave differently from their parent firms? The explanations given are twofold. First, wages in the developing countries considered are so low that any increase in labour costs does not represent more than a marginal increase in the total costs involved. Secondly, union and government pressure may be too strong to let foreign mining companies reduce employment in response to higher wages. In general, however, the author concludes that in both developed and developing countries market prices play a significant role in factor allocation in the case of the copper mining industry. This conclusion does not apply to firms engaged in the production of bauxite because the cost of bauxite production amounts to only a very small fraction of the total cost of making aluminium.

Cooper and Kaplinsky also found that pricing played a role in the choice and use of second-hand machinery, but in a different way. The unit costs of production are more sensitive to rates of output than to machine prices themselves. Thus it is not so much the price of second-hand machinery as the uncertainty and risk involved in attaining a certain rate of output that enter into decision making.

Pack, on the other hand, argues that the use of second-hand machines in the textile industry in the developing countries would be optimal, especially because the relative factor prices prevailing in these countries would be favourable to such a decision. He further argues that even if repair and maintenance were greater in the case of used machines it would tend to be a labour-intensive and less expensive activity in less developed countries, since the wages of repair men would be a small fraction of those of their Western counterparts.

However, it is worth noting that the textile study is based on the restrictive assumption that the productivity of the various vintages or types of equipment would be the same in a less developed country as it is in the United Kingdom. While this is plausible *a priori*, the validity of the assumption is not borne out by the Kenyan jute processing study, the authors of which found wide variations in machine output in a single factory. Notwithstanding this, both studies indicate that in some circumstances a careful selection can mean that the use of second-hand equipment is a feature of the optimal technique.

The point about the differences in relative prices between developing and developed countries and their influence on the choice of techniques needs to be compared with relative factor price differences between urban and rural areas within a developing country. This is done by Frances Stewart in her study on cement blocks in Kenya. She demonstrates that on account of lower wage or labour costs in rural areas, among other factors, manual techniques are more suitable there than machines.

There is one study, however, that on sugar processing in India, which comes up with a rather different result: it is not so much the prices of capital and labour as that of the raw material input that determines the choice of technology. This may be particularly true in the case of a processing industry. It also showed that in the choice between large-scale and small-scale production the general level of wage rates made little difference as long as the wage rate paid in large-scale production does not increase in relation to that paid by small producers. More important parameters are the recovery rate of sugar from cane and the average length of the crushing season.

The cases of both cement blocks and sugar processing suggest that differences in relative factor prices between urban and rural areas are associated partly with the different modes of employment that prevail. In the rural areas, small-scale production units operate under family-based systems, given the low value of the supply price of labour.

Multi-factor substitution

The simplistic choice of factor proportions between two inputs, namely physical capital and labour, does not correspond to reality. In practice, possibilities of substitution exist between labour and material inputs, between labour and working capital, between skilled and unskilled labour, etc. This is borne out by a number of case studies in this volume.

The importance of substitution between labour and non-labour inputs (e.g. intermediate products) at a conceptual level was considered in Chapter 1 on the concept and measurement of labour intensity. The empirical studies on copper and aluminium (Chapter 11), the textile industry (Chapter 6), sugar processing (Chapter 7) and the manufacture of cement blocks (Chapter 8) all show that in one way or another the price and quality of raw material inputs deserve separate treatment in the analysis of substitution possibilities. For example, in the case of extractive industries, the quality and type of ore mined largely determined the elasticity of substitution. On the basis of a comparison between two Zambian mines, A and B, Della Valle concludes that because of the lower quality of copper ore, labour will be replaced five times as quickly in mine A as in mine B.

The importance of intermediate material costs in affecting the wage rates is explicitly recognised by Howard Pack in Chapter 6. Allowance needs to be made for the marginal rate of substitution of intermediate inputs for labour, together with the marginal rate of substitution of capital for labour. In the conventional analysis on factor proportions only the latter is considered.

Alternative techniques may have differential material requirements, thus complicating the problem of technical choice. It is often argued that in textiles and in other manufacturing industries as well, the use of new equipment instead of old, or the use of capital-intensive methods in general, ensures economies in the use of raw materials. There is no unequivocal evidence to support such arguments. Pack demonstrates that in the case of the textile industry the evidence is rather mixed: while newer spinning processes do reduce waste considerably, the more recent loom designs result in higher wastage levels than for the Lancashire looms. It is worth noting, however, that different levels of wastage occur in weaving different types of material.

Baron notes that in sugar processing large capital-intensive mills have an advantage over their small labour-intensive counterparts in so far as they require less sugar cane per unit of sugar produced. Nevertheless, although the small plants are "cane-wasters" this disadvantage may often be offset by the fact that the sugar mills require additional capital.

Stewart, in her study on the manufacture of cement blocks, has estimated non-labour costs, in particular fuel costs, for labour-intensive and capital-intensive techniques respectively. She shows that fuel costs per block of cement are much higher in the case of large stationary vibrating machines than in that of small machines. In general, the manual techniques save on non-labour costs. Stewart's findings about fuel costs are specially relevant in today's circumstances in which oil-importing developing countries are hard hit by increases in oil prices in particular and by the energy crisis in general. Until recently, relatively cheap imported oil encouraged the use of machines instead of labour, and provided little incentive to develop local sources of energy. In oil-short countries at least, wherever possible, efforts may be needed to explore alternative techniques that substitute labour and local sources of energy for fuel and other materials in short domestic supply.

Intermediate products are also relevant to the choice of technique in another sense. We noticed in Chapter 1 that the degree of mechanisation and capital intensity may move in opposite directions if due account is taken of the working capital requirements of alternative techniques: a technique requiring hardly any machines may yet be very capital-intensive. Therefore there may well be a case for treating capital as heterogeneous in the sense that fixed capital and working capital should be treated as separate factor inputs.

Empirical evidence suggests that, in general, the working capital (which represents stocks of raw materials and finished products) required by labour-intensive techniques is larger than that needed for their capital-intensive counterparts. However, the study on sugar processing in India in this volume suggests rather different results, and the author argues that working capital is probably higher for the mills than for small producing units. The reason for this unusual situation is that the Government regulates the sale of sugar by the mills in order to ensure that the supply of sugar is evenly distributed throughout the year. On the other hand, the small units are not subject to such control, with the result that they can sell their sugar in the course of the crushing season. Interestingly enough, government intervention leads to different time lags between production and sales, which influence working capital requirements in the two techniques.

Like capital, labour is also a heterogeneous factor input. It is closer to reality to disaggregate labour, at least broadly, into skilled and unskilled categories. There may be complementarity or substitutability between physical capital and skilled labour. Similarly, skilled labour may be either a substitute for unskilled labour or a complement to it.

A case study undertaken as part of the ILO-organised employment mission to Kenya [1], suggests that the transition from automated to semi-automated techniques may involve factor substitution not only between capital and labour but also between semi-skilled labour (used with automated techniques) and skilled supervision plus unskilled labour (used with semi-automated techniques). In other words, this implies a complementary relationship between supervisors (one type of skill) and unskilled labour. Unfortunately, the study on the manufacture of cans in Kenya, Tanzania and Thailand, reported in Chapter 4 of this volume, is unable to substantiate the hypothesis that supervision costs militate against the use of labour-intensive techniques. Though rather inconclusive, the data suggest that supervision costs per unit "are lowest with more 'intermediate' technologies, but rise with greater labour and capital intensity". The can manufacture study in this volume further indicates that the substitution process is much more complicated than was originally considered in the Kenya employment mission report. For example, in the movement from capital-intensive to labour-intensive methods, there is not only a substitution of supervision for skill but a change in the nature of supervisory skills re-

---

[1] ILO: *Employment, incomes and equality: A strategy for increasing productive employment in Kenya*, op. cit., Technical Paper 7: "A case study of choice of techniques in two processes in the manufacture of cans".

quired. This is indeed a very interesting point that emerges from the analysis of can manufacture in Thailand. The type of supervision needed on automated production lines calls for workers with a high level of technical ability to manage machines, whereas that on labour-intensive production lines requires ability to organise or manage a large number of workers. It might well be that the type of supervision needed for labour-intensive methods (skills needed for organising workers) is more readily available and cheaper in developing countries than the type of supervision needed for capital-intensive methods. This unusual hypothesis needs to be substantiated by further research on the cost of supervision and labour management. At present there are very few studies, if any, dealing specifically with the effects of these factors on technical choice at the level of the firm or the industry. The influence of such non-economic factors as discipline and cultures also seems to be virgin territory. Is the cost of supervision lower in cultures with a high worker discipline than in others? Can one type of abundant skill be substituted for a more scarce one to perform given tasks of supervision? These questions deserve much more attention in the discussions on technical choice in industry than they have received in the past.

Product choice

So far we have assumed that technical choice is confined to identical products with given quality and other characteristics. However, the particular product mix to be manufactured at both intra-industry and inter-industry levels deserves serious consideration. First, it is of crucial importance for determining the employment implications of any industrialisation strategy. Secondly, whether products are well specified or not makes a difference to the issue of choice of techniques: the case studies in this volume, particularly the ones on sugar processing, can making and cement blocks, demonstrate that if the product mix is varied technical choice between one kind of method and another becomes much more complicated in practice. For instance, the study on sugar processing demonstrated that there are in fact three types of sugar, namely white sugar, *gur* and *khandsari*, which meet the same consumer needs. However, although *gur*, a traditional product, is more nutritive than white sugar, the urban population prefers white sugar, while *gur* is consumed mainly in rural areas. *Gur* is produced much more labour-intensively. If government intervention through price control leads to the substitution of white sugar for *gur* in consumer demand, employment would be sacrificed. Unless cross-elasticities of demand for sugar and *gur* are known, it is of course difficult to be precise about the employment effects of product substitution, which itself also depends on the nature of rural-urban income distribution.

While the study on cement blocks reinforces this point, it brings out an additional but very significant point relevant to the problem of product choice: the consideration of the range of products (or materials in this case) requires that the productive process be regarded as a chain in which the choice of techniques at one stage helps to determine the choice at other stages. This interdependence of choices at different stages of production may exist even if the same product is considered. While it is true that each product or material is associated with a technique, so that variation of them would widen the technical choice, it is also true that, given the product or material, there may be scope for labour-capital substitution in relation to one stage or process but not another. This is shown in the study of metalworking: of the 88 metal-machining processes, only 55 permitted technical choice.

There may or may not really be a positive or clear-cut association between labour intensity and product characteristics. It has been argued that products consumed by the rich are more capital-intensive than those consumed by most of the poor in developing countries. Even if this were so—and the study on cement blocks points out exceptions—it need not follow that capital-intensive products are necessarily "inappropriate" for low-income consumers. Take a case in which these products are more durable and cheaper than the labour-intensive ones, e.g. plastic shoes compared with leather shoes. On distributional and equity grounds, such capital-intensive products may well qualify as "appropriate" for even low-income consumers.

Although studies undertaken on several developing countries, e.g. Argentina, Colombia, India and Venezuela, suggest that intra-industry moves towards more labour-intensive goods could generate additional employment, the question whether products "that are appropriate in the factor use sense are also appropriate in the consumption sense" still remains open-ended.

Tasks and processes

We observed in Chapter 1 that much of the literature on technological choice is concerned with the production of final output. However, the manufacture of most products involves a series of processes, each consisting of a different activity or task which requires different inputs of unskilled labour, skills and equipment.

A detailed disaggregation of production into a large number of tasks, as is carried out in the study on metalworking in this volume (Chapter 10), implies an examination of factor proportions in each task or activity separately. For some tasks very little scope will exist for substituting labour for capital, whereas for others a range of efficient possibilities may exist, as is shown in the study of cement block manufacture in Kenya (Chapter 8) and of metalworking in Mexico (Chapter 10).

There is another reason for looking at processes and tasks separately. As demonstrated in Chapter 8, although in theory a very labour-intensive method in one process (e.g. mixing cement on the ground) could be combined with a more mechanised one in another (e.g. the use of a block-maker), in practice the choice of technique at one stage partly determines the choice at the next stage. This interdependence is explained partly by speed and quality requirements. To quote Frances Stewart—

> Rapid speed of production at one stage, if it is to be uninterrupted, requires previous processes to keep pace. This may be achieved by having a number of machines going simultaneously at the previous processes, or by having each machine producing equally rapidly.... The quality requirements of the inputs of one process determine the quality necessary for the output of previous processes.... This type of argument applies to all vertically linked processes of production, whether they are conducted by the same firm or not. [1]

In practice, managers or private entrepreneurs make decisions much more for the type of machines required for different processes than for the manufacture of the final product (see Chapter 10). The introduction of processes and tasks into the analysis implies that the range of technical choice is extended.

One methodological point which is intriguing is the precise manner in which highly disaggregated micro-process analysis can suitably be aggregated into over-all factor proportions for a given product. The manner clearly depends on the possibilities of variation in each process. While this problem of different levels of aggregation has been recognised in the chapter on metalworking, it is not clear whether any meaningful generalisations can be achieved on the basis of very disaggregated analysis. No attempt at aggregation has been made in the studies on cement block manufacture and metalworking. In principle, however, isoquants for each process (each of which will have different points depending on flexibility or rigidity of factor proportions) can be developed into an aggregate isoquant for the group of processes considered.

Market structures

The choice of technology is significantly influenced by market structures and organisation, as well as by the social, economic, institutional and political environment. This is all the more so in market economies where different private investors making technological decisions are often faced with different sets of

---

[1] See Frances Stewart: "Choice of technique in developing countries", op. cit., pp. 115-116.

factor and product prices, thanks to the existence of structural economic dualism.

The monopolistic advantages of a firm or an industry may encourage it to use capital-intensive techniques. The use of international trade names, and the capacity of the monopolists to bear the heavier burden of finance capital, may partly account for this situation. In a study on Indonesia, Louis Wells acquired the impression that international firms not giving much importance to brand names or trade marks tended to use much more labour-intensive techniques than they used in their home countries. [1] Moreover, protected firms making windfall profits may have less incentive to innovate than competitive firms.

Issues of risk and uncertainty, related to the organisation and structures of markets, have an important bearing on the choice of technology, and seem to constitute one of the least explored areas in empirical research. For instance, do firms view returns to training as non-appropriable? If they do, would it result in investment in physical capital instead of in human capital? What effect does uncertainty about government policies or future market demand have on the current technical choices of the firms? It may be that firms maintain greater excess capacity and higher inventory stocks under conditions of uncertainty, thus leading to a higher degree of capital intensity. One hypothesis put forward by Wells is that capital-intensive plants seem to act as an insurance against risks. This may happen in two ways, namely *(a)* the manager of a capital-intensive firm may enjoy a greater flexibility in responding to unexpected demand fluctuations, and *(b)* the capital-intensive plant may make it easier to cope with a future liquidity crisis. To give but one example, it is easier to operate an additional shift in response to increased demand in a capital-intensive plant than in a labour-intensive one which would require large numbers of additional skilled and unskilled workers.

The case study on the second-hand equipment in this volume (Chapter 5) touches on the problems of risk and uncertainty which occur partly because of the special organisational features of the market for second-hand equipment. Transactions between buyers and sellers are irregular and take place under conditions of imperfect information about output performance, maintenance, availability of spare parts, transport, re-assembly and re-installation. Buyers in general may be in a weaker bargaining position. Jute machinery, with which the study is concerned, may however be a peculiar case of particularly weak bargaining power of the buyers since Dundee in the United Kingdom is perhaps the only source of supply. The result of this supplier's monopoly position is that entrepreneurs may prefer to buy new instead of used machinery even though the

---

[1] Louis T. Wells: *Economic man and engineering man: Choice of technology in a low-wage country*, op. cit.

latter saves on capital. It is argued that in certain circumstances machinery dealers might improve the flow of information to buyers and thus reduce their uncertainty.

The study on can making (Chapter 4) brings out the influence of market conditions particularly well. It is demonstrated that the small-scale, more labour-intensive techniques may be preferred until the market gradually builds to a level which would justify investment in a capital-intensive technique.

The risk element and uncertainty are linked in this study to imperfect knowledge about other technical possibilities: in other words the difficulty lies not so much in a lack of alternatives as in uncertainty on the part of the private decision makers about the reliability of certain types of equipment that may, ironically enough, be even cheaper than the more capital-intensive types but may still not be used.

The influence of market structures is further brought out in the study on the capacity of the engineering industry in Colombia (Chapter 9). It is suggested that import-substituting industrialisation and the accompanying protective policies inhibited labour-intensive innovations and the development of an indigenous capital-goods sector. The process of adaptation of imported machine designs and innovations in addition to the importation of capital equipment began only in the export promotion phase when Colombian industry was exposed to greater competition in domestic and foreign markets.

NATURE OF ACTION CALLED FOR

It has been established in this volume that possibilities of substitution exist in both the core processes and the ancillary operations in the manufacturing sector. On its own, however, the identification of technological alternatives is necessary but not sufficient: it is essential that once flexibility in factor proportions has been established, ways and means should be found of applying the efficient labour-intensive technologies that have been identified. The responsibility for taking such action lies in both the public and the private sector. In the former, governments can take action directly in public sector industries by making project appraisals on the basis of social cost-benefit analysis, explicitly introducing employment into the objective function. In some industries less capital-intensive techniques may be efficiently used even without any social evaluation, and thus without calling for a subsidy.

Indirectly, governments can play a major role in influencing private investment decisions through interventions in the market. The ILO-organised comprehensive employment missions, notably to Sri Lanka and the Philippines, pointed out how distortions in the prices of such factors as land, capital and labour, or more generally the whole incentives structure (including fiscal, monetary, wages, and training policies), influence technical choice in the direction of greater capital intensity. For example, the Sri Lanka employment report [1] noted that the private sector was positively encouraged by the tax system to buy capital equipment. Often these subsidies on capital are accompanied by additions to labour costs, so that the hiring of labour is doubly discouraged.

Government interventions in the factor markets need to be such that signals for resource allocation point in the direction of the true scarcity or abundance of factor inputs. No categorical and general recommendations can, however, be made on this point, and this volume does not attempt to do so. The reason for this is that the current state of our empirical knowledge about the influence of fiscal and financial intervention in inducing economy-wide changes in factor combinations towards greater employment from given investment resources is still very limited. [2] Only one systematic quantitative study has been reported that isolates the effect of factor price changes on factor combinations through changes in both technique and output mix. [3] That study is, however, more illustrative than predictive, based as it is on rather restrictive and unrealistic assumptions. Its main conclusion is that for a given factor price change the scale of direct factor substitution (variations in choice of technique) in production is greater than indirect substitution (variations in output mix) through foreign and domestic demand. The authors admit that "without further empirical evidence, it is impossible to say whether or not these results represent realistic orders of magnitude".

In the last resort, the effect of various policy instruments (fiscal or monetary) on relative factor prices and technical choice will depend on the nature and functioning of the particular economic systems; so will the influence of factor prices on product mix.

Notwithstanding the above caveats, private firms can be encouraged to employ more labour-intensive methods in several ways, namely by general measures to make the use of equipment less attractive by comparison with that of labour, by selective measures to attain the above objective in individual enterprises and by making capital-intensive products more expensive than labour-intensive products, thus shifting consumer demand to the latter. [4]

---

[1] ILO: *Matching employment opportunities and expectations: A programme of action for Ceylon*, op. cit., Report, pp. 72-73.

[2] For an interesting review of works on this question, see S. N. Acharya: *Fiscal/financial intervention, factor prices and factor proportions: A review of issues*, op. cit.

[3] H. B. Chenery and W. J. Raduchel: "Substitution in planning models", in H. B. Chenery (ed.): *Studies in development planning* (Cambridge, Mass., Harvard University Press, 1971).

[4] See ILO: *Fiscal measures for employment promotion in developing countries* (Geneva, 1972).

Some of the case studies in this volume, particularly the one on sugar processing, suggest that governments have not always given consideration to the possibility of taking the above practical measures to encourage labour-intensive products and techniques. Government control of sugar supply means distortions of the true market price and the true demand for sweetening agents. As a basis for appropriate government intervention, information is needed on cross-elasticities of demand for alternative or near-substitute products.

Another point about incentives structure and pricing policies is that although the removal of factor price distortions is important for ensuring more appropriate technical choices, it is not sufficient on its own and should constitute one of several elements in a set of measures which may also include the establishment of appropriate institutions of information collection and dissemination, action relating to transport and installation costs in addition to machine prices, and the establishment of adequate planning, organisation and implementation machinery. In the chapter on sugar processing techniques it is shown that one of the reasons for the failure of policy makers to consider expanding small-scale sugar production has been the absence of a "single body responsible for evaluating investment possibilities in the two technologies together".

The above point seems to apply particularly to the buying of second-hand equipment by the developing countries. In Chapter 5 on the use of second-hand equipment in jute processing, considerable emphasis is placed on the need for more technological and economic information about the ageing of such equipment and its deterioration through lifting and re-installation. Nevertheless, even in the absence of further information, the collection of which can often be time consuming, decisions have to be made. It is therefore important to consider whether a policy of using second-hand machinery might contribute towards a technological capacity to create new "appropriate" methods of production by creating greater demands on local machine shops. In fact, the ILO-organised employment mission to Kenya [1] had recommended the possibility of using the large East African Railway Workshops as well as small machine shops to give technical support to industries in the "informal" sector, which often used second-hand machinery.

Existing policies regarding the import of second-hand machinery by the developing countries are rather extreme: on the one hand, some of these countries totally ban the import of this equipment; others tend to follow an open door policy regardless of the risks associated with transactions in second-hand equipment. A discriminating policy based on knowledge about the orga-

---

[1] See ILO: *Employment, incomes and equality: A strategy for increasing productive employment in Kenya,* op. cit., Technical Paper 9: "Industrial research and engineering facilities".

nisation of the market for second-hand machinery and the development of the requisite skills is to be preferred.

This leads to another lesson for policy formulation, namely that even in cases in which efficient technological alternatives do exist the flow of information about them to the potential end users is often very imperfect. The study on can production (Chapter 4) indicates that the use of relatively capital-intensive techniques resulted not so much from the non-availability of less capital-intensive and more appropriate techniques as from the lack of information about existing alternatives. Thus it is the process of selection, and not the nature of the existing techniques, that seemed to be inappropriate.

A final point about the means of correcting factor price distortions is that they tend to influence current prices more than future ones. Indeed, current policy decisions are made on the basis of the current set of prices and from out of a currently known set of technological possibilities. Most of the empirical studies are concerned with a static view of technical choice with the existing set of production possibilities. The current factor prices, even when corrected through proper policy instruments, do not give indications of the potentialities of future increases in technical knowledge and of the development of new appropriate technologies. This dynamic aspect of technological choice and innovations is considered in Chapter 4 on can production and in Chapter 9 on the engineering industry.

There are thus two possible improvements in technological choice from the point of view of policy formulation. First, improvements can be introduced into the decision-making process in order to facilitate more appropriate choices from within the existing range of techniques that have evolved over time. Such an improvement may involve fiscal action on factor prices and the promotion of a better dissemination of information on existing technological alternatives. Secondly, wherever technologies appropriate to the factor endowments of developing countries do not exist, they could be created through massive research and development.

Some efforts are being made in a few developing countries to allocate research and development expenditure specifically to a search for appropriate technologies. But the record in terms of amounts spent or of results achieved is generally poor, partly because the share of the developing countries in the total world expenditure on research and development is as low as 2 per cent, the remainder being devoted almost entirely to the requirements of advanced countries. No doubt most developing countries do have national science and technology institutions, but apart from limited resources, these institutions have the major weakness that science and technology is very rarely related to local production conditions. They also cater to the needs of a very small clientèle of modern industry, little effort being made to harness the forces of science and technology for the "informal" sector. Yet it is precisely the latter kinds of employment-creating innovations that are primarily needed to improve the welfare of the common man.

Thus technological gaps exist not just between the rich and the poor countries but also within the latter between the "formal", or modern, and "informal" sectors. If success is defined in terms of the eradication of poverty, the objective of a successful technology policy cannot simply be a reduction in technological gaps between the rich and the poor countries, but a reduction or elimination of technological dualism within national boundaries; this is perhaps equally important, if not more so. [1] Measures to redress this imbalance include the allocation of a higher proportion of research and development resources to institutes concerned mainly with small-scale and traditional production, the location of these institutes in areas where relevant productive activity is concentrated, and a combination of problem-solving service with research and development activity. The experience of China is relevant here. Indigenous innovations are encouraged both on the farm and on the shop floor: scientists and technologists go down to the communes and the factory floor before defining research priorities; successful innovations and innovators receive public acclaim and recognition. Thus China seems to have institutionalised technological innovation at the local level, linking research and development to applied social needs.

The above measures, namely choosing appropriate technologies from known ones and the development of new appropriate technologies, are inter-related and not mutually exclusive. The acquisition of new technological knowledge cannot be isolated from the current state of knowledge and current techniques of production. For the development of new "appropriate" or "intermediate" technologies we may, as a point of departure, need to look into old designs of equipment and blueprints, or existing ones that are not currently in use.

Moreover, decisions made on the selection of known technologies from the existing range are also likely to influence the new production possibilities.

In view of the interdependence mentioned above, resources need to be optimally allocated between developing new technologies and identifying what already exists but is not widely known. The latter operation is equally important, especially since the development of new technologies, if successful, would yield prototypes for commercial use only after a time lag of several years. From the point of view of an employment policy, gains forgone or postponed in this way until new technologies enter a commercial phase need to be taken into account also.

---

[1] See A. S. Bhalla: "Implications of technological choice in African countries", in *Afrika Spectrum* (Hamburg), No. 1, 1973.

SOME NEW PERSPECTIVES

Recent developments such as the rise in raw material prices, the energy crisis and concern with the environment have put the issue of technology and employment in industry in a new perspective, and the choice has been further complicated by the introduction of new, yet unpredictable, variables. However, the influence of increases in prices of raw materials (including oil) simply means that such scarce factors should be further economised wherever feasible. Hitherto, a relatively cheap supply of raw materials and of imported oil meant easy replacement of labour by machines. This is no longer economically feasible with high-cost energy. Wherever manual techniques can be substituted for machine-intensive ones, a saving of this scarce input could be achieved. Nevertheless, there are fields in which the only alternative is to develop new local sources of energy from the sun, wind and tides, for example. Here is thus a new area for research and development work for the second half of the 1970s, which reinforces the recent interest in the question of choice of "appropriate" products in general.

Given the energy resources, both the possibilities of labour-energy substitution and the development of new and cheap sources of energy need to be explored. One direction in which new research and development work needs to concentrate is a shift to indigenous raw materials, small hand-operated machinery and the use of various waste materials for the generation of energy in individual industrial plants.

Similarly, concern for environmental factors implies a modification of policy regarding alternative technologies. Research will have to be concentrated on industrial technologies that either are non-polluting or cause less pollution than others; this is essential since pollution, in a sense, means negative output. For instance, techniques using natural fibres, it is argued, are less polluting than those using artificial inputs such as synthetic fibres. This fact has major implications for the choice and location of processing industries; to the extent that such industries are located in less developed countries and are more labour-intensive, the objectives of increased employment and pollution control would be achieved simultaneously. There are two main unsolved problems with regard to technology, the environment and employment. The first is the identification of the technologies that are appropriate for avoiding pollution. The second lies in the fact that the technologies that are needed to prevent pollution may well be very capital-intensive and labour-displacing. ILO research on technology and employment is now beginning to be concerned with these new problems.

It is clear that even in the absence of new developments such as inflation, the energy crisis and pollution, the resources of developing countries for

323

research and development are extremely limited. In fact, they have become even more so, in particular for the oil-short developing countries, since new demands are imposed for additional research work on the subjects just mentioned. Partnership between the developed and the developing countries in research and development will therefore be much more essential in the future than it has ever been in the past. This is indeed implicit in the recent resolution of the General Assembly of the United Nations on the New International Economic Order, which re-asserts the need for assistance from developed to developing countries in programmes of research and development, for the creation of suitable indigenous technology, and for the promotion of international co-operation in research and development. This international co-operation may include an increase in the resources devoted by the advanced countries to the development of technology appropriate for developing countries, the financing of international and national institutions to promote appropriate technology, and inducements to multinational enterprises to give more attention to the modification of modern technology in the light of local conditions in developing countries, and wherever feasible, to locate labour-intensive processes under mutually acceptable and equitable arrangements in developing countries where labour costs are low. [1]

At present there are very few developing countries with any comprehensive and consistent national science or technology plans and policies; yet the formulation of such policies in terms of development plans and objectives is essential. At the national level, these policies need to embrace the choice of appropriate products and industries which, in the context of fuller employment and more equal income distribution, must be based on consumption plans for the poorer income groups; the choice of appropriate technologies, which would require enlisting the support of equipment manufacturers and research institutes; the reorientation of research and development efforts towards informal sector activities, which would call for the encouragement of local machine fabrication facilities; and incentives to ensure that the appropriate technologies identified or created are actually used.

---

[1] Recommendations along these lines were included in United Nations: *The impact of multinational corporations on the development process and on international relations*, Report of the Group of Eminent Persons to Study the Role of Multinational Corporations on Development and on International Relations, doc. E/5500/Add. 1 (Part I) (New York, 1974; mimeographed).